Manual of Microbiological Methods

MANUAL OF

Microbiological Methods

BY THE

SOCIETY OF AMERICAN BACTERIOLOGISTS

COMMITTEE ON BACTERIOLOGICAL TECHNIC

M. J. Pelczar, Jr., *Chairman*

R. C. Bard
G. W. Burnett
H. J. Conn, *Editor*
R. D. DeMoss
E. E. Evans
M. W. Jennison
A. P. McKee
A. J. Riker
J. Warren
O. B. Weeks
F. A. Weiss

McGRAW-HILL BOOK COMPANY

New York Toronto London

1957

MANUAL OF MICROBIOLOGICAL METHODS

Library of Congress Catalog Card Number 57-8629

13 14 15 16 17 18 19 20 - MAMM - 7 5 4 3 2

59556

Committee Members and Other Contributors

ALLEN, O. N. Department of Bacteriology, University of Wisconsin, Madison, Wis.

ARK, P. A. Division of Plant Pathology, University of California, Berkeley, Calif.

*BARD, R. C. Smith Kline & French Laboratories, Philadelphia 1, Pa.

BARTHOLOMEW, J. W. Department of Bacteriology, University of Southern California, Los Angeles 7, Calif.

*BURNETT, G. W. Dental Division, Walter Reed Army Institute of Research, Washington, D.C.

†COHEN, BARNETT *Formerly*, Johns Hopkins Medical School, Baltimore, Md.

*CONN, H. J. *Emeritus*, Division of Food Science and Technology, N.Y. Agricultural Experiment Station, Geneva, N.Y.

*DEMOSS, R. D. Department of Bacteriology, University of Illinois, Urbana, Ill.

*EVANS, E. E. Department of Microbiology, Medical Center, University of Alabama, Birmingham 3, Ala.

HILDEBRAND, E. M. Horticultural Field Station, U. S. Department of Agriculture, Beltsville, Md.

HILDEBRANDT, A. C. Department of Plant Pathology, University of Wisconsin, Madison, Wis.

*JENNISON, M. W. Department of Bacteriology, Syracuse University, Syracuse 10, N.Y.

LINDBERG, ROBERT B. Department of Bacteriology, Walter Reed Army Institute of Research, Washington, D.C.

MCCLUNG, L. S. Department of Bacteriology, Indiana University, Bloomington, Ind.

*MCKEE, A. P. Department of Bacteriology, College of Medicine, State University of Iowa, Iowa City, Iowa

*PELCZAR, M. J., Jr. Department of Microbiology, University of Maryland, College Park, Md.

*RIKER, A. J. Department of Plant Pathology, University of Wisconsin, Madison, Wis.

*WARREN, Joel Division of Biological Standards, National Institutes of Health, Bethesda 14, Md.

*WEEKS, O. B. Department of Bacteriology, University of Idaho, Moscow, Idaho

*WEISS, F. A. American Type Culture Collection, 2029 M Street, N.W., Washington, D.C.

*Members of Committee on Bacteriological Technic, 1955–7

†Deceased

Preface

This manual is intended for use in those types of microbiological work which involve the study of microbial cultures or of viruses, either for identification or for learning the properties of the organisms investigated. This book takes the place of the loose-leaf publication issued during the period of 1923–1956 under the name of "Manual of Methods for Pure Culture Study of Bacteria." The present manual covers a wider scope but still includes the subject of "pure culture study," the meaning of which is discussed.

The methods given here are not to be regarded as official. The committee has always taken the stand that official methods should not be adopted in the case of research work, because it is continually necessary to modify research methods in order to keep them up to date. The standardization of methods tends to hinder the development of new technics, while the chief function of this committee is to stimulate its development.

The methods in this manual, therefore, are merely claimed to be those regarded as satisfactory by the committee at the time of publication. Whenever practical, the methods have been tested by the committee in comparison with other procedures.

Of the chapters in this book, V, VIII, IX, X, and XI are almost entirely new. The others are revisions, more or less complete, of leaflets of the old manual. In the case of these revised chapters, the names of the contributors at the chapter heads are the latest revisers, not the original authors. In the case of the five entirely new chapters, however, the actual authors are cited at the chapter heads.

H. J. Conn
M. J. Pelczar, Jr.

Contents

CHAPTER I

Introductory

SCOPE OF THE MANUAL

There has sometimes been misunderstanding as to the sense in which the Committee on Bacteriological Technic uses the expression "pure culture study of bacteria." It is occasionally thought that such an expression would cover nearly the whole field of bacteriological technic. On the other hand, the definition of pure culture study of bacteria which has been drawn up by the committee is the study of bacterial cultures with the object of learning their characteristics and behavior or determining their identity, or both. Such a study may be regarded as including isolation methods, methods for the cultivation and the storage of various kinds of bacteria, microscopic study of pure cultures either stained or unstained, determination of cultural characteristics of an organism, a study of its physiological characteristics, the chemical methods necessary in making the last-mentioned study, the determination of pathogenicity and study of pathological effects, the serological characterisics of an organism when used as a means of description.

It is clear from such a statement that pure culture study of bacteria is fairly comprehensive but that there are many fields of bacteriological technic not included within it, e.g., methods for the enumeration of bacteria in their natural habitats, the diagnosis of disease and many other phases of medical bacteriology, methods employed in the study of food spoilage and controlling the processes of fermentation, etc. Such a list might be extended almost indefinitely, for the field of pure culture study, although fairly broad, is actually a small part, though basic, of bacteriological technic.

The scope of the present manual has been widened to include virological methods and procedures for the maintenance and preservation of bacteria, and it may in the future be expanded to include methods for yeasts and molds. Nevertheless its subject matter still does not cover those topics listed in the preceding paragraph.

RELATION TO TAXONOMY

Clearly, one of the objects of pure culture study is to determine the identity of any bacterial culture under investigation. This brings the subject very close to the field of bacterial taxonomy, i.e., the naming and classifying of bacteria. Inasmuch as bacteria cannot be classified without studying their characteristics in pure culture, it is an obvious conclusion that pure culture study is a necessary prelude to bacterial taxonomy.

It must be recognized, nevertheless, that one can consider pure culture study without regard to taxonomy and that one can study the taxonomy of bacteria without paying special attention to the methods of pure culture study. Since this distinction can be made and the committee editing this series of publications is a committee on technic, care has always been taken to maintain the distinction so as not to interfere with the functions of other committees that have been appointed and to deal with matters of nomenclature and classification.

PUBLICATIONS OF THE COMMITTEE ON TECHNIC

Descriptive charts. The first descriptive chart actually adopted by the Society of American Bacteriologists was in 1907. The chart has been revised from time to time and at present there are two forms: one known as the Standard Descriptive Chart and the other as the Descriptive Chart for Instruction. The latter is very much simpler than the former. The former is printed on both sides of an 8½- by 11-in. sheet of light cardboard; the latter on a sheet of heavy paper of the same size.

The object of the descriptive chart is to provide a space for recording the most important characteristics of a single culture. The Standard Chart is the most complete and is intended especially for advanced work in bacteriology. Unfortunately, however, it does not meet modern research needs at all perfectly because each group of bacteria requires its own set of tests and no form can be drawn up sufficiently detailed to cover all of them. The Chart for Instruction, on the other hand, is so much simpler and contains so much blank space that it sometimes is found to be more satisfactory in research work than the Standard Chart. It is, however, intended primarily for students to use in characterizing cultures furnished them in connection with their classwork.

Manual of Microbiological Methods. The origin of the present manual traces back to a committee report which was printed in the *Journal of Bacteriology* and was distributed in reprint form by the committee (1918). It was only 14 pages long and covered only the methods used in carrying out the determinations called for on the descriptive chart of those days. After one or two minor revisions it was converted in 1923 into the "Manual of Methods for Pure Culture Study of Bacteria," which as remarked above was published in loose-leaf form until 1956. The first edition was only 48 pages in length and, like its predecessor, was confined wholly to the methods needed in using the chart. Gradually, however, it was expanded until it included 10 leaflets, and it has come to include a variety of methods other than those called for in the use of the descriptive chart. By 1953 several other subjects had been selected as desirable to include in future editions, and plans were made for converting the manual into a larger publication.

While the old manual was in loose-leaf form, it was kept up to date by periodic revision of its leaflets, one or two at a time, and by means of a continuation service owners were enabled to secure the latest editions to insert in their copies. This feature now, unfortunately, has to be given up because of the increased size of the book and of certain practical difficulties involved in loose-leaf publication. The committee regrets the necessity of abandoning the old system, but there seems to be no other course than to convert it into a regular book and to hope that revisions to bring the contents up to date can be accomplished by means of periodic new editions.

HISTORICAL

The first efforts toward producing a descriptive chart for characterizing bacteria were made by two different individual investigators, H. W. Conn and S. de M. Gage. The work of these two investigators called the matter to the attention of bacteriologists in general, and it was finally brought before the Society of American Bacteriologists by F. D. Chester at the Philadelphia meeting in December, 1903, and then again at the 1904 meeting, when he explained his idea of a "group number" which would be descriptive of the salient characters of an organism. On his recommendation the society appointed a Committee on Methods for the Identification of Bacterial Species of which Professor Chester was made chairman. This committee drew up the first descriptive chart with which the Society of American Bacteriologists had any connection.

This chart was put before the society at its 1905 meeting. It was presented at this time as a preliminary effort, and no endorsement of it was given by the society, nor apparently was such endorsement requested. The committee was instructed to continue its work, and a second chart was prepared during 1906 and presented at the society meeting in December of that year. At this meeting it was decided that the chart should call for more complete data concerning bacteria than provided for by either of the two charts already submitted, so the committee was instructed to do further work along this same line.

The committee at this time was composed of F. D. Chester, F. P. Gorham, and E. F. Smith, but Professor Chester was largely responsible for the first two charts presented at society meetings. Before the committee undertook a further revision, however, he had left bacteriological work and hence was no longer active on the committee. During 1907, therefore, Dr. Smith acted as chairman of the committee, and under his supervision the committee drew up another chart which was presented to the society at its meeting in December of that year. This chart was officially endorsed by the society and was put on sale by the secretary of the society.

For several years following no changes were made in the chart. The next step in its development was brought about by H. A. Harding (1910), who published a paper in which he outlined the complete history of the chart, with copies of the early charts, and discussed improvements that might be made. This paper is available for those desiring more detail concerning this early history than is given here.

As the society felt that further modifications were now needed, a new committee was appointed in 1911 consisting of F. P. Gorham, C. E. A. Winslow, Simon Flexner, H. A. Harding, and E. O. Jordan. This committee gave a report at the 1913 meeting, presenting a chart which was put on sale by the society but was not officially endorsed. As this committee was unable to continue the work, an entirely new one was appointed at this time consisting of H. A. Harding, H. J. Conn, Otto Rahn, W. D. Frost, and L. J. Kligler. This committee soon lost Dr. Rahn, who left the country in 1914, and M. J. Prucha was added in his place. The committee was called the Committee on Revision of the Chart for the Identification of Bacterial Species.

The new committee was instructed by the society to make a conservative revision of the chart and at the same time to draw up a manual of methods to be used in connection with it. At the 1914 meeting of the society, therefore, a chart was presented for approval, much like the 1907 chart except for its more logical arrangement of data. This chart was given the society's endorsement and was issued during 1915.

The 1914 chart was printed on a sheet with its back entirely blank, the glossary previously on the back having been omitted. The committee gave as the reason for this that the glossary would be included in the manual on methods shortly to be published. The publication of this manual was delayed, however, pending investigation of the methods to be included in it. This investigation of methods was to be undertaken not only for the sake of the manual but also as a preliminary step toward radical revision of the chart, which was felt to be badly needed. Early in 1917, however, and before this program could be carried out, the chairman of the committee was forced by pressure of other duties to drop the work. As he wished to remain on the committee, however, no change in membership was made, but H. J. Conn was asked to become chairman.

The committee then undertook the first step toward the preparation of a manual on methods. A report was presented at the 1917 meeting, giving the methods recommended at that time for use with the chart. The report was printed in the *Journal of Bacteriology*, March, 1918, and was subsequently sold by the society in the form of reprints. This report was considered a preliminary manual on methods.

The committee proposed at the same time a much simplified chart in the form of a four-page folder, which it recommended for use in instruction until the official chart could be given the revision it needed. This chart was not endorsed by the society but was printed and sold by the society for two or three years.

This same committee (but now called the Committee on the Descriptive Chart) issued another report on methods which appeared in the *Journal of Bacteriology*, March, 1919, dealing with the gram stain, production of acid, and the reduction of nitrates. At the 1919 meeting it issued a further report which appeared in the *Journal of Bacteriology* in two parts, March and May, 1920. The first part of the report was a revision of the one which had been published in March, 1918, and was sold as a revised manual of methods until the reprints were exhausted in 1922.

At the 1920 meeting the Committee on the Descriptive Chart was discharged with the understanding that its functions would be taken over by a committee of broader scope then appointed and called the Committee on Bacteriological Technic. This committee was appointed with the understanding that its membership should fluctuate from year to year in order to keep on it men actively interested in the work.

The new committee made a further revision of the chart, which was presented at the 1920 meeting and endorsed by the society. Later editions of this chart have been drawn up by the committee but have not been submitted to the society for official endorsement. In order to avoid committing the society in favor of any of the methods concerned, recent editions of the chart have merely been presented by the committee and permission asked to put them on sale.

The committee issued four further reports in the *Journal of Bacteriology* (1921, 1922*a*, *b*, and *c*) before the manual was prepared. One of these reports (1922*b*) proposed certain revisions of methods in the case of the gram stain, fermentation, nitrate reduction, indole and hydrogen sulfide production. The committee presented this report at the 1922 meeting of the society with the recommendation that the revised material be published as part of a "Manual of Methods for Pure Culture Study of Bacteria." The committee was thereupon instructed by the society to publish this manual, using the loose-leaf form of binding, with the understanding that new folders be issued from time to time to keep it up to date. This was done, and the system continued till 1956, when, as explained above, it proved necessary to convert the manual into book form and its name was changed to "Manual of Microbiological Methods."

The Committee on Bacteriological Technic has seen the following changes in personnel:

1920	H. J. Conn,[1] K. N. Atkins, I. J. Kligler, J. F. Norton, G. E. Harmon.
1921	H. J. Conn,[1] K. N. Atkins, G. E. Harmon, Frederick Eberson, Alice Evans.
1922	H. J. Conn,[1] K. N. Atkins, G. E. Harmon, Frederick Eberson, F. W. Tanner, and S. A. Waksman.
1923	H. J. Conn,[1] K. N. Atkins, J. H. Brown, G. E. Harmon, G. J. Hucker, F. W. Tanner, and S. A. Waksman.
1924–5	H. J. Conn,[1] K. N. Atkins, J. H. Brown, Barnett Cohen, G. J. Hucker, F. W. Tanner.
1926–7	H. J. Conn,[1] Barnett Cohen, Elizabeth F. Genung, W. L. Kulp, W. H. Wright; with G. J. Hucker and S. Bayne-Jones as a subcommittee on serological methods.
1928	H. J. Conn,[1] Victor Burke, Barnett Cohen, Elizabeth F. Genung, W. L. Kulp, W. H. Wright.
1929–30	H. J. Conn,[1] Victor Burke, Barnett Cohen, Elizabeth F. Genung, I. C. Hall, W. L. Kulp, W. H. Wright (deceased, May, 1929).
1931–4	H. J. Conn,[1] Barnett Cohen, Elizabeth F. Genung, Victor Burke, I. C. Hall, J. A. Kennedy.
1935	H. J. Conn,[1] Victor Burke, Barnett Cohen, M. W. Jennison, J. A. Kennedy.
1936–42	H. J. Conn,[1] J. H. Brown, Victor Burke, Barnett Cohen, C. H. Werkman, M. W. Jennison, J. A. Kennedy, A. J. Riker.
1943–5	H. J. Conn,[1] Victor Burke, Barnett Cohen, C. H. Werkman, M. W. Jennison, J. A. Kennedy, L. S. McClung, A. J. Riker.
1946–7	H. J. Conn,[1] G. H. Chapman, Barnett Cohen, I. C. Gunsalus, M. W. Jennison, L. S. McClung, A. J. Riker, C. E. ZoBell.
1948	M. W. Jennison,[1] G. H. Chapman, Barnett Cohen, H. J. Conn, I. C. Gunsalus, J. A. Kennedy, L. S. McClung, A. J. Riker, C. A. Stuart, C. E. ZoBell.
1949	M. W. Jennison,[1] G. H. Chapman, H. J. Conn, I. C. Gunsalus, L. S. McClung, C. A. Stuart, A. J. Riker, C. E. ZoBell.
1950	M. W. Jennison,[1] R. C. Bard, G. H. Chapman, H. J. Conn, I. C. Gunsalus, L. S. McClung, C. A. Stuart, A. J. Riker, C. E. ZoBell.
1951–2	M. W. Jennison,[1] R. C. Bard, G. W. Burnett, H. J. Conn, H. C. Lichstein, L. S. McClung, A. P. McKee, M. J. Pelczar, A. J. Riker, C. A. Stuart, C. E. ZoBell.

[1] Chairman.

1953–4 M. J. Pelczar,[1] R. C. Bard, G. W. Burnett, H. J. Conn, E. E. Evans M. W. Jennison, H. C. Lichstein, L. S. McClung, A. P. McKee, A. J. Riker, J. Warren, O. B. Weeks, F. A. Weiss.

1955–7 M. J. Pelczar,[1] R. C. Bard, G. W. Burnett, H. J. Conn, R. D. DeMoss, E. E. Evans, M. W. Jennison, A. P. McKee, A. J. Riker, J. Warren, O. B. Weeks, F. A. Weiss.

GENERAL CONSIDERATIONS

Pitfalls to Be Avoided by the Student

In studying microbial cultures with the object of identifying them or describing them, the student is apt to run into certain pitfalls. Some of these apply specifically to certain types of work and are therefore best taken up in the various chapters of this book where they seem properly to fit. Others are more general; some, in fact, are well known even to beginning students in bacteriology. However, as others are less fully appreciated, a few words concerning some of these pitfalls seem called for here—even at the risk of repeating cautions that may seem too elementary. These pitfalls arise primarily from three sources: (1) the danger of impure cultures, (2) confusing results because of variation of bacterial species, (3) differences in methods of study.

The danger in impure cultures is, of course, thoroughly understood. Unfortunately, however, the second consideration just mentioned makes it more important to emphasize the danger of impure cultures today than was the case before 1920. In those days bacteriologists quite generally accepted the idea of monomorphism, and whenever a culture was observed to be noticeably abnormal in either morphology or physiology, it was promptly discarded as a contaminant. When, however, it began to be learned that even the most strictly guarded pure cultures might show changes in morphology during their life history, and then later when it was realized that the same organism might occur in two or more phases showing distinctly different cultural and physiological characteristics, the old ideas of monomorphism were decidedly upset. As a result of the changing point of view, it is very easy for a careless student today to believe that he is observing two phases of the same pure culture when, actually, one of his "phases" is a contaminant. This makes constant checking as to purity of cultures even more important than it was before dissociation into phase variants was generally accepted by bacteriologists.

Accepting the idea of dissociation presents other difficulties to the student. Without exhaustive study, it is sometimes very easy to describe two phases of the same species as though they were different organisms. It is also easy to prepare a description of some culture which is an illogical jumble of the characteristics of two or more phases, due to

[1] Chairman.

the fact that it was first studied in an unstable form and dissociation was taking place during the course of the study. On the other hand, some of the methods employed in the hopes of inducing phase variation may actually cause contamination and be incorrectly interpreted. Some of these points are very adequately discussed by Frobisher (1933).

The third source of error mentioned above (variation in methods) also needs emphasis. When a species is described in such terms as one frequently encounters in published descriptions, e.g., "produces acid (without gas) from glucose and lactose but not from sucrose; does not reduce nitrates," one has to guess at the answers to such questions as these: What basal medium was used in each instance? What indicator of acid production was employed? How thorough a study was made to show the absence of any acid from sucrose or of any reduction of nitrate? Or, in the last instance, is it safe to assume that the author of the species merely failed to find nitrite in some nitrate medium? Unless such questions are answered correctly, the description is meaningless; the attempt to identify an unknown culture with such a description may well give misleading results.

With all these pitfalls to avoid, it is easy to see how the same set of data, no matter how carefully prepared, can be differently interpreted by two different bacteriologists. As a result extreme caution is urged, both in determining the identity of a culture and in deciding whether or not to pronounce it a new species.

Practical Hints

Determining the characteristic of a culture. One should always, if possible, make a complete study of a culture promptly after its first isolation while it is in a condition to display its true characteristics. When a culture has been carried in the laboratory for a long period of time, it may change in some respects from the original. When practical, such cultures should be exposed to conditions which might bring them back to the "normal." When this is done, however, the possibility should always be recognized that by such manipulation dissociation may be induced so that the phase subsequently studied may be quite different from the original isolation. Whenever distinct evidence of dissociation is observed, each phase should be studied and recorded separately, and efforts should be made to reverse the change or to obtain the same change with other strains until the possibility of impure cultures seems to be out of the question. No importance should ever be attached to a single determination unless supported by replications giving the same results. In describing morphology, one should not be contented with one or two observations but should study several transfers and should follow up each of them day by day for about a week. When changes are observed, a careful study should be made to learn whether they indicate morphologic variation, dissociation, or merely contamination. In making special staining tests, like the gram stain, several determinations should be made on separate transfers of the culture and at different ages, because there are species that vary in their staining reactions and such variation cannot be detected by single determinations. As a check on the technic, a known positive and a known negative culture should be included in the study. For example, when making a gram stain, it is good practice to place on the slide, beside the culture under study, a smear containing a mixture of a known gram-positive and a known gram-negative organism (which differ markedly in morphology). Then it is possible to observe if the expected results are obtained with the known cultures and thus to have some degree of control on the technic.

Identification. After recording the characteristics of an organism, the next step is identification, if possible, with a previously described species. This should never be attempted until at least six representative strains of the unknown organism isolated from more than one source, if possible, have been studied. No rules can be given for identifying the culture. Descriptions of bacteria are scattered so widely through the literature and vary so greatly in their form that identification is often extremely different. Bergey's "Manual of Determinative Bacteriology" is a great help, but it is usually necessary to go back to original descriptions and often to secure transfers of authentic strains before certain identification can be made. Difficult as this procedure is, no one is justified in naming a new species of bacteria until a comprehensive search through the literature of species already described has been made. Frequently it is necessary to refer in some publication to a previously described species on the basis of such an identification as this. In this case it is important to state in the publication whether or not an authentic strain of the species has been obtained for comparison; if so, from where obtained; if not, what published description of the species was followed in making the identification. As to a name to use for such a species, one may follow the original author's nomenclature or may give it the name employed in some modern system (e.g., Bergey). Whatever name is chosen, no confusion will result if it is accompanied by the name of the original author of the specific name and by that of the one making the combination of generic and specific names. Thus, whether one says "*Bacillus coli* Migula" or "*Escherichia coli* (Migula) Castellani and Chalmers," it is entirely clear what species is intended.

Naming a new species. When it proves impossible to identify a culture with any species described in the literature, it is often desirable to publish a description of it as a new species. When publishing such a description, there are five important points to be kept in mind:

1. The description should be based on at least six representative isolations of the organism.
2. If variations are found to occur among these strains, a critical study must be made to be sure that they are not the result of contamination.
3. In naming any characteristic of the species, especially if it is a negative character (e.g., "nitrates not reduced"), the technic by which it is determined must be stated.
4. Before giving the results of any test as positive or negative, comparisons must be made with a control culture known to be positive and one known to be negative.
5. Before actually assigning a name one should consult a specialist in bacterial taxonomy, both as to the necessity for a new name and as to the validity of the name selected. The Board of Editor-Trustees of Bergey's Manual, for example, are always very glad to offer such advice.

If these hints were followed by all who are trying to identify species or to publish descriptions of them, much of the confusion in bacterial nomenclature would be eliminated.

REFERENCES

Committee on Bacteriological Technic. 1922*a*. An investigation of American stains. *J. Bacteriol.*, **7**, 127–248.

———. 1922*b*. Methods of pure culture study. *J. Bacteriol.*, **7**, 519–528.

———. 1922*c*. An investigation of American gentian violets. *J. Bacteriol.*, **7**, 529–536.

Committee on Descriptive Chart. 1918. Methods of pure culture study. *J. Bacteriol.*, **3**, 115–128.

———. 1919. Methods of pure culture study. Progress report for 1918. *J. Bacteriol.*, **4**, 107–132.

———. 1920*a*. Methods of pure culture study. Revised. *J. Bacteriol.*, **5**, 127–143.

———. 1920*b*. Progress report for 1919. *J. Bacteriol.*, **5**, 315–319.

Frobisher, M. 1933. Some pitfalls in bacteriology. *J. Bacteriol.*, **25**, 565–571.

Harding, H. A. 1910. The constancy of certain physiological characters in the classification of bacteria. *N.Y. State Agr. Expt. Sta. Tech. Bull.* **13**.

CHAPTER II

Staining Methods

H. J. Conn in collaboration with J. W. Bartholomew and M. W. Jennison

GENERAL PRINCIPLES

The staining of bacteria depends in general upon the same properties of dyes as does the staining of animal or plant tissue for histological purposes. Short discussions of the nature of dyes, with special reference to staining are given elsewhere (Conn, 1953), and only the briefest summary of the subject need be given here.

All bacterial dyes are synthetic products—anilin dyes, or coal-tar dyes, as they are generally called. Although the synthetic dyes vary greatly in their chemical nature and staining properties, they are for practical purposes often divided into two general groups, the acid dyes and the basic dyes. These terms do not mean that the dyes in question are free acids or free bases. The free color acids and bases, when obtainable, are colored, to be sure, but they are often insoluble in water and rarely have appreciable staining action; i.e., the colors do not "stick." The salts of these compounds, on the other hand, are more soluble, penetrate better, and stain more permanently; they are the true dyes.

An acid dye is the salt of a color acid; a basic dye the salt of a color base. In other words, acid dyes owe their colored properties to the anion, basic dyes to the cation. The actual reaction of an aqueous solution of a dye, however, depends on several factors, and an acid dye may well be basic in reaction, while a basic dye may be acid. This is because the reaction of such a solution depends on the relative strengths of the dye ion and of the anion or cation with which it is combined in the dye salt.

Basic dyes have greatest affinity for the nuclei of cells, probably because of the acid nature of the nuclear material. Acid dyes have a stronger tendency to combine with the cytoplasm. Bacteria do not show typical cell structure, and they tend to stain fairly uniformly with nuclear, i.e., the basic, dyes. Hence, the stains in common use by the bacteriologists are rarely acid dyes.

Preparation of Smears

Pure cultures of bacteria can ordinarily be prepared for staining by the simple process of making an aqueous suspension and drying a drop of it on a slide or cover glass, without any fixation other than gentle heat. The use of this simple procedure depends upon the fact that most bacteria, because of their small size or their stiff walls, can be dried without great distortion. For this reason it is not always necessary, as with higher organisms, to coagulate the tissues before microscopic preparations can be made, although for cytological studies and for accurate determinations of size and shape of the cells, some fixation other than heat is needed.

The best bacterial smears are usually made by removing a small amount of surface growth from some solid medium and mixing it with distilled water. It is often possible to use a drop of a culture growing in a liquid medium, but such a smear is not always so satisfactory, since certain constituents of the medium may prevent the bacteria from adhering to the slide or may interfere with the staining.

The suspension used should always be sufficiently dilute. Ordinarily, only a faint turbidity should be visible to the naked eye, for it is always best to avoid the occurrence on the slides of solid masses of bacteria, piled one on top of the other. If a smear after staining does not show any portions where the bacteria are well separated one from another, a new, more dilute smear should be made. This is particularly important in the case of the gram stain or flagella staining.

The usual method of fixing the suspension to the slide or cover glass is to pass it rapidly after drying through a bunsen flame two or three times. Another very satisfactory method is to allow the drop of material to dry on a slide lying on a flat, moderately hot surface, such as a plate of some nonrusting metal resting on a boiling-water bath. With many bacteria an aqueous suspension of the surface growth from agar can be dried in the air at room temperature and stained without any fixing; this method is not universally successful, however.

For some cytological procedures special methods of making bacterial preparations are necessary, sometimes calling for fixing solutions rather than heat. It must be seen that the technic described for staining dried smears is too crude for accurate measurements of cells or for studying cytological details. One cytological method is given on page 30.

It is also beyond the scope of this publication to give staining methods for other than pure culture work.

In using any of the methods it must be remembered that blind adherence to a staining technic is no guarantee that the result will be satisfactory. Even experienced workers sometimes discover to their dismay that they took too much for granted as to the purity of their reagents, cleanliness of slides and covers, or proper compounding of the staining solutions. A technic should, therefore, be checked upon known organisms as controls. It is, furthermore, important to know that the solutions and water used for dilution are reasonably free from bacteria and spores.

Staining Formulas

There has always been a surprising amount of inaccuracy in the literature concerning staining solutions. This is due to a variety of causes: indefiniteness in the original publication, mistakes of copying by later authors, modifications of the original which are not described as modifications and come later to be ascribed to the original author, failure of authors to cite references when giving their methods. For such reasons it has proved necessary in this publication to give in many instances both the original (rather indefinite) formula and an emended formula as interpreted by the committee. The committee, however, *assumes no responsibility for the identity of the two and offers the emendation merely to prevent the perpetuation of formulas which are clearly ambiguous or indefinite as to their ingredients.* Recent cooperation among this committee, the Biological Stain Commission, and the National Formulary Committee of the American Pharmaceutical Association has resulted in the virtual adoption of these emended formulas.

Staining schedule. *Tap vs. distilled water.* When washing slides after applying any stain, tap water is ordinarily more convenient to use than distilled water, and in the staining schedules that follow, tap water is specified in those instances where its use is considered to be ordinarily unobjectionable. It must be remembered, however, that the *use of distilled water is never contraindicated* for such purposes, and many bacteriologists perfer it for all steps where washing is called for, because it is not subject to variation in composition, buffer content, etc.

GENERAL BACTERIAL STAINS—RECOMMENDED PROCEDURES[1]

Ziehl's Carbol-fuchsin

Old statement of formula		*Emended statement of formula*	
		Solution A	
Sat alc sol basic fuchsin	10 ml	Basic fuchsin (90 % dye content)	0.3 g
5 % sol carbolic acid	100 ml	Ethyl alcohol (95 %)	10 ml
		Solution B	
		Phenol	5 g
		Distilled water	95 ml
		Mix solutions A and B	

[1] In these discussions, small-size type is used for all formulas and directions for preparing them, text-size type for all other directions concerning recommended procedures, and small-size type for similar matters concerning alternate procedures.

Ammonium Oxalate Crystal Violet (Hucker's)

Solution A		*Solution B*	
Crystal violet (90% dye content)[1]	2 g	Ammonium oxalate	0.8 g
Ethyl alcohol (95%)	20 ml	Distilled water	80 ml

Mix solutions A and B

Crystal Violet in Dilute Alcohol

Crystal violet (90% dye content)	2 g
Ethyl alcohol (95%)	20 ml
Distilled water	80 ml

Loeffler's Alkaline Methylene Blue

Original statement of formula		*Emended statement*	
		Solution A	
Conc sol methylene blue in alcohol	30 ml	Methylene blue (90% dye content)	0.3 g
Sol KOH in distilled water (1:10,000)	100 ml	Ethyl alcohol (95%)	30 ml
		Solution B	
		Dilute KOH (0.01% by weight)	100 ml
		Mix solutions A and B	

Methylene Blue in Dilute Alcohol

Methylene blue (80% dye content)	0.3 g
Ethyl alcohol (95%)	30 ml
Distilled water	100 ml

Carbol Rose Bengal

Rose bengal (90% dye content)	1 g
Phenol (5% aqueous solution)	100 ml
$CaCl_2$	0.01–0.03 g

(The amount of $CaCl_2$ added determines the intensity of staining.)

Staining schedule: Follow the general procedure given under "Preparation of Smears," page 11, allowing 5–60 sec for application of the stain. Overstaining rarely occurs except with carbol fuchsin; understaining does not have to be feared except with rose bengal.

Results: The results depend on which of the above staining fluids is selected. They are listed in the order of intensity of action; i.e., carbol fuchsin gives the most intense stain and is not indicated when selective staining is desired or when much debris is present on the slide. The crystal violet solutions are very good for routine purposes. The methyl-

[1] It is not necessary that dry stains of the exact dye content specified be used in this or in the preceding and following formulas. Samples of higher or lower dye content may be employed by making the proper adjustment in the quantity used.

ene blue solutions are much more selective, with special affinity for metachromatic granules. The rose bengal solution is much less commonly used; it is specially valuable when mucus or colloidal organic material is present, as such material is not ordinarily stained by it.

GENERAL BACTERIAL STAINS—ALTERNATE PROCEDURES

Kinyoun's Carbol Fuchsin

Basic fuchsin (dye content not specified; probably 90%)	4 g
Phenol crystals	8 g
Ethyl alcohol (95%)	20 ml
Distilled water	100 ml

This formula is preferred in some quarters to the Ziehl carbol fuchsin. It is attributed to Kinyoun, but the reference to its original publication has not been located.

Carbol Crystal Violet (*Nicolle*)

Original statement of formula		*Emended statement*	
		Solution A	
Sat alc gentian violet	10 ml	Crystal violet (90% dye content)	0.4 g
1% aqu sol phenol	100 ml	Ethyl alcohol (95%)	10 ml
		Solution B	
		Phenol	1 g
		Distilled water	100 ml
		Mix solutions A and B	

This formula is sometimes preferred either as a general stain or in the gram technic. If properly prepared it is permanent, but it has a tendency to gelatinize if the amount of dye is too great. To prevent this sort of deterioration the quantity of dye in the above amended formula has been reduced to 0.4 g from the 1.0 g recommended previously. Even when the solution is so prepared as to be permanent, however, it seems to have no advantage over the ammonium oxalate crystal violet given above.

Anilin "Gentian Violet" (*Ehrlich*)

Original statement of formula		*Emended statement*	
		Solution A	
Sat alc sol gentian violet	5–20 ml	Crystal violet (90% dye content)	1.2 g
Anilin water (2 ml anilin shaken with 98 ml water and filtered)	100 ml	Ethyl alcohol (95%)	12 ml
		Solution B	
		Anilin	2 ml
		Distilled water	98 ml
		Shake and allow to stand for a few minutes, then filter.	
		Mix solutions A and B	

This formula is given largely for its historic interest. It is a quite unstable solution and has no special value today. It was, however, one of the first important bacterial staining fluids and was formerly regarded as the standard formula for the gram stain. It is not, however, certain what was the "anilin gentian violet" originally employed in the gram stain, even though ascribed to Ehrlich. As a matter of fact Ehrlich seems to be properly credited only with the idea of using anilin water in the formula, as he apparently did not recommend any one definite formula.

NEGATIVE STAINING OF BACTERIA—RECOMMENDED PROCEDURES

Dorner's Nigrosin Solution

Nigrosin, water soluble (nigrosin B Grübler recommended by Dorner; American nigrosins certified by Commission on Standardization of Biological Stains ordinarily satisfactory)	10 g
Distilled water	100 ml
Immerse in boiling water bath for 30 min, then add as preservative	
Formalin	0.5 ml

Filter twice through double filter paper and store in serological test tubes, about 5 ml to the tube.

This staining solution is used for the negative demonstration of bacteria, in place of the Burri india ink. For its use in Dorner's spore stain, see page 20.

Staining schedule:

1. Mix a loopful of the bacterial suspension on the slide with an equal amount of the staining solution. (If prepared from growth on solid media, the suspension must not be too heavy.)
2. Allow the mixture to dry in the air, and examine under microscope.

Results: Unstained cells in a background which is an even dark gray if the preparation is well made.

Benians' Congo Red

Congo red (80% dye content)	2 g
Distilled water	100 ml

Staining schedule:

1. Place a drop of the above staining fluid on a slide.
2. Mix culture with the drop and spread out into a rather thick film.
3. After film has dried, wash with 1 per cent HCl.
4. Dry, either in the air or by blotting.

Results: Cells unstained in a blue background. Good results are not to be expected from broth cultures or from cultures in salt solutions unless the cells are first removed by centrifuging.

THE GRAM STAIN—RECOMMENDED PROCEDURES

There are numerous modifications of the gram stain, many of which have been listed by Hucker and Conn (1923, 1927). The two modifications given below have proved especially useful to the committee. The Hucker modification is valuable for staining smears of pure cultures; that of Kopeloff and Beerman (1922) for preparations of body discharges such as gonorrhoeal pus, also for pure cultures of strongly acid-forming organisms. The latter is itself a variation of the modification by Burke (1921).

Hucker Modification

Ammonium Oxalate Crystal Violet

(*See page 13*)

Gram's Modification of Lugol's Solution

Iodine	1 g
KI	2 g
Distilled water	300 ml
Counterstain	
Safranin O (2.5% solution in 95% ethyl alcohol)	10 ml
Distilled water	100 ml

Staining schedule:

1. Stain smears 1 min with ammonium oxalate crystal violet. This formula has sometimes been found to give too intense staining, so that certain gram-negative organisms (e.g., the gonococcus) do not properly decolorize. If this trouble is encountered, it may be avoided by using less crystal violet.
2. Wash in tap water for not more than 2 sec.
3. Immerse 1 min in iodine solution.
4. Wash in tap water, and blot dry.
5. Decolorize 30 sec with gentle agitation, in 95 per cent ethyl alcohol. Blot dry.
6. Counterstain 10 sec in the above safranin solution.
7. Wash in tap water.
8. Dry, and examine.

Results: Gram-positive organisms, blue; gram-negative organisms, red.

Burke and Kopeloff-Beerman Modifications

Alkaline Gentian Violet

Solution A		*Solution B*	
Gentian or crystal violet[1]	1 g	$NaHCO_3$	1 g
Distilled water	100 ml	Distilled water	20 ml

Burke's Iodine Solution

Iodine, 1 g; KI, 2 g; distilled water, 100 ml

Kopeloff and Beerman's Iodine Solution

Iodine	2 g
Normal NaOH (40.01 g per liter)	10 ml

After the iodine is dissolved, make up to 100 ml with distilled water.

Burke's Counterstain

Safranin O (85% dye content), 2 g; distilled water, 100 ml

[1] The authors specify either crystal violet or methyl violet 6B. Probably any of the gentian violets now sold under the commission certification are satisfactory, i.e., either crystal violet or one of the bluer grades of methyl violet (e.g., methyl violet 2B).

Kopeloff and Beerman's Counterstain

Basic fuchsin (90% dye content), 0.1 g; distilled water, 100 ml

Staining schedule:

1. Dry thinly spread films in the air without heat.
2. Flood with solution A; mix on the slide with 2–3 drops (or more, depending on size of flooded area) of solution B, and allow to stand 2–3 min.

   Kopeloff and Beerman mix the two solutions in advance, 1.5 ml of solution A to 0.4 ml of solution B, and allow to stay on slide 5 min or more.
3. Rinse with either of the above iodine solutions. (The committee indicates no preference between the two; some workers prefer one, some the other.)
4. Cover with fresh iodine solution, and let stand 2 min or longer.
5. Rinse with tap water; then blot water from surface of smear, *without drying*. (Kopeloff and Beerman omit the washing.) The amount of drying is important in this step. One must get rid of all free water but not allow the cells to dry.
6. Follow the blotting very quickly with decolorization in ether and acetone (1 vol of ether to 1–3 vol of acetone), adding to the slide drop by drop until practically no color comes off in the drippings (usually less than 10 sec). In this step the speed of decolorization can be varied by varying the ratio of ether to acetone; the more acetone, the more rapid the process. It is sometimes desirable to slow down the process by using a ratio of 1:1.
7. Dry in the air.
8. Counterstain 5–10 sec in one of the above given counterstains. Burke's (i.e., safranin) is preferred. The Kopeloff and Beerman counterstain is too powerful to be used when the shorter staining time recommended by Burke is followed.
9. Wash in tap water.
10. Dry, and examine.

Results: Gram-positive organisms, blue; gram-negative organisms, red. This technic is claimed to have the advantage of not giving false positives due to vacuolar bodies that resist decolorization by other gram-staining procedures.

Interpretation of the Gram Stain

A word of caution is necessary as to the interpretation of the gram stain. The test is often regarded with unjustified finality because organisms are generally described as being either gram-positive or gram-negative. Many organisms, however, actually are gram-variable. Hence, one should never give the gram reaction of an unknown organism on the basis of a single test. He should repeat the procedure on cultures having different ages and should use more than one staining technique in order to determine the constancy of the organism toward the stain. Two phenomena deserve consideration. (1) Henry and Stacey (1943) and Bartholomew and Umbreit (1944) have shown that gram-positive organisms can be made gram-negative by treatment with ribonuclease and that their gram-positive reaction can be restored subsequently by treatment with magnesium ribonucleate. (2) Some organisms have granules which resist decolorization and which may cause misinterpretation. Such observations show that the gram stain does not always give a clear-cut reaction and that the results must be interpreted with care.

ACID-FAST STAINING—RECOMMENDED PROCEDURES

Ziehl-Neelsen Method

Ziehl (1882); Neelsen (1883)

Staining schedule:

1. Stain dried smears 3–5 min with Ziehl's carbol fuchsin (page 12), applying enough heat for gentle steaming.
2. Rinse in tap water.
3. Decolorize in 95 per cent ethyl alcohol, containing 3 per cent by volume of conc HCl, until only a suggestion of pink remains.
4. Wash in tap water.
5. Counterstain with one of the methylene blue solutions given on page 13.
6. Wash in tap water.
7. Dry, and examine.

Results: Acid-fast organisms, red; others, blue.

Gross' "Cold" Method

Of recent years an effort has been made (see Darrow, 1948; Gross, 1952) to eliminate the necessity of applying heat during the fuchsin staining so as to simplify the technic and to avoid "messy" preparations. Such procedures seem to have justified themselves and can be recommended for pure culture work; whether or not they are reliable for diagnostic purposes would require detailed comparison in actual use, and to the committee's knowledge no such comparison has been made. Gross' method is as follows:

Preparation of basic fuchsin solution:

Add 25 ml of a stock 4 per cent alcoholic basic fuchsin solution to 75 ml of 6 per cent aqueous phenol.

To this add 3–4 drops of Tergitol No. 7 (a Carbide & Chemical Corp. product), and stir thoroughly.

Preparation of methylene blue solution:

Add 30 ml of a stock 1.5 per cent alcoholic methylene blue solution to 100 ml of 0.01 per cent aqueous KOH.

Staining schedule:

1. Stain 5–10 min, without heating, in the above basic fuchsin solution.
2. Rinse in warm water.
3. Agitate for 30–60 sec in acid alcohol (3 ml of conc HCl in 97 ml ethyl alcohol).
4. Rinse with cold water.
5. Counterstain 3–5 min in the above methylene blue solution.

Note: The only essential difference between this method and Darrow's is that the latter author states that equally good results were obtained with a weaker (0.3 per cent) fuchsin solution in phenol. In the hands of one the committee's collaborators, however, Gross' 1 per cent solution has proved more satisfactory.

ACID-FAST STAINING—ALTERNATE PROCEDURES

Fluorescence Method
Richards and Miller (1941)

Although this method is not of special importance in pure culture work, special mention should be made of it because of the amount of attention now given to it in diagnostic work. Its real advantage is that it can be used with relatively low magnification, and the large fields that can be examined assure positive diagnoses in cases where the numbers of tubercle organisms are few.

Solution A		*Solution B*	
Auramine O (90% dye content)	0.1 g	Ethyl alcohol (70%)	100 ml
Liquefied phenol	3 ml	Conc HCl	0.5 ml
Distilled water	97 ml	NaCl	0.5 g

Staining schedule:

1. Stain dried smears 2–3 min in solution A.
2. Wash in tap water.
3. Destain 3–5 min in solution B, freshly prepared.
4. Dry, and examine under a monocular microscope, using 8 mm dry objective and a 20× ocular; illumination should be a low-voltage, high-amperage microscope lamp, supplied with a blue (ultraviolet-transmitting) filter, a complementary yellow filter having been provided for the ocular.

Results: Acid-fast bacteria, bright yellow, fluorescent; other organisms, not visible; background, nearly black.

Much's Method
Much (1907)

Much's method 2, which is now quite widely used, employs carbol gentian violet of essentially the formula given on page 13 for carbol fuchsin except that in the place of basic fuchsin the author calls for methyl violet BN. Preparations are stained cold for 24 hr or by gentle application of heat until steaming. They are then washed in water and treated with Lugol's iodine (see page 16) from 1 to 5 min. After a second washing they are treated with 5 per cent nitric acid for 1 min followed by 3 per cent hydrochloric acid for 10 sec. They are then decolorized 1 min in equal parts of acetone and 95 per cent ethyl alcohol. Weiss (1909) has modified this procedure by staining with a mixture of 3 parts of carbol fuchsin to 1 part of carbol gentian violet and counterstaining with 1 per cent aqueous safranin (5 to 10 sec) or with Bismarck brown (1 min). The counterstain is applied immediately after the decolorization, the acetone-alcohol being removed merely by blotting. In some laboratories this method of counterstaining is employed following the Much technic with carbol gentian violet alone for the primary stain.

Cooper's Method
Cooper (1926)

The Cooper method calls for staining in Ziehl's carbol fuchsin to which 3 per cent of a 10 per cent aqueous sodium chloride solution is added just before use. Smears are stained either by steaming 3–4 min, then allowing them to cool until a precipitate forms, or else by standing overnight in a 37° incubator and cooling in an icebox for 20 min to allow precipitation to occur. After the precipitation, the smears are washed with tap water and decolorized 1–10 min in acid alcohol (5 ml of nitric acid, sp gr 1.42, to 95 ml of 95 per cent ethyl alcohol); washed again with water and finally for 1 min with 95 per cent ethyl alcohol. They are counterstained with 1 per cent brilliant green or, if the smear is heavy, with a greater dilution of this same stain; washed with water; dried; and examined.

SPORE STAINING—RECOMMENDED PROCEDURES

Dorner's Method
Dorner (1922, 1926)

Staining schedule:

1. Make a heavy suspension of the organism in 2–3 drops of distilled water in a small test tube.
2. Add equal quantity of freshly filtered Ziehl's carbol fuchsin (page 13).
3. Allow the mixture to stand in a boiling water bath 10 min or more.
4. On a cover slip or slide mix one loopful of the stained preparation with one loopful of Dorner's nigrosin solution (page 15).
5. Smear as thinly as possible and do not dry too slowly.

Note: If even backgrounds for exhibiting or photographing are required, especially in the case of slime-producing bacteria, the following procedure is recommended:

1. Make the suspension in 0.5 ml of nutrient broth or water.
2. Add 1 ml of 10 per cent gelatin solution.
3. Add 1 ml of carbol fuchsin, and stain as in steps 1 and 2 above.
4. Wash out the colloids with warm tap water, with the help of centrifuge or sedimentation.
5. Mix with nigrosin, and proceed as above.

Results: Spores, red; vegetative cells, unstained; background, gray.

Dorner's Method—Snyder's Modification
Snyder (1934)

Staining schedule:
1. Prepare a dried smear on a slide, and cover with a small piece of blotting paper.
2. Saturate blotting paper with freshly filtered Ziehl's carbol fuchsin.
3. Allow to steam 5–10 min, keeping paper moist by adding more staining fluid.
4. For neat preparations, decolorize instantaneously with 95 per cent ethyl alcohol (but omit this step if the organisms do not hold color well).
5. Wash with tap water.
6. Apply a drop of saturated aqueous nigrosin (or Dorner's fluid), and spread evenly.
7. Allow slide to dry quickly with gentle heat, without prior washing.

Results: Same as with original method, but this modification proves applicable to some bacteria (e.g., *Bacillus subtilis*) that are difficult to stain by Dorner's technic.

Conklin's Modification of Wirtz Method
Wirtz (1908); Conklin (1934)

Staining schedule:
1. Make smears as usual and fix by heat.
2. Flood slide with 5 per cent aqueous malachite green, and steam for 10 min, keeping slide flooded by addition of fresh staining fluid.
3. Wash 30 sec in running water.
4. Counterstain 1 min with 5 per cent aqueous mercurochrome.
5. Wash in running water.
6. Blot dry, and examine.

Results: Spores, green; rest of cell, red. Trouble is sometimes experienced with the green fading after the slides have stood a few days. Apparently this is the result of an alkaline reaction and can be prevented by treating the slides in acid before making the smears. (The alkalinity may be due to an invisible film of soap or washing powder.)

SPORE STAINING—ALTERNATE PROCEDURE

Bartholomew and Mittwer's "Cold" Method

Just as recent work is showing that heat is not necessary in making an acid-fast stain, it is proving that it may also be eliminated from spore staining, in which a very similar principle is involved. The following modification of the Wirtz method by Bartholomew and Mittwer (1950) is a good illustration:

Staining schedule:

1. Fix the smear by passing through a flame 20 times.
2. Stain 10 min with saturated aqueous malachite green (i.e., about 7.6 per cent), without heat.
3. Rinse with tap water for about 10 sec.
4. Stain 15 sec in 0.25 per cent aqueous safranin.
5. Rinse, blot, and dry.

Results are the same as with the Conklin modification.

STAINING THE DIPHTHERIA ORGANISM—RECOMMENDED PROCEDURES

Various special procedures have been devised for staining the diphtheria organism in such a manner as to render it distinctive in appearance by differentiation of its characteristic metachromatic granules.

Staining with Methylene Blue

Staining schedule:

1. Prepare smear as usual, and fix with gentle heat.
2. Stain for a few seconds with either of the methylene blue solutions (i.e., Loeffler's or dilute alcoholic) given on page 13.
3. Wash in tap water.
4. Dry, and examine.

Results: Metachromatic granules, dark blue to violet; bacteria without such granules, evenly stained. The picture varies a little according to which of the two methylene blue solutions is employed. The Loeffler formula gives purplish shades of staining because of the oxidation of methylene blue caused by the alkali. Some users consider the polychrome effect thus obtained to give better differentiation; others think the metachromatic granules show more sharply with the clear blue of the unpolychromed dye.

Albert's Diphtheria Stain
Albert (1920)

Toluidine blue	0.15	g
Methyl green	0.20	g
Acetic acid (glacial)	1	ml
Ethyl alcohol (95%)	2	ml
Distilled water	100	ml

Laybourn's Modification

Laybourn (1924) has modified the Albert stain by replacing the methyl green with an equal amount of malachite green.

Staining schedule:

1. Make smears as usual, and fix with gentle heat.
2. Stain 5 min in either Albert's staining fluid or Laybourn's modification

of it. The latter is claimed to give deeper staining of both granules and body of the cells without lessening the contrast between them.

3. Drain without washing.
4. Treat 1 min in a modified Lugol's solution (iodine, 2 g; KI, 3 g; distilled water, 300 ml).
5. Wash briefly in tap water.
6. Blot with filter paper, and examine.

Results: Metachromatic granules, black; bars of diphtheria cells, dark green to black; body of cells, light green.

Ljubinsky Stain
(from Blumenthal and Lipskerow, 1905)

Original formula			*Emended formula*		
Solution A			*Solution A*		
Pyoktanin (Merck)	0.25	g	Methyl violet 2B or crystal violet (85% dye content)	0.25	g
5% acetic acid	100	ml	Glacial acetic acid	5	ml
			Distilled water	95	ml
Solution B			*Solution B*		
Vesuvin	0.1	g	Bismarck brown Y	0.1	g
Distilled water	100	ml	Distilled water	100	ml

Staining schedule:

1. Make smears as usual and fix with gentle heat.
2. Stain 30 sec to 2 min in solution A.
3. Wash in tap water.
4. Stain 30 sec with solution B.
5. Wash in tap water.
6. Dry, and examine.

Results: Metachromatic granules, dark blue or black; rest of cell, reddish or yellowish.

STAINING THE DIPHTHERIA ORGANISM—ALTERNATE PROCEDURES

Neisser's Diphtheria Stain
Neisser (1903)

Solution 1		*Solution 2*	
Methylene blue (dye content not specified; probably 90%)	1 g	Crystal violet (dye content not specified; probably 85%)	1 g
Alcohol (e.g., 95%)	20 ml	Alcohol (e.g., 95%)	10 ml
Acetic acid (glacial)	50 ml	Distilled water	300 ml
Distilled water	1000 ml	*Solution 3*	
Mix, and agitate until dye is dissolved		Chrysoidin	1 or 2 g
		Hot water	300 ml
		Filter after dissolving	

Dried films are stained 10 sec in a mixture of 2 parts of solution 1 and 1 part of solution 2. Wash. Stain 10 sec in solution 3. Wash briefly in water, or not at all. Blot dry.

Ponder's Diphtheria Stain
Ponder (1912); Kinyoun (1915)

	Original formula	As modified by Kinyoun
Toluidine blue	0.02 g	0.1 g
Azure I		0.01 g
Methylene blue		0.01 g
Glacial acetic acid	1 ml	1 ml
Ethyl alcohol (see below)	2 ml	5 ml
Distilled water	100 ml	120 ml

Dissolve the dyes in the alcohol; add the water, then the acid; and let stand 24 hr before using. Do not filter. After prolonged standing, action may be intensified by adding 1 or 2 drops of glacial acetic acid.

According to Kinyoun, smears are fixed with heat, allowed to cool, and stained 2–7 min.

In the source of the original formula above cited, absolute alcohol is specified; Kinyoun calls for 95 per cent alcohol. On theoretical grounds, indeed, absolute alcohol is not indicated, and the 95 per cent strength may well be substituted even in the original formula. Although the committee has had no personal experience with either formula, information is at hand indicating the superiority of the Kinyoun modification.

FLAGELLA STAINING—RECOMMENDED PROCEDURES

Flagella staining is a difficult technic, and there have been numerous methods proposed for the purpose. It has long been realized that flagella are actually below the visual limit in size, but of recent years the electron microscope has given a definite idea how small they really are—around 0.02–0.03 μ in diameter. Electron micrographs, in fact, often show many more flagella than do stained preparations. Until the electron microscope, however, has become a routine laboratory instrument, one must have resort to the principle introduced by Loeffler of mordanting the preparations before staining to increase the apparent size of the flagella.

A second difficulty in staining flagella is the ease with which bacteria shed these delicate appendages unless the cultures are properly handled. To prevent this one ordinarily employs specially cleaned slides and specially prepared smears on the slides.

Methods for preparing slides. Ordinary cleaning of glassware is not sufficient for the purpose. Various methods have been proposed, but the following directions seem to give as good results as any:

Use new slides if possible, preferably of Pyrex glass or similar heat-resistant properties. (This is because under the drastic method of cleaning to remove grease, old slides have a greater tendency to break.) Clean first in a dichromate cleaning fluid, wash in water, and rinse in 95 per cent alcohol; then wipe with a clean piece of cheesecloth. (Wiping is not always necessary but is advisable unless fresh alcohol is used after every few slides.) Pass each slide back and forth through a flame for some time, ordinarily until the appearance of an orange color in the flame; some experience is necessary before the proper amount of heating can be accurately judged.

Unless heat-resistant slides are used, cool slides gradually in order to minimize breakage. An ordinarily satisfactory method of doing this is to place the flamed slides on a metal plate (flamed side up) standing on a vessel of boiling water and then to remove the flame under the water so as to allow gradual cooling. (Too rapid cooling may result in breakage, sometimes as long as 2 weeks after the heating.)

Methods of handling cultures. Of various methods proposed, it is not possible to recommend any one as uniformly the best. As any laboratory worker becomes familiar with one particular method, he soon finds he can get better results with that than with any other. The following method, however, can be given as one of the most satisfactory, especially for students who have not had previous experience with some other method:

Use young and actively growing cultures (e.g., 18–22 hr old) on agar slants. Before proceeding, check the culture for motility in hanging drop. If motile, wash off the growth by gentle agitation with 2–3 ml of sterile distilled water. Transfer to a sterile test tube, and incubate at optimum temperature for 10 min (30 min for those producing slime). At this point, again check motility under a microscope. Transfer a small drop from the top of the suspension (where motile organisms are most numerous) by means of a capillary pipet to one end of the slide prepared as above described. Tilt the slide, and allow the drop to run slowly to the other end. (Two or three such streaks can be placed on a slide.) Place the slide in a tilted position, and allow it to dry in the air.

Staining Procedure

Good results can be obtained with any of the following methods, especially after familiarity has been obtained with it. Special recommendation must be given to the last of the four procedures (modified Bailey method). Although seeming a little more complicated on first reading, it has been found to give the most uniformly satisfactory results in inexperienced hands.

Casares-Gil Flagella Stain[1]
As Published by Plimmer and Paine (1921)

Mordant:

Tannic acid	10	g
$AlCl_3 \cdot 6H_2O$	18	g
$ZnCl_2$	10	g
Basic fuchsin[2]	1.5	g
Alcohol (60%)	40	ml

The solids are dissolved in the alcohol by trituration in a mortar, adding 10 ml of the alcohol first, and the rest slowly. This alcoholic solution may be kept several years. For use, mix with an equal quantity of water (Thatcher, 1926) or dilute with 4 parts of water (Casares-Gil), filter off precipitate, and collect filtrate on the slide.

Staining schedule:

1. Prepare smears of young cultures, on scrupulously cleaned slides as above directed.
2. Filter mordant onto slide as above directed (preferably using Thatcher's 1:1 dilution); allow to act for 60 sec without heating.
3. Wash in tap water.
4. Flood slide with freshly filtered Ziehl's carbol fuchsin (page 13), and allow to stand 5 min without heating.
5. Wash with tap water.
6. Air-dry, and examine. Sometimes considerable search may be needed before finding a satisfactorily stained part of the smear.

Results: Fagella well stained (red) in the case of those bacteria (e.g., colon-typhoid group, aerobic sporeformers) that do not have extremely delicate flagella.

Gray's Flagella Stain
Gray (1926)

Mordant:

Solution A

$KAl(SO_4)_2 \cdot 12H_2O$ (sat aqu solution)	5 ml
Tannic acid (20% aqu solution)	2 ml
(A few drops of chloroform must be added to this if a large quantity is made up)	
$HgCl_2$ (sat aqu solution)	2 ml

Solution B

Basic fuchsin (sat alc solution)	0.4 ml

Mix solutions A and B less than 24 hr before using. Both solutions separately may be kept indefinitely, but deteriorate rapidly after mixing.

Staining schedule:

1. Prepare smears from young cultures as above directed.
2. Flood slide with freshly filtered mordant, and allow to act 8–10 min.

[1] See Galli-Valerio (1915).

[2] The authors specify rosanilin hydrochloride. There are, however, other basic fuchsins more universally available which ought to prove equally satisfactory.

3. Wash with a gentle stream of distilled water, and follow steps 4–6 of above schedule (Casares-Gil method).

Results: Same as with Casares-Gil method.

***Leifson's Stain*[1]**
Leifson (1930)

$KAl(SO_4)_2 \cdot 12H_2O$, or $NH_4Al(SO_4)_2 \cdot 12H_2O$ (sat aqu solution)	20 ml
Tannic acid (20% aqu solution)	10 ml
Distilled water	10 ml
Ethyl alcohol, 95%	15 ml
Basic fuchsin (sat solution in 95% ethyl alcohol)	3 ml

Mix ingredients in order named. Keep in tightly stoppered bottle, and the stain may be good for a week.

Staining schedule:

1. Prepare slides as for the preceding methods.
2. Flood slides with the above solution, and allow to stand 10 min at room temperature in warm weather or in an incubator in cold weather.
3. Wash with tap water. (If a counterstain is desired, borax methylene blue may be applied, without heat, followed by another washing. See page 29.)
4. Dry and examine.

Results: When no counterstain is used, same as with the two above procedures; with methylene blue counterstain, flagella red, cells blue.

Bailey Method
Bailey (1929)
Modified by Fisher and Conn (1942)

This method is specially recommended for bacteria on which flagella are difficult to stain (as is frequently the case with soil and water non-sporeformers and with plant pathogens) because of slime production, unusually fine flagella or flagella that are readily lost.

Mordant:	
Solution A	
Tannic acid (10% aqu solution)	18 ml
$FeCl_3 \cdot 6H_2O$ (6% aqu solution)	6 ml
Solution B	
Solution A	3.5 ml
Basic fuchsin (0.5% in ethyl alcohol)	0.5 ml
HCl, concentrated	0.5 ml
Formalin	2.0 ml

Staining schedule:

1. Prepare smears of young cultures, following carefully the procedure recommended on page 25 under "Methods of handling cultures."
2. Filter the above solution A onto the slide and allow it to remain 3½ min without heating.

[1] This stain, already mixed, is now available commercially in powder form.

3. Pour off solution A, and without washing add solution B, also through a filter, and allow it to stand 7 min without heating.
4. Wash with distilled water.
5. Before the slide dries, cover with Ziehl's carbol fuchsin (page 13), allowing it to stand 1 min on a hot plate heated just enough for steam to be barely given off.
6. Wash in tap water.
7. Dry in the air, and examine.

Results: Similar to the preceding methods, but the background precipitate is usually finer and less conspicuous, thus interfering less with the demonstration of unusually fine, delicate flagella.

Staining flagella of anaerobes. O'Toole (1942) calls attention to certain difficulties in staining the flagella of anaerobes and gives a modification of the above Bailey stain which is intended to overcome them. The method is not unlike that of Fisher and Conn, who had the O'Toole procedure in mind when working out their modification.

CAPSULE STAINS—RECOMMENDED PROCEDURES

Bacterial capsules are more easily confused with artifacts than any other structure pertaining to the organisms. Inasmuch as capsules sometimes show merely as unstained areas around the cells, there is a temptation to call any such surrounding area a capsule; very often, however, they merely represent the tendency of a lightly stained surrounding medium to retract from the cells on drying. For this reason the best way to demonstrate capsules is actually to stain them by some procedure which differentiates them from the cell itself. Several of the flagella stains accomplish this, notably those of Bailey and Leifson, given above. Much simpler is the procedure of Anthony described below. The Anthony method can be recommended because of both its simplicity and its dependability. Any of the other methods which follow give satisfactory results. The student is specially urged, however, not to pronounce any organism capsulated, as a result of any of these staining procedures, until he has carefully compared it with other organisms generally recognized as having capsules.

Leifson Method
Leifson (1930)

This method is described in detail above (page 27) and does not need to be repeated here. The special methods of handling slides and cultures, outlined for flagella staining, do not need to be observed, but the following is essential:

After step 3:

4. Stain 5–10 min, without heating, in borax methylene blue (methylene blue, 90 per cent dye content, 0.1 g; borax 1 g; distilled water 100 ml).
5. Wash in tap water.
6. Dry, and examine.

Results: capsules, red; cells, blue.

Anthony's Method
with Tyler's Modification
Anthony (1931)

Original formula		*Tyler's modification*[1]	
Crystal violet (85% dye content)	1 g	Crystal violet (85% dye content)	0.1 g
Distilled water	100 ml	Glacial acetic acid	0.25 ml
		Distilled water	100 ml

Staining schedule:

1. Prepare smears, and dry them in the air.
2. Stain 2 min in the above aqueous crystal violet or, according to Tyler, 4–7 min in the above acetic crystal violet.
3. Wash with 20 per cent aqueous $CuSO_4 \cdot 5H_2O$.
4. Blot dry, and examine.

Results: capsules, blue violet; cells, dark blue.

Hiss's Method
Hiss (1905)

Original statement of formula		*Emended formula*	
Sat alc basic fuchsin or gentian violet	5–10 ml	Basic fuchsin (90% dye content)	0.15–0.3 g
Water	to make 100 ml	Distilled water	100 ml
		or	
		Crystal violet (85% dye content)	0.05–0.1 g
		Distilled water	100 ml

Staining schedule:

1. Grow organisms in ascitic fluid or serum medium, or mix with drop of serum and prepare smears from this mixture.
2. Dry smears in the air, and fix with heat.
3. Stain with one of the above solutions a few seconds by gently heating until steam rises.
4. Wash off with 20 per cent aqueous $CuSO_4 \cdot 5H_2O$.
5. Blot dry, and examine.

Results: Capsules, faint blue; cells, dark purple.

[1] See Park and Williams (1933), p. 84.

STAIN FOR FAT DROPLETS

Burdon's Method

Burdon (1946)

Staining solution: 0.3 g of Sudan black B (commission certified) in 100 ml of 70 per cent ethyl alcohol. After the bulk is dissolved, shake at intervals and allow to stand over night.

Staining schedule:

1. Prepare smears as usual from 18- to 24-hr cultures, and fix by heat.
2. Flood the entire slide with the above staining solution and allow it to stand undisturbed at room temperature for 5–15 min. (Exact time is unimportant, as good results are often obtained after only 1 or 2 min; on the other hand no harm results if the slides stain until the solution is completely dry.)
3. Drain, and blot slide completely dry.
4. Cover with xylene by pouring from a dropping bottle or dipping several times in a staining jar. Blot till dry.
5. Counterstain 5–10 sec with 0.5 per cent aqueous safranin, taking care not to overstain.

Note: For acid-fast organisms, Ziehl's carbon fuchsin diluted 1:10 with distilled water may be applied for 1–3 min, instead of safranin.

6. Wash in tap water, blot, and dry.

Results: Fat droplets blue-black or blue-gray; rest of cell pink.

ROBINOW'S STAIN FOR NUCLEAR APPARATUS

Slightly modified from Robinow (1944)

Among several methods given by Robinow for staining nuclear material the following seems as generally applicable as any:

Staining solution: Add 1 drop of Giemsa stain to 1 ml of Sorensen's buffer of pH 6.9–7.0. Robinow specifies Gurr's R66. Giemsa stain as certified by the Biological Stain Commission, however, is equally good and requires less prolonged staining; the staining time given below is that called for by the American-type Giemsa.

Fixation and smearing:

1. Incubate petri-dish cultures 2–5 hr.
2. Remove a block of agar, and fix it from a few seconds to several hours in the vapor of 2 per cent osmic acid.
3. Make an impression smear on a cover slip or glass slide.
4. Store in 70 per cent alcohol till needed.

Staining procedure:

1. Remove preparation from alcohol, and wash in water.
2. Place for 5–10 min in normal HCl at 60°C.

3. Remove, and wash three times in tap water.
4. Stain 1–15 minutes, at 37°C, in the above diluted Giemsa stain.
5. Mount in water for oil immersion examination.

Note: If it is desired to mount in balsam, the staining time must be increased to several hours.

Results: The deeper colors (blue and violet) tend to be localized in the chromatinic material comprising the nuclear structures.

STAINS FOR SPIROCHAETES—RECOMMENDED PROCEDURE

Fontana Stain

Preparation of ammoniacal silver nitrate:

Dissolve 5 g of $AgNO_3$ in 100 ml of distilled water. Remove a few milliliters, and to the rest of the solution add drop by drop a concentrated ammonia solution until the sepia precipitate which forms redissolves. Then add drop by drop enough more of the silver nitrate solution to produce a slight cloud which persists after shaking. It should remain in good condition for several months.

Staining schedule:

1. Prepare smear, and fix with heat.
2. Pour on a solution of 5 per cent tannic acid in 1 per cent phenol, and allow to steam 30 sec.
3. Wash 30 sec in running water.
4. Cover with a drop of the above ammoniacal silver nitrate, heat gently over a flame, and allow it to stand 20–30 sec after steaming begins.
5. Wash in tap water.
6. Blot dry, and examine.

Results: Spirochaetes, dark brown or black, in a dark maroon field.

STAINS FOR SPIROCHAETES—ALTERNATE PROCEDURE

Tunnicliff's Stain

Tunnicliff has employed carbol gentian violet (3 to 4 sec) followed by Lugol's iodine (see page 16) for the same period in staining bacterial smears. With a slight modification this proves a good spirochaete stain. The modification is:

Carbol crystal violet (1 vol of 10 per cent alc crystal violet to 10 vol of 1 per cent aq phenol) 30 sec; wash with water; the Lugol-Gram iodine solution 30 sec; wash with water; safranin 30 sec; wash with water and dry.

STAIN FOR RICKETTSIAE

Macchiavello's Method

Staining solution: 0.25 g of basic fuchsin (90 per cent dye content) dissolved in 100 ml of distilled water, buffered to pH 7.2–7.4 with the proper phosphate buffer mixture.

TABLE 1. DYE SOLUBILITIES AT 26°C

Color index number	Name of dye	Per cent soluble in Water	Per cent soluble in 95% alcohol
1027	Alizarin	nil	0.125
1034	Alizarin red S	7.69	0.15
40	Alizarole orange G	0.40	0.57
36	Alizarole yellow GW	25.84	0.04
184	Amaranth	7.20	0.01
847	Amethyst violet	3.12	3.66
655	Auramin O	0.74	4.49
12	Aurantia	nil	0.33
146	Azo acid yellow	2.17	0.81
88	Azo Bordeaux	3.83	0.19
448	Benzopurpurin 4B	......	0.13
280	Biebrich scarlet	......	0.05
332	Bismarck brown R	1.10	0.98
331	Bismarck brown Y	1.36	1.08
252	Brilliant croceine	5.04	0.06
29	Chromotrope 2R	19.30	0.17
21	Chrysoidin R	0.23	0.99
20	Chrysoidin Y	0.86	2.21
370	Congo red	......	0.19
89	Crystal ponceau	0.80	0.06
681	Crystal violet (chloride) } gentian violets	1.68	13.87
	Crystal violet (iodide) } gentian violets	0.035	1.78
	Cresyl violet (National Aniline)	0.38	0.25
715	Cyanole extra	1.38	0.44
771	Eosin B (Na salt)	39.11	0.75
768	Eosin Y† (Na salt)	44.20	2.18
	Eosin Y† (Mg salt)	1.43	0.28
	Eosin Y† (Ca salt)	0.24	0.09
	Eosin Y† (Ba salt)	0.18	0.06
130	Erika B	0.64	0.17
254	Erythrin X	6.41	0.06
773	Erythrosin† (Na salt)	11.10	1.87
	Erythrosin† (Mg salt)	0.38	0.52
	Erythrosin† (Ca salt)	0.15	0.35
	Erythrosin† (Ba salt)	0.17	0.04
770	Ethyl eosin	0.03	1.13
	Fast green FCF	16.04	0.35
176	Fast red A	1.67	0.42
16	Fast yellow	18.40	0.24
766	Fluorescein (color acid)	0.03	2.21
	Fluorescein (Na salt)	50.20	7.19
	Fluorescein (Mg salt)	4.51	0.35
	Fluorescein (Ca salt)	1.13	0.41
	Fluorescein (Ba salt)	6.54	0.56

TABLE 1. DYE SOLUBILITIES AT 26°C (*Continued*)

Color index number	Name of dye	Per cent soluble in	
		Water	95 % alcohol
	Fuchsin, basic:		
676	Pararosanilin (chloride)	0.26	5.93
	Pararosanilin (acetate)	4.15	13.63
	Rosanilin (chloride)	0.39	8.16
678	New fuchsin (chloride)	1.13	3.20
	Gentian violet (see crystal or methyl violet)		
666	Guinea green B	28.40*	7.30
1180	Indigo carmine	1.68	0.01
133	Janus green	5.18	1.12
670	Light green SF yellowish	20.35	0.82
657	Malachite green (oxalate)	7.60	7.52
9	Martius yellow, Na salt	4.57	0.16
	Martius yellow, Ca salt	0.05	1.90
138	Metanil yellow	5.36	1.45
142	Methyl orange	0.52	0.08
	Methyl orange (acid)	0.015	0.015
680	Methyl violet (gentian violet)	2.93	15.21*
922	Methylene blue ($ZnCl_2$ double salt)	2.75	0.05
	Methylene blue (chloride)	3.55	1.48
	Methylene blue (iodide)	0.09	0.13
924	Methylene green	1.46	0.12
10	Naphthol yellow G	8.96	0.025
152	Narcein	10.02	0.06
825	Neutral red (chloride)	5.64	2.45
	Neutral red (iodide)	0.15	0.16
826	Neutral violet	3.27	2.22
927	New methylene blue N	13.32*	1.65
728	New Victoria blue R	0.54	3.98
520	Niagara blue 4B	13.51	nil
914	Nile blue 2B	0.16	0.62
73	Oil red O	nil	0.39
150	Orange I	5.17	0.64
151	Orange II	11.37	0.15
27	Orange G	10.86	0.22
714	Patent blue A	8.40	5.23
774	Phloxine† (Na salt)	50.90*	9.02
	Phloxine† (Mg salt)	20.84	29.10
	Phloxine† (Ca salt)	3.57	0.45
	Phloxine (Ba salt)	6.01	1.17
7	Picric acid	1.18	8.96
28	Ponceau 2G	1.75	0.21
186	Ponceau 6R	12.98	0.01
741	Pyronin B (iodide)	0.07	1.08
739	Pyronin Y	8.96	0.60

TABLE 1. DYE SOLUBILITIES AT 26°C (*Continued*)

Color index number	Name of dye	Per cent soluble in	
		Water	95% alcohol
148	Resorcin yellow	0.37	0.19
749	Rhodamine B	0.78	1.47
750	Rhodamine G	1.34	6.31
779	Rose bengal† (Na salt)	36.25	7.53
	Rose bengal† (Mg salt)	0.48	1.59
	Rose bengal† (Ca salt)	0.20	0.07
	Rose bengal† (Ba salt)	0.17	0.05
831	Safranin	5.45	3.41
689	Spirit blue	nil	1.10
24	Sudan I	nil	0.37
248	Sudan III	nil	0.15
258	Sudan IV	nil	0.09
920	Thionin	0.25	0.25
925	Toluidine blue O	3.82	0.57
690	Victoria blue 4R	3.23	20.49
659	Victoria green 3B	0.04	2.24
8	Victoria yellow	1.66	1.18

* These figures are grams per hundred grams of saturated solution (the others being grams per hundred milliliters).

† The color acids of these dyes (not listed here) are practically insoluble in water.

NOTE: These figures are ordinarily for recrystallized dyes. Commercial samples are generally less soluble often by as much as 30 per cent.

SOURCE: Based on data obtained at the Color Laboratory of the U. S. Department of Agriculture. See Conn (1953), pp. 289–290.

Staining schedule:

1. Smear a bit of tissue on a slide.
2. Dry in the air, and fix with gentle heat.
3. Pour the above staining fluid onto the slide through a coarse filter paper. Allow to stand 4 min.
4. Rinse very rapidly with 0.5 per cent aqueous citric acid.
5. Wash quickly and thoroughly with tap water.
6. Counterstain about 10 sec with 1 per cent aqueous methylene blue.
7. Rinse in tap water.
8. Dry, and examine.

Results: Rickettsiae, red; cell nuclei, deep blue; cytoplasm, light blue.

REFERENCES

Albert, Henry. 1920. Diphtheria bacillus stains with a description of a "new" one. *Am. J. Public Health,* **10,** 334–337.

———. 1921. Modification of stain for diphtheria bacilli. *J. Am. Med. Assoc.,* **76,** 240.

Anthony, E. E. 1931. A note on capsule staining. *Science,* **73,** 319.

Bailey, H. D. 1929. A flagella and capsule stain for bacteria. *Proc. Soc. Exptl. Biol. Med.,* **27,** 111–112.

Bartholomew, J. W., and Tod Mitwer. 1950. A simplified bacterial spore stain. *Stain Technol.,* **25,** 153–156.

Bartholomew, J. W., and W. W. Umbreit. 1944. Ribonucleic acid and the Gram stain. *J. Bacteriol.,* **48,** 567–578.

Benians, T. H. C. 1916. Relief staining for bacteria and spirochaetes. *Brit. Med. J.* 1916 (2), 722.

Blumenthal, J. M., and M. Lipskerow. 1905. Vergleichende Bewertung der differentiellen Methode zur Färbung des Diphtheriebacillus. *Centr. Bakteriol.,* I Abt., Orig., **38,** 359–366.

Burdon, Kenneth L. 1946. Fatty material in bacteria and fungi revealed by staining dried, fixed slide preparations. *J. Bacteriol.,* **52,** 665–678.

Burke, Victor. 1921. The Gram stain in the diagnosis of chronic gonorrhea. *J. Am. Med. Assoc.,* **77,** 1020–1022.

———. 1922. Notes on the Gram stain with description of a new method. *J. Bacteriol.,* **7,** 159–182.

Conklin, Marie E. 1934. Mercurochrome as a bacteriological stain. *J. Bacteriol.,* **27,** 30.

Conn, H. J. 1953. "Biological Stains," 6th ed. Biotech Publications, Geneva, N. Y.

———, and Mary A. Darrow. 1943–1945. "Staining Procedures." Biotech Publications, Geneva, N. Y.

Cooper, F. B. 1926. A modification of the Ziehl-Neelsen staining method for tubercle bacilli. *Arch. Pathol. Lab. Med.,* **2,** 382–385.

Darrow, Mary A. 1948. Staining of tubercle organism in sputum smears. *Stain Technol.,* **24,** 93–94.

Dorner, W. C. 1922. Ein neues Verfahren für isolierte Sporenfärbung. *Landwirtsh. Jahrb. Schweiz.,* **36,** 595–597.

———. 1926. Un procédé simple pour la coloration des spores. *Lait,* **6,** 8–12.

———. 1930. The negative staining of bacteria. *Stain Technol.,* **5,** 25–27.

Fisher, P. J., and Jean E. Conn. 1942. A flagella staining technic for soil bacteria. *Stain Technol.,* **17,** 117–121.

Fontana, Artur. 1912. Verfahren zur intensiver und raschen Färbung des Treponema pallidum und anderer Spirochäten. *Derm. Wochschr.,* **55,** 1003–1004.

Galli-Valerio, B. 1915. La méthode de Casares-Gil pour la coloration des cils des bactéries. *Cent. Bakteriol.,* I Abt. Orig., **76,** 233–234.

Gray, P. H. H. 1926. A method of staining bacterial flagella. *J. Bacteriol.,* **12,** 273–274.

Gross, Milton. 1952. Rapid staining of acid fast bacteria. *Am. J. Clin. Pathol.,* **22,** 1034–1035.

Henry, H., and M. Stacey. 1943. Histochemistry of the Gram-staining reaction for micro-organisms. *Nature,* **151,** 671.

Hiss, P. J., Jr. 1905. A contribution to the physiological differentiation of Pneumococcus and Streptococcus, and to methods of staining capsules. *J. Exptl. Med.,* **6,** 317–345.

Hucker, G. J. 1922. Comparison of various methods of Gram staining. (Preliminary Report.) *Abstr. Bacteriol.,* **6,** 2.

———, and H. J. Conn. 1923. Methods of Gram staining. *N.Y. State Agr. Expt. Sta Tech. Bull.* 129.

———. 1927. Further studies on the methods of Gram staining. *N.Y. State Agr. Expt. Sta. Tech. Bull.* 128.

Kinyoun, J. J. 1915. A modification of Ponder's stain for diphtheria. *Am. J. Public Health,* **5,** 246–247.

Kopeloff, N., and P. Beerman. 1922. Modified Gram stains. *J. Infectious Diseases,* **31,** 480–482.

Laybourn, R. L. 1924. A modification of Albert's stain for the diphtheria bacilli. *J. Am. Med. Assoc.,* **83,** 121.

Leifson, Einar. 1930. A method of staining bacterial flagella and capsules together with a study of the origin of flagella. *J. Bacteriol.,* **20,** 203–211.

Loeffler, F. 1884. Untersuchungen über Bedeutung der Mikroorganismen für die Entstehung der Diphtherie beim Menschen, bei der Taube und beim Kalbe. *Mitt. Gesundheitsamte,* **2,** 421–499. (See p. 439.)

Much, H. 1907. Über die granuläre, nach Teil nicht färbbare Form des Tuberkulosevirus. *Beitr. Klin. Tuberk.,* **8,** 85–99.

Neelsen, F. 1883. Ein casuistischer Beitrag zur Lehre von der Tuberkulose. *Centr. Med. Wisse.,* **21,** 497–501. (See p. 500.)

Neisser, M. 1903. Die Untersuchung auf Diphtheriebacillen in centralisierten Untersuchungsstationen. *Hyg. Rundschau.,* **13,** 705–717.

Nicolle, Ch. 1895. Pratique des colorations microbiennes. *Ann. inst. Pasteur.,* **9,** 664–670.

O'Toole, Elizabeth. 1942. Flagella staining of anaerobic bacilli. *Stain Technol.,* **17,** 33–40.

Park, W. H., and Anna W. Williams. 1933. "Pathogenic Microorganisms." 10th ed. Lea & Febiger, Philadelphia.

Plimmer, H. G., and S. G. Paine. 1921. A new method for the staining of bacterial flagella. *J. Pathol. Bacteriol.,* **24,** 286–288.

Ponder, C. 1912. The examination of diphtheria specimens. A new technique in staining with methylene blue. *Lancet,* **2,** 22–23.

Richards, O. W., and D. K. Miller. 1941. An efficient method for the identification of *M. tuberculosis* with a simple fluorescence microscope. *Am. J. Clin. Pathol.,* **11,** 1–7.

Robinow, C. F. 1944. Cytological observations on *Bact. coli, Proteus vulgaris* and various aerobic spore-forming bacteria, with special reference to the nuclear structures. *J. Hygiene,* **43,** 413–423.

Snyder, Marion A. 1934. A modification of the Dorner spore stain. *Stain Technol.,* **9, 71.**

Thatcher, Lida M. 1926. A modification of the Casares-Gil flagella stain. *Stain Technol.,* **1,** 143.

Tunnicliff, Ruth. 1922. A simple method of staining Gram-negative organisms. *J. Am. Med. Assoc.,* **78,** 191.

Wirtz, R. 1908. Ein einfache Art der Sporenfärbung. *Centr. Bakteriol.,* I Abt. Orig., **46,** 727–728.

Ziehl, F. 1882. Zur Färbung des Tuberkelbacillus. *Deut. med. Wochschr.,* **8,** 451.

CHAPTER III

Preparation of Media

G. W. BURNETT, M. J. PELCZAR, JR., AND H. J. CONN

INTRODUCTORY

Scope

The presentation of the data in this chapter is an attempt to describe growth media which have proved to be of general value to bacteriologists. The media to be discussed are usually employed for the isolation and maintenance of pure cultures and for the identification of species according to physiological properties. No group of bacteria will receive particular attention except to the extent that data will naturally be more complete concerning media for groups rather thoroughly described. The information presented herein is intended primarily as an introduction to the description of growth media for those unfamiliar with the problems of bacterial cultivation. It is realized that the experienced specialist may employ media considerably different from those described below, but the fundamental ideas for a given medium would probably be similar to those presented here.

For the purposes of this chapter, the media included are classified as follows: "Cultivation and Storage Media," "Enrichment Media," "Differential Media," "Media for Determination of Physiological Properties," "Media for Specific Bacteriological Procedures," "Media for Special Purposes." It is obvious that all previously described media will not be included; some time ago Levine and Schoenlein (1930) compiled a few thousand of the published formulas. Suggestions for media of particular value in clinical diagnostic laboratory procedures may be found in Hitchens (1945), Gradwohl (1948), Marshall *et al.* (1947), Schaub and Foley (1952), Simmons and Gentzkow (1955), Stitt, Clough, and Branham (1948), Wadsworth (1947), the laboratory manuals of the U.S. Army, Navy, and other such sources. Very valuable information is available also in the manuals and catalogues of the companies which prepare dehydrated media; these catalogues are available free of charge directly from the companies.

Bacterial Growth Requirements

All living organisms require a utilizable source of energy in order to grow. Those using radiant energy are known as *phototrophs*, whereas those utilizing the chemical energy liberated from oxidation-reduction reactions are referred to as *chemotrophs*. In addition to an energy source, all living organisms require suitable carbon and nitrogen sources, as well as inorganic salts. The *autotrophic* organisms can utilize CO_2 as the sole carbon source, whereas *heterotrophic* organisms, although they may need CO_2, also require carbon sources more complex than CO_2, either for the carbon skeleton proper or for the hydrogen atoms linked to this skeleton, or both. The requirement for nitrogen may be satisfied in the form of NH_4^+, NO_3^-, or N_2, although many organisms need complex organic nitrogenous compounds for this purpose. Such elements as Na, K, Ca, Mg, Mn, Fe, Zn, Cu, S, P, Cl, etc., are required for growth and are utilizable generally in the form of inorganic salts. In addition, many organisms require growth factors—organic substances which the organisms cannot synthesize at a significant rate and which are usually required in small amounts.

Many bacteria require atmospheric oxygen for growth (*obligate aerobes*), whereas others fail to grow in the presence of oxygen (*obligate anaerobes*); some can grow under either set of conditions (*facultative organisms*), whereas still others can grow only under low oxygen tension (*microaerophiles*). The relationship of a given organism to oxygen is a manifestation of the oxidation-reduction potential range commensurate with the physiological activity of that organism (see Chap. IV). The requirement for oxygen by some bacteria can be satisfied by oxidizing agents such as NO_3^-, SO_4^-, etc. On the other hand, obligate anaerobes can be grown in the presence of oxygen provided a sufficiently low O/R potential is obtained by the addition of reducing agents or by permitting the formation of reduced products (see Chap. V). Thus, although all the physiological phenomena associated with the oxygen relationship have not been explained completely, this relationship must be considered in order to obtain vigorous bacterial growth.

Most bacteria grow well only within a limited pH range. To maintain this range, at least during the initial growth buffers are added to the medium; these buffers may be of the types described in Chap. IV, or for the neutralization of acids, $CaCO_3$ may be included in the medium. Finally, to be available to the organism, all the components necessary for growth must be in aqueous solution.

Bacteria apparently display all possible variations of the major nutritional requirements. The autotrophs and many heterotrophs can be grown in chemically simple media of defined composition (*synthetic media*). By this means, qualitative and quantitative assay of each ingredient of such media can be made, leading to the microbiological assay of vitamins, amino acids, carbohydrates, inorganic ions, etc. Because of our imperfect knowledge of the exact nutritional requirements of most bacteria, the comparatively few truly synthetic media have been devised chiefly for the cultivation of autotrophs or in connection with specific research problems involving the growth of certain heterotrophs. However, the large number of heterotrophic bacteria familiar to most bacteriologists continue to be grown in rather complex media. To a great extent this complexity is associated merely with the empirical manner in which bacteriology has been practiced since the days of Pasteur and probably can be eliminated as exact nutritional information becomes available. A sort of tradition has developed in regard to the complex growth media made from natural plant and animal materials, which having been employed for many years remain as the familiar means of growing bacteria.

The purpose of this brief review of bacterial nutrition is to stimulate reflection on the part of the laboratory investigator regarding the cultivation of bacteria. Growth media selected arbitrarily or according to tradition may be less adequate for many purposes than media whose composition has been determined according to the nutritional rationale; indeed, the latter approach will more certainly lead to a clearer understanding of the nature of the microorganism being cultivated.

General Procedures in Media Preparation

In selecting the components of a medium, it is desirable to employ substances of defined composition, purity, or mode of preparation. In the preparation of synthetic media, known chemical compounds of chemically pure (C.P.) or reagent grade should be used. It is less easy to define the composition of some of the complex substances used in certain formulas.

Peptones and gelatin should conform to the minimal standards given in the United States Pharmacopeia XV. Gelatin, in addition, should meet the additional specifications as set forth in the Military Medical Purchase Description, ASMPA 1-212-000. Tentative specifications for bacteriological-grade agar have been proposed. The specifications referred to above for agar, gelatin, and some peptones are included in the appendix to this chapter. In addition to chemical analysis, the growth response of selected test organisms is used in designating a product as satisfactory for bacteriological use. Distilled water is used generally to dissolve the components of the medium, although it is obviously unnecessary when crude plant or animal extracts are employed. Many media, especially those prepared from natural products, are turbid or become turbid upon heating and require filtration through paper or other agents prior to sterilization.

The reaction of most bacteriological media is usually adjusted to a hydrogen-ion concentration near neutrality. Since sterilization by wet heat usually causes a drop of pH of about 0.2–0.4 unit, it is necessary to adjust the pH of the medium *before* sterilization to a value higher by the amount indicated. Although solutions of natural extracts often do not require pH adjustment, each batch of medium should be tested to determine if the desired pH is obtained. Detailed instructions for testing and adjusting the reaction of media are presented in Chap. IV. For ordinary purposes, satisfactory results will be obtained by adjusting the medium to slightly above (pH 7.2–7.4) the neutral point with bromthymol blue, the aqueous solution of the sodium salt or the alcoholic (95 per cent ethanol) solution of the unneutralized indicator (Table 2) being employed. A comparator block containing either standard solutions of known pH values and colors or a comparator consisting of colored glass standards for the corresponding pH and indicator color values should be used to determine pH. For deeply colored media, a glass electrode pH meter should be used.

TABLE 2. ACID-BASE INDICATORS

Indicator	Concentration recommended, %*	Sensitive pH range	Full acid color	Full alkaline color
Bromphenol blue	0.04	3.1–4.7	Yellow	Blue
Bromcresol green	0.04	3.8–5.4	Yellow	Blue
Methyl red	0.02	4.2–6.3	Red	Yellow
Bromphenol red	0.04	5.2–6.8	Yellow	Red
Bromcresol purple	0.04	5.4–7.0	Yellow	Purple
Bromthymol blue	0.04	6.1–7.7	Yellow	Blue
Phenol red	0.02	6.9–8.5	Yellow	Red
Cresol red	0.02	7.4–9.0	Yellow	Red
Thymol blue (alk. range)	0.04	8.0–9.6	Yellow	Blue
Phenolphthalein	0.10	8.3–10.0	Colorless	Red

* Stock solutions in 95 per cent ethanol for the indicator acids or in water for the indicator salts. See Chap. IV, Table 6, for details of preparation.

Buffers are often required in medium to facilitate growth. This is particularly true of media composed of simple compounds or in which acid-producing bacteria are cultivated. Mixtures of sodium and potassium phosphates are generally employed (see Chap. IV), although dibasic phosphate is also used singly. Buffer requirements will be indicated in the media described below.

To determine pH changes during growth, indicators may be included in the medium. For this purpose, an indicator covering the desired range (Table 2) is selected and 1-2 ml of a 1–2 per cent alcoholic solution is added to each liter of medium before sterilization. Although litmus is an insensitive pH indicator, it has the advantage of also indicating major changes in oxidation-reduction potential and is thus useful to show these changes, particularly in milk media.

General directions for media sterilization will not be presented here, since these are available in many bacteriological laboratory guides. A few critical points will, however, be emphasized. Most tubed media are safely sterilized in an autoclave, using 121°C (250°F) for 15 min, care being taken not to pack containers tightly; media in large flasks or bottles require a longer sterilization time (15–30 min). The autoclave should reach this temperature rapidly (within 10 min) and should cool down just as rapidly after interrupting the steam flow. It is to be noted that *temperature* is the important factor in heat sterilization, pressure being merely the means of obtaining water at elevated temperatures. Thus, during sterilization the temperature must be observed (a pressure gauge is not reliable as an indicator of temperature) and used as the criterion of an adequate level of heat. It should be realized that overheating a medium may lead to modifications of its composition. In general, maintaining most media at 121°C for 15 min will not cause important changes. The chemical composition of many substances, particularly carbohydrates, is changed, however, even by this limited heat treatment (Davis and Rogers, 1939). For critical study, such substances should be sterilized separately by filtration through bacteriological filters (Seitz or sintered glass) and added aseptically to the remainder of the medium previously sterilized by heat. If filtration facilities are lacking, concentrated stock solutions (20–25 per cent) of such substances can be sterilized by heat in the autoclave and then dispensed aseptically to the desired final concentration. In cases where conclusions based on comparative studies of the sterilization modifications described above are not available, the effect of heat upon the medium must be considered in each new investigation.

Dehydrated media refer to powdered, water-soluble commercial products which yield a growth medium. Usually all that is required in the preparation is to dissolve the proper amount of the powder (according to the directions accompanying the medium), to dispense as desired, and to sterilize. A wide variety of types, each suitable for a specific purpose, is available, and these have been found to be adequate for bacteriological use. Indeed, in most instances, the greater uniformity of these products over that attained by preparation of individual batches prepared in the laboratory from separate ingredients indicates their desirability for comparative work. No attempt will be made here to describe each of these dehydrated media (in some instances formulas for preparation of media will be given even though satisfactory dehydrated products are available), but those persons desiring to use these products should seek information from the manufacturing companies. Companies in this country which specialize in dehydrated media preparation include (1) Albimi Laboratories, Inc., 16 Clinton Street, Brooklyn 1, New York, (2) Baltimore Biological Laboratories, 1640 Gorsuch Avenue, Baltimore 18, Maryland, and (3) Difco Laboratories, Inc., 920 Henry Street, Detroit 1, Michigan. Products such as peptone, beef or yeast extract, agar, etc., are also available from other supply houses not specializing in dehydrated media preparation.

CULTIVATION AND STORAGE MEDIA

The media to be described in this section will include the formulas for various complex, nonsynthetic media which may be used for the general cultivation of bacteria, either from a natural sample or after a pure culture has been obtained. No attempt will be made to designate any one medium as the standard for a particular purpose, but it may be noted that for certain purposes (for example, estimation of organisms present in water and milk) "standard media" have been designated by other organizations (American Public Health Association, 1946, 1948).

The peptone[1] listed as an ingredient in some of the formulas is a product derived by digestion of proteinaceous materials of either plant or animal origin, by use of acid, alkali, or added or natural proteolytic enzymes (Asheshov, 1941; Brewer, 1943; Gladstone and Fildes, 1940; Leifson, 1943; Mueller and Johnson, 1941). Since the composition of peptone varies with the origin and the method of preparation, not all types may be suitable in all instances; any type of bacteriological grade found to give best results for a particular purpose may be used. Data available at present indicate that peptone prepared by pancreatic digestion of casein (for example, B.B.L. "trypticase" and Difco "casitone") will often contain growth-promoting substances, required by many organisms, which are not found in some other peptones. In the formulas listed below in which peptone is specified, a particular type is indicated in a few instances; when this procedure is not followed, the worker may choose the type giving more satisfactory results.

Agar,[1] a complex carbohydrate refined from marine algae, is the usual agent for solidification of media. This material should be free from starch and debris and capable of producing a clear solution when hot; the exact concentration to be used to give the desired degree of solidity may vary with the degree of purification, although usually 1.5 per cent is sufficient. Agar media should not be adjusted to a pH lower than 6.0 prior to sterilization, since the agar is hydrolyzed under these conditions. When such agar media are required, the pH is adjusted after sterilization by the aseptic addition of acid.

All laboratory-prepared *infusions*, especially those from meat, should be checked microscopically to assure freedom from bacterial growth, especially if the infusion is held for even a few hours at temperatures which will permit microbial reproduction.

Beef-extract peptone broth (often called *nutrient broth*) ordinarily has the following composition: beef extract, 3 g; peptone, 5.0 g; distilled water, 1,000 ml. The water is heated to 60°C to promote solution of ingredients; after cooling, the pH is adjusted to 7.0–7.2. After dispensing in tubes or other containers, it is autoclaved at 121°C for 15 min. This medium, still used quite widely for the general cultivation of aerobic organisms and as a basal medium for a variety of physiological tests, is now recognized to be nutritionally inadequate for many types of fastidious organisms. In many instances addition of 5 g of yeast extract will support growth of such types. For *beef-extract agar* (*nutrient agar*), 1.5 per cent of agar is added to the above medium before dispensing and autoclaving.

[1] See appendix to this chapter for specifications of some peptones and agar.

Thioglycollate broth, originally used for the growth of anaerobic bacteria, is now being used also for other types with regard to oxygen relationships. The liquid medium may be prepared as follows: Dissolve 15 g of peptone (preferably a pancreatic digest of casein), 5 g of glucose, 5 g of yeast extract, 0.75 g of L–cystine, 2.5 g of NaCl, 0.75 g of agar, and 0.1–0.5 g of sodium thioglycollate in 1 liter of water; adjust pH to 7.2. Sterilize for 15 min at 121°C. The small amount of agar which is necessary to maintain a low oxidation-reduction potential does not affect appreciably the fluidity of the medium. This medium is unsatisfactory for stock cultures unless $CaCO_3$ is added. If a dye to indicate the O/R potential is desired, add 0.002 g of methylene blue or 0.001 g of resazurin per liter. If the medium to be used contains an indicator dye, for obligate anaerobes, the dye should show the medium to be reduced except in the upper layer. Therefore, unless the dye indicates that the medium has been reoxidized to a considerable extent, it is unnecessary to follow the usual practice of heating the medium for exhaustion of oxygen immediately prior to inoculation. This medium should not be stored in the refrigerator after preparation, as it absorbs more oxygen at lower temperatures. For special purposes, such as the sterility testing of biological products in which inactivation of mercurials presents a problem, dehydrated media are available which meet the specifications of the National Institutes of Health and other agencies.

Sodium caseinate agar. This medium is often used for the enumeration of bacteria, including the Actinomycetes, in soil. It is prepared as follows: sodium caseinate, 1.0 g; glucose, 1.0 g; $MgSO_4$, 0.2 g; K_2HPO_4, 0.2 g; $FeSO_4$, trace; distilled water, 1,000 ml; agar, 15 g. It is adjusted to pH 7.0. Before pouring plates or dispensing in other containers, it is shaken gently to disperse precipitate.

Infusion broths may be prepared by extraction of either lean skeletal muscle or heart muscle in the cold or by heat. In the former case the following is typical: Add 1,000 ml of water to 400–600 g of lean veal or beef tissue which has been finely ground after removing as much fat as possible. Allow to infuse overnight at refrigerator temperature, remove scum of fat, squeeze infusion through muslin cloth, and restore volume to 1,000 ml. Add 5 g of peptone, and heat for 20 min at 100°C; filter through paper; adjust pH to desired value. For infusions prepared by heat, use the same proportion of meat to water and boil over free flame for 15 min, with or without a previous overnight infusion period in a refrigerator. Add peptone, and if desired as the basal medium for blood agar, add 5 g NaCl to 1,000 ml, and continue as above. For aerobic organisms adjust the pH to 7.0–7.2, but for anaerobic types adjust to 7.6 and tube the liquid deep over a 1- to 2-cm column of desiccated tissue particles saved from the infusion. Autoclave for 20 min at 121°C. The use of infusions is diminishing owing to their replacement by media prepared from peptones or other materials of more uniform composition, better growth-promoting properties, and less complicated preparation details.

Beef-liver infusion, used principally for anaerobic types, is prepared as follows: Remove fat from 500 g of fresh beef liver, grind, and heat, with occasional stirring, in 1,000 ml of tap water for 1 hr in flowing steam. Cool, and strain through cheesecloth. Restore filtrate to original volume, and add 1 per cent of peptone and 0.1 per cent of K_2HPO_4. Dry the tissue (at 55°C if possible) rapidly. Tube broth over several chunks of tissue. Use the broth (before addition of peptone and phosphate) in the original strength or diluted five times. Sterilize 20 min at 121°C. Avoid longer heating of medium, for this diminishes its value with respect to initiation of growth from small inocula.

Brain medium is used to stimulate spore production by many clostridia and hence is a valuable stock culture medium. The blackening reaction produced by certain species has some diagnostic value (Hall and Peterson, 1924). The medium is prepared as follows: Secure fresh sheep (or calf) brains which are as free as possible from injury. Using forceps, remove blood and membranous material from brain tissue. Add distilled water, in the ratio of 100 ml of water to 100 g of brain, and boil slowly for ½ hr. Put brains through potato ricer. Add 1.0 per cent of peptone and 0.1 per cent of glucose to the resulting mixture, and heat slightly to dissolve peptone. Tube in deep columns while the mixture is stirred in order to effect an even distribution of the brain tissue. It is sometimes recommended that reduced iron ("iron reduced by hydrogen," Merck & Co., Rahway, New Jersey), a thin strip of iron, or iron wire be added to the tube before tubing the liquid mixture. Sterilize in autoclave for 30 min at 121°C and *check sterility* by incubation at 37°C for a minimum of 24 hr. The finished medium should have approximately an equal amount of liquid broth above the brain particles. Proteolysis is indicated by putrefactive odors, a disintegration of the particles, and a blackening reaction.

Yeast infusion may be prepared by several procedures. A satisfactory one follows: Obtain fresh yeast, preferably starch-free, and add 10 per cent by weight to several liters of tap water. Autoclave for 3 hr or more. Allow cells to settle by standing for several days at room temperature. Remove liquid infusion by siphon or with the Sharples centrifuge. Sterilize the liquid, after removal from the cells, in screw-capped bottle, and store indefinitely. If fresh yeast is not available, or if a simpler procedure is preferred, a similar medium may be prepared by adding 0.5 per cent of dehydrated yeast extract to distilled water. Care should be observed in selection of the yeast extract, since not all are equal in growth-promoting properties and some contain intact yeast cells which may be misleading in microscopic preparations.

Semisolid agar. Some organisms, especially microaerophiles, are cultivated more successfully in semisolid media in which agar, in concentrations varying from 0.2 per cent to 0.5 per cent according to the purpose for which the medium is intended, is added to a suitable base medium. Such a medium may be useful also for determination of motility and for fermentation reactions. To aid in reducing the degree of oxidation of the medium the lower concentration of agar is sufficient; for determination of motility, concentrations approaching 0.5 per cent are necessary in order to stiffen the medium sufficiently to show distinctly the hazy zone characteristic of motile bacteria (Tittsler and Sandholzer, 1936). Since the concentration for this purpose is critical and the percentage to be used will vary with the purity of the agar, the exact concentration to be used must be tested for each batch of agar and each lot of medium checked with a known motile culture.

Silica gel. It is sometimes desired to cultivate bacteria on inorganic gels to avoid unknown and possibly undesirable chemical contaminants in agar, to avoid liquefaction of agar by certain types, and to eliminate agar in the cultivation of certain autotrophic bacteria. Sterges (1942*a* and *b*) published full details (which cannot be condensed here) of preparation of such media based on the reaction between sodium silicate and hydrochloric acid. A simplified technic recommended by Ingelman and Laurell (1947) follows: Mix 1 vol of ortho-silicic acid tetraethyl ester, $Si(OC_2H_5)_4$, with 1 vol of ethanol; add 6 vol of boiled water a little at a time with stirring. Remove turbidity by centrifuging, and dispense in tubes or plates. Sterilize at 120°C for 30–40 min in autoclave, during which time gel formation occurs. Cool slowly, and remove ethanol by flooding with sterile water. Remove water, and replace with suitable nutrient solution. Reautoclave if necessary for sterility. Temple (1949) recommends a commercially available, highly purified colloidal silica preparation such as Ludox (duPont); for use a 30 per cent aqueous solution is diluted to 10 per cent, nutrients are added, pH adjusted, it is dispensed in petri dishes and autoclaved.

Storage media include those media in which bacteria are stored in "stock culture" condition for indefinite periods to provide a source of viable cultures as needed. The medium to be used will vary with the species to be maintained. A critical factor to be considered is whether or not the presence of a fermentable carbohydrate reduces viability; frequently 2 per cent $CaCO_3$ (sterilized separately by dry heat) is added to a medium to neutralize the acids produced by an organism which will not grow well in the absence of a fermentable carbohydrate. After incubation in the chosen medium at the optimum temperature for a period allowing approximately half-maximal growth, stock cultures should be stored at refrigerator temperature between transfer periods, which may vary from two to several weeks according to the longevity of the types under study. Screw-capped tubes or small bottles may be used to minimize contamination and drying of medium during the storage period. Some workers (see McClung, 1949) prefer to cover the growth with a layer of mineral oil (sterilized in shallow layers for 2 hr at 160°C with dry heat on three successive days). In some laboratories, stock cultures are prepared as agar slant cultures, but usually the stab technique is employed. Spore-producing types may be kept in stock condition in the form of spore suspensions; often these are dried on sterile soil. McClung (1949) gives references on this technic and on details for lyophilization of all types.

For those organisms which will grow on nutrient agar, peptone (1–2 per cent) agar, or yeast-infusion-glucose (0.5 per cent) agar, these media are generally used. For maintenance of *Acetobacter* species, Vaughn (1942) recommends agar containing yeast infusion, glucose, calcium carbonate, or liver-infusion broth prepared as above but diluted with an equal volume of water before the addition of peptone and phosphate. For aerobic nitrogen-fixing types (*Rhizobium*, *Azotobacter*), the yeast-extract-mannitol agar of Fred, Baldwin, and McCoy (1932) is prepared as follows: mannitol, 10.0 g; K_2HPO_4, 0.5 g; $MgSO_4$, 0.2 g; NaCl, 0.2 g; $CaCl_2$, 3.0 g; yeast infusion (pH 6.8), 100 ml; distilled water, 900 ml; agar, 15.0 g. The following medium is recommended for *Neisseria* by Vera (1948) and can be used for a variety of genera: peptone (pancreatic digest of casein), 20.0 g; cystine, 0.5 g; Na_2SO_3, 0.5 g; NaCl, 3.0 g; agar, 3.5 g; distilled water, 1,000 ml; pH 7.3.

In addition to the above, many formulas will be found in the literature which are of special value for particular types; dehydrated media are available for this purpose also.

ENRICHMENT MEDIA

The enrichment culture technic constitutes a means for the isolation of a wide variety of bacteria by adjusting the nutritional environment in such a manner as to enhance selectively the growth of a certain bacterial type within a given mixed inoculum. Use of this simple methodological approach, so ably exploited by Beijerinck (1921–1940), has assured the ready isolation of nearly all types of bacteria (van Niel, 1949) and constitutes a powerful tool for the bacteriologist in the isolation and identification of pure cultures from an initially mixed population. Even fastidious pathogenic bacteria can be isolated in this manner, using for this purpose the animal body as a selective medium. In addition, selection of the optimum temperature and optimum pH for growth aids in the cultivation of certain bacterial types, as does the proper adjustment with regard to oxygen relationship or O/R potential.

In this discussion, a few examples will be presented of the media used for enrichment cultures; detailed information concerning many other media similarly employed will be found in the literature dealing with specific groups. It is to be noted that certain of the formulas given in the section on "Cultivation and Storage Media" and on "Differential Media" are of media which also may be used as enrichment media. In contrast, many of the media listed in the latter section contain compounds which inhibit the growth of certain types which might be expected in the initial sample in addition to the organism sought. In other instances, the composition of the medium allows the demonstration of the growth characteristics of the desired organism to make presumptive identification possible.

The sulfur oxidizing bacteria. The chemoautotrophic growth of *Thiobacillus thiooxidans* is accomplished by inoculating a shallow layer of the following medium with about 1 g of mud or soil and incubating at 30°C: powdered sulfur, 10 g (or $Na_2S_2O_3 \cdot 5H_2O$, 5 g); $(NH_4)_2SO_4$, 0.4 g; KH_2PO_4, 4 g; $CaCl_2$, 0.25 g; $MgSO_4 \cdot 7H_2O$, 0.5 g; $FeSO_4$, 0.01 g; water, 1 liter (Starkey, 1935). When good development has occurred, a transfer is made to fresh medium of the same composition. From the latter, isolations are made on the following solid medium: $Na_2S_2O_3 \cdot 5H_2O$, 5 g; K_2HPO_4, 0.1 g; $NaHCO_3$, 0.2 g; agar, 20 g; water, 1 liter. *Thiobacillus* colonies are recognized by the deposition of free sulfur: $2Na_2S_2O_3 + O_2 \rightarrow 2Na_2SO_4 + 2S$.

The nonsulfur photosynthetic bacteria (*Athiorhodaceae*) utilize organic compounds as H-donors and require growth factors for their development. Some are strict anaerobes, growing only photosynthetically, but others are facultative aerobes and can grow well heterotrophically in the dark under aerobic conditions, obtaining energy by the oxidation of organic substrates. Here, general enrichment media for cultivation under anaerobic conditions will be described, the source of energy being light (van Niel, 1944). The basal medium consists of $(NH_4)_2SO_4$, 1 g; K_2HPO_4, 0.5 g; $MgSO_4$, 0.2 g; NaCl, 2 g; $NaHCO_3$, 5 g; water, 1 liter. This medium is supplemented by the addition of a single organic substance (ethanol, glycerol, mannitol, formate, acetate, succinate, malate, alanine, asparagine) in a final concentration of 0.15–0.2 per cent, after which the reaction is adjusted to pH 7.0 with H_3PO_4. The various media are dispensed into glass-stoppered bottles, inoculated with a small amount of surface water or mud, and the completely filled and stoppered bottles incubated in a light cabinet at 25–30°C under continuous illumination with electric bulbs (25–40 watts). For more rapid and abundant growth, peptone and yeast extract may be substituted for the single organic compound. Isolations are made by preparing shake cultures of successive dilutions, using a medium of yeast extract, 5 g; $NaHCO_3$, 2 g; Na_2S, 0.1 g (to provide anaerobiosis); agar, 20 g; water, 1 liter; pH 7.0. Streaked plates of this medium (less the Na_2S) may also be used for the isolation of facultatively aerobic strains, in which case growth will occur in the dark or in the light.

Lactic acid bacteria. Enrichment cultures of the extensively fermentative types such as the lactic acid bacteria depend upon the use of poorly buffered media rich in organic nitrogenous compounds and growth factors and containing fermentable carbohydrates which serve as energy sources. Lactic acid bacteria will predominate in such a nutritional environment because they withstand the high concentrations of acid produced by the breakdown of the carbohydrates, whereas most other bacteria are killed or inhibited. In general, lactic acid bacteria can be isolated from fermenting plant juices, dairy products, buccal and vaginal cavities, raw sewage, etc. Here, a few common enrichment media and methods will be indicated which will permit growth of a wide variety of types, although additional study is required for their isolation and identification. Incubation temperatures are commonly 30, 37, or 45°C. Milk, raw or pasteurized, can be used both as a source of lactic acid bacteria and as an enrichment medium. Glass-stoppered bottles are filled with raw milk (with or without previous heating at 60°C for 10 min) and incubated. Similarly, glass-stoppered bottles completely filled with a medium consisting of 0.5 per cent yeast extract and 2 per cent of glucose or 10 per cent of sucrose are inoculated with raw sewage. Shredded cabbage or other plant tissue, pressed tightly and covered with water, or ground grain mash can serve as enrichment media. Further types can be isolated by streaking throat swabs on blood agar plates (described below) and incubating at 37°C. When growth has occurred in the enrichment media, plates of yeast-extract-glucose (or sucrose) agar containing 1 per cent $CaCO_3$ (sterilized separately by dry heat) are streaked, and portions of isolated colonies tested for catalase (lactic acid bacteria are catalase negative) with 5 per cent H_2O_2 before streaking again to obtain pure cultures. With certain types which show some sensitivity to atmospheric oxygen in primary cultures, poured, seeded plates may be more successful than streaked plates. Plates should be incubated at temperatures used for the corresponding enrichment. It is apparent that many variations of enrichment media may be employed and that the predominant occurrence of given lactic acid bacterial types depends upon the source used.

The coliform group. Owing to the vast amount of work performed with the *Enterobacteriaceae*, highly specialized media have been developed for their rapid isolation and identification. Enrichment media for the nonpathogenic *Escherichia* and *Aerobacter* include nutrient broth plus 1 per cent of lactose. Gas production gives presumptive evidence of the group, and the necessary confirmation is accomplished by use of one of a variety of media listed under "Differential Media." Similarly the pathogenic *Salmonella* may be enriched from feces, urine, water, sewage, contaminated foodstuffs, etc., by use of tetrathionate and selenite (Leifson, 1936) broths. These media are available in dehydrated form.

For obligately anaerobic bacteria, a primary requirement is the initial attainment of a low O/R potential, either by eliminating oxygen from the nutritional environment or by counteracting its effect by the addition of chemicals. The material in Chap. VI outlines methods for the cultivation of members of the genus *Clostridium.* For the enrichment of the sporeforming anaerobic bacteria a variety of media are available, and choice among them will depend upon the species desired and the sample material. For the pathogenic species the thioglycollate medium or beef heart infusion listed under "Cultivation and Storage Media" is suitable. For certain soil types (organisms producing butyric acid, or butyl alcohol, and some others) the liver infusion medium (same section), corn liver medium, or potato infusion is useful. In the last two, the fermentation of starch may be observed, and some bacterial types give a characteristic "head" (a slimy mass of unfermented cellulosic material). The starch-fermenting types usually sporulate readily on these natural infusions. The *corn liver medium* is prepared as follows: Add 50 g of ground (white or yellow) corn meal and 10 g of dried liver powder to 1 liter of tap water. Heat in flowing steam for 1 hr with occasional stirring. Remove from steam, cool almost to room temperature, and dispense in tubes, flasks, or bottles. Sterilize 45 min at 121°C. The resulting medium, on cooling, should be semisolid, with the coarser particles of corn settling to the bottom leaving a 2- to 3-cm layer of starchy material at the top. For determination of pigment production, omit the liver powder. The *potato infusion* is prepared as follows: White potatoes, 200 g; glucose, 5 g; $(NH_4)_2SO_4$, 1.0 g; $CaCO_3$, 3 g; tap water to make 1 liter. Peel potatoes, and add water. Steam for 1 hr, or boil slowly until soft, and put through potato ricer. Add other ingredients, and bring up to original volume. Cool, and tube, with stirring, so as to obtain an even distribution of the potato particles.

As an example of the enrichment of an anaerobic bacterium, which has been isolated (Baker and Taha, 1942; Bornstein and Barker, 1948) by use of a medium constituting a simple and chemically defined nutritional environment, *Clostridium kluyveri* may be isolated from black mud using a medium containing the following components: ethanol, 8 g; sodium acetate hydrate, 8 g; KH_2PO_4–Na_2HPO_4 buffer, 1 M, pH 7.0, 25 ml; $(NH_4)_2SO_4$, 0.5 g; Na_2CO_3, 0.1 g; $MgSO_4 \cdot 7H_2O$, 0.2 g; $CaSO_4 \cdot 2H_2O$.

10 mg; $FeSO_4 \cdot 7H_2O$, 5 mg; $MnSO_4 \cdot 4H_2O$, 2.5 mg; $NaMoO_4 \cdot 2H_2O$, 2.5 mg; biotin, 3 μg; p-aminobenzoic acid, 50 μg; sodium thioglycollate, 0.5 g; distilled water, 1 liter. The thioglycollate may be replaced by 0.2 g of $Na_2S \cdot 9H_2O$ which is best added after sterilization of the medium; the growth factors may be replaced by 0.5–1.0 g of yeast extract. The procedure is: Add heavy inoculum (5 per cent of black mud from fresh-water or marine sources) and incubate culture at 35°C in completely filled glass-stoppered bottles in order to exclude oxygen. To isolate pure cultures, use a solid medium of the same composition in an anaerobic jar.

Use of bacteriostasis. A principle sometimes invoked in the enrichment culture of a particular type of organism from a mixed population is to utilize the inhibitory (bacteriostatic) property of a specific chemical without which the medium would be suitable for many species in the sample. For example, addition of crystal violet in a final concentration of 1:100,000 will inhibit most gram-positive types without affecting the gram-negative group. Similarly, the antibiotics penicillin and streptomycin may be used for selective inhibition. Sodium azide (0.03 per cent) has bacteriostatic action for gram-negative bacteria, the aerobic gram-positive sporeforming bacilli, and certain other aerobes (Lichstein and Soule, 1944).

DIFFERENTIAL MEDIA

The formulas presented in this section include media employed, commonly for original isolation, to determine differential reactions which may permit presumptive identification of bacterial species. In some instances the media constitute selective growth environments, often as a result of the inclusion of specific compounds which inhibit the growth of those organisms not under investigation.

Blood agar. This medium is used frequently in the study of streptococci and other groups exhibiting hemolytic properties. The medium is valuable also in the routine cultivation of fastidious pathogenic species which may or may not show hemolytic reactions. To prepare medium: Add aseptically 5 per cent of sterile rabbit, sheep, horse, or human blood to a satisfactory carbohydrate-free agar base medium containing 0.5–0.85 per cent of NaCl (to maintain isotonicity). Blood agar seeded with a light inoculum yields more typical hemolytic reactions than do heavily streaked plates. *Chocolate agar* is prepared as above except that the blood is added to the base medium at 80°C; this causes a distinct darkening of the added blood (protein coagulation).

Egg-yolk agar. This medium is used as a presumptive identification medium for certain pathogenic clostridia and for sporeforming aerobes. Plates of an agar-base medium to which sterile egg yolk has been added are streaked, and the characteristic reactions obtained are due principally to production of lecithinase. To prepare the medium: After washing shell with disinfectant, asceptially withdraw first the white and then the yolk of fresh hen's egg. Dilute yolk with equal volume of 0.85 per cent NaCl, and add 1 ml of yolk suspension to each 9 ml of the following medium which is sterilized before addition of egg: peptone, 40 g; Na_2HPO_4, 5.0 g; KH_2PO_4, 1.0 g; NaCl, 2.0 g; $MgSO_4$, 0.1 g; glucose, 2 g; agar, 25 g, 1 liter water, pH 7.6. Streak plates in such a manner as to yield well-isolated colonies. Consult McClung and Toabe (1947) for details of species characters displayed by clostridia and Colmer (1948) and McGaughey and Chu (1948) for information on the aerobic bacilli.

Media for gram-negative nonsporeforming bacteria. Many differential media have been developed for the presumptive identification of the gram-negative rods occurring in the normal and diseased intestines and in samples contaminated with fecal material. Not all these media can be discussed, and although those which are mentioned may be prepared in the laboratory, readily available dehydrated products are generally employed because of their greater uniformity. On *Endo's agar* other lactose nonfermenting types, such as *Salmonella typhosa*, produce clear, colorless colonies which do not affect the faint pink color of the medium, whereas colonies of coliform organisms which ferment lactose are surrounded by a dark red zone. On *eosin-methylene-blue agar* (*E.M.B.*), colonies of lactose nonfermenting organisms are translucent; of the lactose fermenting types, *Escherichia coli* colonies are small, dark and have a greenish metallic sheen, while those of *Aerobacter aerogenes* are large, moist, and gray-brown in color with a pronounced tendency to coalesce. On the desoxycholate agar of Leifson (1935) growth of gram-positive organisms is inhibited. Colonies of the lactose nonfermenting *Salmonella* and *Shigella* are colorless, those of *E. coli* are red, and those of *Aerobacter* are pale with pink centers (Paulson, 1937).

Medium for isolation of staphylococci. Chapman (1948) suggests the following medium for selective isolation of chromogenic staphylococci associated with food poisoning: D-mannitol, 10 g; peptone, 10 g; gelatin, 30 g; yeast extract 2.0 g; NaCl, 55 g; K_2HPO_4, 5 g; $(NH_4)_2SO_4$, 75 g; 10 per cent NaOH, 6 ml; water, 1,000 ml; agar, 15 g. Following sterilization, shake medium to disperse precipitate.

Media for Brucella. Although liver-infusion agar has been used widely for cultivation and differentiation of *Brucella*, it is being replaced, owing to batch variations, by one of the following media which are available in dehydrated form: (1) *Difco tryptose agar* ("tryptose," 20 g; glucose, 1 g; NaCl, 5 g; agar, 15 g). (2) *B.B.L. trypticase soy agar* ("casein peptone," 15 g; "soy peptone," 5 g; NaCl, 5 g; agar, 15 g). Incubate cultures at 37°C in an atmosphere of 10 per cent CO_2 and in original isolations, if gram-positive contaminants are suspected, add crystal violet in concentration of 1:700,000. For differentiation of *Brucella* species on the basis of bacteriostatic action of dyes, add 1:100,000 concentration of thionin and 1:100,000 concentration of basic fuchsin. *Brucella melitensis* and *B. suis* grow in the presence of thionin, but *B. abortus* in inhibited, whereas on agar containing basic fuchsin, *B. suis* is inhibited by *B. melitensis* and *B. abortus* is not inhibited.

Media for cultivation of Corynebacterium diphtheriae. Perhaps the most widely used medium for the cultivation of *C. diphtheriae*, which yields cells of characteristic morphology and metachromatic granule staining reactions, is *coagulated blood serum*. Although it may be prepared from fresh serum (horse, cow, sheep, or pig) by adding 1 vol of 1 per cent dextrose broth to 3 vol of serum, users of small quantities of this medium will find the dehydrated product more satisfactory. Special precautions are necessary in the successful sterilization of this medium. It may be sterilized by inspissation or by the following method: Add desired quantity of dehydrated powder to warm (50°C) water, and maintain this temperature for 45 min during which time stir medium gently to avoid bubbles. Tube in amounts such that a deep butt will not be produced when tube is slanted. Place tubes in slanted position in autoclave equipped with manually operated air-escape valve. Cover tubes with several layers of newspaper, or pack in metal box between layers of nonabsorbent cotton. Close door and air-escape valve of autoclave, and admit steam, raising pressure of air and steam mixture quickly to 15 lb. After 20 min open air-escape valve cautiously to permit escape of air while maintaining steam pressure. When all air is removed, close escape valve and continue heating at 121°C for 20 min. Reduce pressure slowly, and open door when pressure is at zero.

Other media, often containing potassium tellurite (0.03–0.05 per cent), are valuable in the differentiation of this organism. The details of these cannot be condensed here, but those interested should consult the original papers, for example, Kellogg and Wende, 1946; Frobisher *et al.*, 1948; Petran, 1948; Whitley and Damon, 1949; Buck, 1949.

Media for the cultivation of Neisseria gonorrhoeae. The problems involved in the successful isolation of this organism from samples of clinical material will not be discussed here. For suggestions concerning the media to be used, consult Carpenter *et al.* (1949).

Media for the cultivation of Mycobacterium tuberculosis. Many media have been devised for the cultivation of this organism from sputum, urine, pleural exudates, gastric contents, etc. The media of Dorset, Loewenstein, Petragnani, Petroff, etc., have been used widely; descriptions of such media are available in manuals for laboratory clinical pathology, particularly Gradwohl (1948). Recently, Dubos and his associates described growth of this organism in liquid semisynthetic medium, containing surface-active agents (esters of long-chain fatty acids and polyhydric alcohols), which supported comparatively rapid cell multiplication with a small inoculum and yielded cultures of uniform turbidity without granule or pellicle formation (see Frobisher, 1949).

MEDIA FOR DETERMINATION OF PHYSIOLOGICAL PROPERTIES

In the study of pure cultures, the so-called cultural characters and physiological reactions assume importance in the classification of many groups of heterotrophic bacteria. The latter reactions include the dissimilation of various carbohydrates and polyalcohols, the breakdown of simple and complex nitrogenous compounds, and the production of specific compounds as end products of the metabolism of the bacteria under study. The media for some of the more commonly used reactions will be included in this section. (See also Chap. VII, "Routine Tests for the Descriptive Chart.")

Carbohydrate indicator media. The common procedure is to add the carbohydrate or polyalcohol to be studied to a basal medium (either liquid or agar) to which an indicator has been added to detect changes in pH which develop during growth. Growth of some organisms, particularly the sporeforming anaerobes, may result in a marked reduction of the indicator, in which case the indicator must be added after rather than before growth; use of a spot plate or other methods of pH determination is then made at the time of observation. Early observation of fermentation results generally eliminates the difficulty due to reduction of the indicator, since acidity changes usually precede reduction. Production of gas is detected in liquid media by placing Durham tubes (small inverted vials which will fill with liquid during sterilization) in the tubes at the time the medium is dispensed. The tubes are unnecessary if a solid or semisolid medium is used. Semisolid agar is prepared by adding 0.3–0.5 per cent of agar to a satisfactory liquid medium and making stab inoculations in the column of medium with a straight inoculating needle. In such a medium (or in full-strength agar) gas production will be denoted by the appearance of gas bubbles and cracks in the medium; the same semisolid medium may also be used to determine motility. Full-strength (1.5 per cent) agar should be cooled in slanting position and inoculated on the surface of the slant.

The basal medium to be employed for fermentation tests should provide the necessary nutrients for the organism to be studied and must be free of fermentable carbohydrates. If good growth can be obtained on 2 per cent casein or gelatin-peptone solutions (or agar), these media are preferred. According to Vera (1949), some samples of beef and yeast extracts contain fermentable carbohydrate. In all instances, *control tubes* to which carbohydrate has not been added must be inoculated to check changes of pH due to the breakdown of carbohydrates or other substances.

The synthetic medium of Ayers, Rupp, and Johnson (1919) may be used if a peptone-free medium is desired and if the organism being studied can utilize ammonium salts as a source of nitrogen. It is prepared as follows: $NH_4H_2PO_4$, 1.0 g; KCl, 0.2 g; $MgSO_4 \cdot 7H_2O$, 0.2 g; water, 1,000 ml; carbohydrate, 10.0 g; adjust the pH by addition of 1 *N* NaOH.

The soluble carbohydrates or polyalcohols are added to the basal medium at a level of 0.5–1.0 per cent. (For precautions taken in sterilization see page 39.) The indicator (commonly 1 ml of a 1.6 per cent alcoholic solution per 1,000 ml of medium) is added before sterilization. Although litmus and Andrade's indicator (acid fuchsin decolorized with alkali) have been used widely, they do not give accurate results in terms of H-ion concentration; thus, except for special purposes, (see Chap. VII, page 165) it is recommended that sulfon-phthalein indicators be employed. Select the appropriate indicator from the list given in Table 2, governing choice by following considerations: phenol red, with pH range 6.9–8.5, is useful for indication of changes on the alkaline side of neutrality and slight changes to acid; bromthymol blue has a sensitive range extending slightly in either direction from neutrality; bromcresol purple, with a pH range of 5.4–7.0, is useful in synthetic media and for pronounced pH changes in more highly buffered media. Bromthymol blue is frequently the most useful of these but must be used with caution in synthetic media, since it indicates even the minor pH changes due to CO_2 absorption.

Double and triple sugar agars are of particular value in the rapid identification of the gram-negative enteric bacteria. These media are tubed in columns sufficiently deep to allow a 1.5-in. butt in addition to a slant. They should be inoculated by smearing the slant and stabbing the butt with a straight inoculating needle. *Russell's double sugar agar* is prepared as follows: to a liter of peptone broth or beef-extract peptone broth add glucose, 1.0 g; lactose, 10.0 g; NaCl, 5.0 g; phenol red, 0.025 g; and agar, 15 g. After incubation, those organisms (*Salmonella typhosa*) which attack glucose but not lactose will show an acid reaction (yellow) in the butt but not in the slant, whereas those types (*Escherichia coli*) which ferment lactose will give an acid reaction throughout the entire medium. Gas production is indicated by formation of gas bubbles or splitting of the medium. *Krumwiede's triple sugar agar* is prepared by adding 10.0 g of sucrose to the above formula. Fermentation of either sucrose or lactose, or both, will give rise to an acid reaction throughout the medium, whereas fermentation of only glucose will produce acid in the butt but not in the slant. The addition of 0.2 g of ferrous sulfate and 0.3 g of sodium thiosulfate to the above formula allows the determination of the production of hydrogen sulfide in the same medium (*triple sugar iron agar*); cultures producing H_2S show an extensive blackening of the agar due to iron sulfide precipitation (Sulkin and Willett, 1940).

Starch agar. Add 0.2 per cent of soluble starch to a suitable nutrient agar basal medium, sterilize, and pour plates. After inoculation and growth, test for starch hydrolysis by flooding the plate, which has been streaked across center line, with dilute (Lugol's) iodine. Absence of the bluish-purple color characteristic of the starch-iodine complex indicates hydrolysis.

Media to demonstrate gelatin liquefaction. *Plain gelatin* may be made by adding 12 per cent of bacteriological grade gelatin to distilled water, adjusting the pH to 7.0 before sterilization, sterilizing 12–15 min at 120°C, and cooling the tubes immediately. *Nutrient gelatin* is prepared by adding the above amount of gelatin to a sugar-free nutrient broth. For most obligate anaerobes, it is necessary to add 0.25 per cent of glucose and to tube the medium in deep columns. If growth does not occur, sodium thioglycollate (0.1 per cent final concentration) is added to serve as a reducing agent. Utilization of gelatin may be determined in an *agar medium* by adding gelatin (0.4 per cent final concentration) to a nutrient agar. A sufficient amount of the sterile medium is poured into a petri dish to avoid a thin spot if center of dish is slightly raised. A single streak of the culture is made across the surface of the hardened medium, and following incubation, the plate is flooded with a gelatin precipitant: acid $HgCl_2$ (mercuric chloride 15.0 g; concentrated HCl, 20.0 ml; distilled water, 100 ml); or saturated ammonium sulfate solution. A white precipitate indicates presence of nonhydrolyzed gelatin; absence of the precipitate in the region of growth indicates gelatin hydrolysis.

Proteolysis. Gelatin hydrolysis (liquefaction) represents enzymatic action upon an incomplete protein, and positive action is not necessarily an indication that the organism can hydrolyze complex proteins; this is a characteristic of particular value in the study of certain obligate anaerobes. The beef-heart infusion with particles of tissue represents one of the media in which muscle protein hydrolysis may be observed. *Coagulated serum* (as slants) represents another type of protein to be tested; organisms with ability to hydrolyze serum proteins will cause partial or complete liquefaction. For action on *coagulated egg albumin* one should include a small cube of hard-boiled egg white in a tube of a suitable base medium; disintegration of coagulated egg white during growth is evidence for proteolytic activity. Another medium for the indication of proteolytic activity is *alkaline egg medium* which is prepared as follows: Mix the yolk of one and the whites of two fresh eggs (preferably in a Waring blender). Add 500 ml of distilled water, and adjust pH to 7.6. Stir well, or mix in blender. Add 1 part of above to 5 parts of nutrient broth, tube in deep columns, and sterilize for 20 min at 121°C. The final medium should be an opaque, whitish liquid. During growth proteolysis is indicated by progressive clearing of the medium.

Nitrate broth. Add 0.1 per cent of KNO_3 to a nutrient broth or agar. For obligate anaerobes, also add 0.1 per cent of glucose and 0.1 per cent of agar to basal medium and tube in deep columns. For organisms not reducing nitrates in a peptone medium the following synthetic medium of Dimmick (1947) is recommended: K_2HPO_4, 0.5 g; NaCl, 0.5 g; $MgSO_4 \cdot 7H_2O$, 0.2 g; $NaNO_3$, 2.0 g; glucose, 10.0 g; agar, 15 g; distilled water 1,000 ml. If the organism being studied requires more calcium, add 0.05 g of $CaCl_2$ to the above; in this case it is important that the final pH be 7.2; to assure this *after sterilization,* adjustment before sterilization should be to about 7.8.

Indole determination. Use 1 per cent concentration of a peptone high in tryptophane, such as those prepared by enzymatic digestion of casein or lactalbumin (see Chap. VII for methods of testing for indole).

H_2S ***production.*** If the lead acetate paper tests recommended in Chap. VII are not desired, use dehydrated media or consult the original papers of Vaughn and Levine (1936), Hunter and Crecelius (1938), and Untermohlen and Georgi (1940) for directions for preparation of media containing lead, bismuth, or iron salts which will precipitate as the sulfides in the presence of H_2S.

Methyl red and Voges-Proskauer reaction. Prepare basal medium for these tests as follows: peptone, 7.0 g; glucose, 5.0 g; K_2HPO_4, 5.0 g. Since all peptones are not suitable, use for peptone an enzymatic digest of casein. For details concerning these reactions consult Chap. VII.

Medium for determination of utilization of citrate. Within the gram-negative nonsporeforming bacilli, differentiation of some types is based on utilization of citrate as the sole source of carbon. Koser's synthetic medium is suitable for this and is prepared as follows: KH_2PO_4, 1.0 g; $MgSO_4$, 0.2 g; $NaNH_4PO_4$, 1.5 g; $Na_3C_6H_5O_7$, 3.0 g; water, distilled, 1 liter. Growth, as evidenced by turbidity, in the water-clear medium indicates utilization of the citrate radical as a carbon source. One should make certain that a small enough inoculum is used in this medium so that it is not noticeably turbid before incubation. This can be accomplished by using a straight needle for inoculation or by using a loopful of a suspension in sterile water.

Litmus milk. Prepare saturated aqueous solution of litmus. Add a sufficient quantity of the solution to give a light lavender color to fresh *skimmed* milk (some grades of dried milk powder may be substituted, but many are unsatisfactory); sterilize 12–15 min at 120°C; cool tubes immediately by immersing in cold water. For anaerobic organisms, Spray's system of classification is based upon use of this medium (tubed in a deep column) to which is added 0.05 g of reduced iron or a thin strip of iron. The reactions determined for aerobic bacteria on this medium are explained in Chap. VII; those for anaerobes in Spray's (1936) original paper.

Pigment production. This character may be observed on a variety of media, and reports describing characteristics of new species should indicate the medium used. Starch agar (as described above) is often satisfactory, and if the organism will grow on *potato slants,* this medium may be used. Potato slants are prepared as follows: Peel white potatoes and cut plug from center, using cork borer of appropriate size. Slice plugs obliquely to make slants. Wash slants overnight in slowly running cold tap water. Place plugs in tube supporting them with small glass rod, stick of wood, or potato slice; add 1 ml of water to keep slants moist during incubation, and sterilize. Pigment production by clostridia may be observed in corn-meal infusion medium (page 49) prepared without addition of liver or in potato infusion.

Lipolysis. For media to be used to detect lipolytic action and for methods of proper preparation of fat emulsions, consult the papers of Castell (1941), Castell and Bryant (1939), Collins and Hammer (1934), Eisenberg (1939), Knaysi (1941), and Starr (1941).

MEDIA FOR SPECIFIC BACTERIOLOGICAL PROCEDURES

In the preceding sections a variety of media have been presented which are suitable for the isolation, cultivation, characterization, and maintenance of bacteria. Many of these media can be employed for multiple purposes, and conversely, in some instances, several different formulations may prove satisfactory for the same purpose. The established performance of a medium and the personal preference of the laboratory worker result in the adoption of one formula in place of another.

However, there are numerous bacteriological procedures which require that media of a designated composition be used. Listed in Table 3 are several such microbiological procedures, together with references which specify the composition of media required for performance of the tests.

MEDIA FOR SPECIAL PURPOSES

Voluminous experimental evidence attests to the fact that the composition of the culture medium has a profound influence on the microbial cell with respect to formation of enzymes, toxins, antibiotics, and other products. Some aspects of this subject have been reviewed by Gale (1951). A few examples are presented below.

Metabolically active cells. In many studies dealing with the physiological activity of cells harvested from culture media, comparatively little attention has been paid to the conditions of growth, a large cell crop being the usual criterion of the adequacy of the medium. Wood and Gunsalus (1942) have pointed out the limitation of this criterion and have studied the effects of the components of the growth medium upon the dehydrogenase activity of suspensions of *Streptococcus mastitidis*. A suitable and easily prepared medium was described consisting of the following ingredients: "tryptone," 10 g; yeast extract, 10 g; K_2HPO_4, 5 g; glucose, 2 g; water, 1 liter. Metabolically active cells were obtained after incubation at 37°C for 12–15 hr. The final pH was about 6.8. Tsuchiya and Halvorson (1947) made similar studies to obtain suspensions of glycolytically-active *Lactobacillus casei* and *L. arabinosus*. The significance of this type of study rests upon the knowledge of the effects of each major component of the medium upon the particular physiological activity being studied, and such an analysis should precede biochemical investigation.

Production of enzyme, toxin, or antibiotic. The report of Bellamy and Gunsalus (1945) on the composition of a pyridoxine-deficient growth medium for *Streptococcus faecalis*, which yielded cells containing large amounts of tyrosine decarboxylase apoenzyme, will serve as an example of a medium for the production of a bacterial enzyme available in a convenient form to study coenzyme function. The composition of the medium used is described in the paper cited above; at this point, the major intention is the statement of the availability of a cultural method of great potential advantage in biochemical investigations.

The production of bacterial toxins has engaged the attention of many bacteriologists. In at least one case (*Clostridium perfringens*), one of the toxins produced by the organism is an enzyme (lecithinase). It may be assumed, therefore, that the effect of the constituents of the medium upon toxin production is similar to the situation with respect to the production of other enzymes and may be studied effectively along parallel lines of investigation. The nutritive requirements for toxin production by the organisms of diphtheria (Mueller and Miller, 1941), botulinus (Lamanna, Eklund, and McElroy, 1946; Lamanna and Glassman, 1947; Lewis and Hill, 1947; Stevenson, Helson and Reed, 1947), tetanus (Mueller and Miller, 1943, 1948), and gas gangrene (Adams and Hendee, 1945; Logan *et al.*, 1945) are available.

The production of antibiotic compounds varies considerably with the nature of the culture medium and other factors. For a discussion of this topic consult Prescott and Dunn (1949).

TABLE 3. LITERATURE REFERENCES CONCERNING SPECIAL MEDIA

Microbiological procedure	*Reference for composition of media required*
Assay of antibiotics	Grove and Randall, 1955 U.S.P. XV, 1955
Assay of vitamins	Association of Vitamin Chemists, 1947 Barton-Wright, 1952 Johnson, 1949 U.S.P. XV, 1955
Assay of amino acids	Barton-Wright, 1952 Dunn, 1947 Horn *et al.*, 1950
Sterility testing of pharmaceutical products	National Institutes of Health Specifications. U.S.P. XV, 1955
Disinfectant testing	A.O.A.C., 1950
Enumeration of bacteria in milk and dairy products	Am. Assoc. Med. Milk Commissions, Inc., 1954–1955 Standard Methods for the Examination of Dairy Products, 1953
Enumeration of bacteria in waters	Standard Methods for the Examination of Water and Sewage, 1955

REFERENCES

Adams, M. H., and E. D. Hendee. 1945. Methods for the production of the alpha and theta toxins of *Clostridium welchii*. *J. Immunol.*, **51,** 249–256.

American Association of Medical Milk Commissions, Inc. 1954–1955. "Methods and Standards for the Production of Certified Milk." New York.

American Public Health Association, American Water Works Association, and Federation of Sewage and Industrial Wastes Associations. 1955. "Standard Methods for the Examination of Water and Sewage," 10th ed., 522 pp. American Public Health Association, New York.

American Public Health Association and Association of Official Agricultural Chemists. 1953. "Standard Methods for the Examination of Dairy Products," 10th ed., 315 pp. American Public Health Association, New York.

Asheshov, I. N. 1941. Papain digest media and standardization of media in general. *Can. J. Public Health,* **32,** 468–471.

Association of Official Agricultural Chemists. 1950. "Official Methods of Analysis of the A.O.A.C.," 7th ed, 910 pp. Association of Official Agricultural Chemists, Washington.

Association of Vitamin Chemists, Inc. 1947. "Methods of Vitamin Assay," 189 pp. Interscience Publishers, Inc., New York.

Ayres, S. H., P. Rupp, and W. T. Johnson. 1919. A study of the alkali-forming bacteria in milk. *U. S. Dept. Agr. Bull.* 782.

Barker, H. A., and S. M. Taha. 1942. *Clostridium kluyverii*, an organism concerned in the formation of caproic acid from ethyl alcohol. *J. Bacteriol.*, **43**, 347–363.

Barton-Wright, E. C. 1952. "The Microbiological Assay of the Vitamin B Complex and Amino Acids," 79 pp. Pitman Publishing Corporation, New York.

Beijerinck, M. W. 1921–1940. "Verzamelde Geschriften," 6 vols. M. Nijhoff, 's-Gravenhage.

Bellamy, W. D., and I. C. Gunsalus. 1945. Tyrosine decarboxylase. II. Pyridoxine-deficient medium for apoenzyme production. *J. Bacteriol.*, **50**, 95–103.

Bornstein, B. T., and H. A. Barker. 1948. The nutrition of *Clostridium kluyveri. J. Bacteriol.*, **55**, 223–230.

Brewer, J. H. 1943. Vegetable bacteriological media as substitutes for meat infusion media. *J. Bacteriol.*, **46**, 395–396.

Buck, T. C., Jr. 1949. A modified Loeffler's medium for cultivating *Corynebacterium diphtheriae. J. Lab. Clin. Med.*, **34**, 582–583.

Carlquist, P. R. 1950. Culture Media in "Diagnostic Procedures and Reagents," 3d ed., pp. 1–48. American Public Health Association, New York.

Carpenter, C. M., M. A. Bucca, T. C. Buck, E. P. Casman, C. W. Christensen, E. Crowe, R. Drew, J. Hill, C. E. Lankford, H. E. Morton, L. R. Peizer, C. I. Shaw, and J. D. Thayer. 1949. Evaluation of twelve media for the isolation of the gonococcus. *Am. J. Syphilis, Gonorrhea, Venereal Diseases*, **33**, 164–176.

Castell, C. H. 1941. *P*-aminodimethylaniline monohydrochloride as an indicator of microbial action on fats. *Stain Technol.*, **16**, 33–36.

———, and L. R. Bryant. 1939. Action of microorganisms on fats. I. The significance of color changes in dyes used for the detection of microbial action on fat. *Iowa State Coll. J. Sci.*, **13**, 313–328.

Chapman, G. H. 1948. An improved Stone medium for the isolation and testing of food-poisoning staphylococci. *Food Research*, **13**, 100–105.

Collins, M. A., and B. W. Hammer. 1934. The action of certain bacteria on some simple tri-glycerides and natural fats, as shown by Nile-blue sulphate. *J. Bacteriol.*, **27**, 473–485.

Colmer, A. R. 1948. The action of *Bacillus cereus* and related species on the lecithin complex of egg yolk. *J. Bacteriol.*, **55**, 777–785.

Davis, J. G., and H. J. Rogers. 1939. The effect of sterilization upon sugars. *Zentr. Bakteriol.*, Abt. II, **101**, 102–110.

Dimmick, I. 1947. Phosphorous deficiency in relation to the nitrate reduction test. *Can. J. Research*, Sect. C., **25**, 271–273.

Dunn, M. S. 1947. Amino acids in food and analytical methods for their determination. *Food Technol.*, **1**, 269–286.

Eisenberg, G. M. 1939. A Nile blue culture medium for lipolytic micro-organisms. *Stain Technol.*, **14**, 63–67.

Fred, E. B., I. L. Baldwin, and E. McCoy. 1932. Root nodule bacteria and leguminous plants. *Univ. Wis. Studies in Sci.*, **5**, 343 pp.

Frobisher, M., Jr. 1953. "Fundamentals of Microbiology," 5th ed., 633 pp. W. B. Saunders Company, Philadelphia.

———, E. I. Parsons, E. L. Yeates, and K. L. Gay. 1948. A comparative study of tellurite plating media for *Corynebacterium diphtheriae. Am. J. Hyg.*, **48**, 1–5.

Gale, E. F. 1951. "The Chemical Activities of Bacteria," 213 pp. Academic Press, Inc., New York.

Gladstone, G. P., and P. Fildes. 1940. A simple culture medium for general use without meat extract or peptone. *Brit. J. Exptl. Pathol.*, **21**, 161–173.

Gradwohl, R. B. H. 1948. "Clinical Laboratory Methods and Diagnosis," 4th ed., vol. **2,** 1297–1721. The C. V. Mosby Company, St. Louis, Mo.

Grove, D. C., and W. A. Randall. 1955. "Assay Methods of Antibiotics," 238 pp. Medical Encyclopedia, Inc., New York.

Hall, I. C., and E. C. Peterson. 1924. The discoloration of brain medium by anaerobic bacteria. *J. Bacteriol.*, **9,** 211–224.

Horn, M. J., D. B. Jones, and A. E. Blum. 1950. Methods for microbiological and chemical determinations of essential amino acids in proteins and foods. *U.S. Dept. Agr. Misc. Publ.* 696.

Hunter, C. A., and H. G. Crecelius. 1938. Hydrogen sulfide studies. I. Detection of hydrogen sulfide in cultures. *J. Bacteriol*, **35,** 185–196.

Ingelman, B., and H. Laurell. 1947. The preparation of silicic acid jellies for the cultivation of microorganisms. *J. Bacteriol.*, **53,** 364–365.

Johnson, B. C. 1948. "Methods of Vitamin Determination," 109 pp., Burgess Publishing Co., Minneapolis, Minn.

Kellogg, D. K., and R. D. Wende. 1946. Use of a potassium tellurite medium in the detection of *Corynebacterium diphtheriae*. *Am. J. Public Health*, **36,** 739–745.

Knaysi, G. 1941. On the use of basic dyes for the demonstration of the hydrolysis of fat. *J. Bacteriol.*, **42,** 587–589.

Lamanna, C., H. W. Eklund, and O. E. McElroy. 1946. Botulinum toxin (type A); including a study of shaking with chloroform as a step in the isolation procedure. *J. Bacteriol.*, **52,** 1–13.

———, and H. N. Glassman. 1947. The isolation of type B botulinum toxin. *J. Bacteriol.*, **54,** 575–584.

Leifson, E. 1935. New culture media based on sodium desoxycholate for the isolation of intestinal pathogens and for the enumeration of colon bacilli in milk and water. *J. Pathol. Bacteriol.*, **40,** 581–599.

———. 1936. New selenite enrichment media for the isolation of typhoid and paratyphoid (*Salmonella*) bacilli. *Am. J. Hyg.*, **24,** 423–432.

———. 1943. Preparation and properties of bacteriological peptones. I. Enzymatic hydrolysates of casein. *Bull. Johns Hopkins Hosp.*, **72,** 179–199.

Levine, M., and H. W. Schoenlein. 1930. "A Compilation of Culture Media for the Cultivation of Microorganisms," 969 pp. The Williams & Wilkins Company, Baltimore.

Lewis, K. H., and E. V. Hill. 1947. Practical media and control measures for producing highly toxic cultures of *Clostridium botulinum*, Type A. *J. Bacteriol.*, **53,** 213–230.

Lichstein, H. C., and M. H. Soule. 1944. Studies of the effect of sodium azide on microbic growth and respiration. I. The action of sodium azide on microbic growth. *J. Bacteriol.*, **47,** 221–230.

Logan, M. A., A. A. Tytell, I. S. Danielson, and A. M. Griner. 1945. Production of *Clostridium perfringens* alpha toxin. *J. Immunol.*, **51,** 317–328.

McClung, L. S. 1949. Recent developments in microbiological techniques. *Ann. Rev. Microbiol.*, **3,** 395–422.

———, and R. Toabe. 1947. The egg yolk plate reaction for the presumptive diagnosis of *Clostridium sporogenes* and certain species of the gangrene and botulinum groups. *J. Bacteriol.*, **53,** 139–347.

McGaughey, C. A., and H. P. Chu. 1948. The egg-yolk reaction of aerobic sporing bacilli. *J. Gen. Microbiol.*, **2,** 334–340.

Marshall, M. S., J. B. Gunnison, A. S. Lazarus, E. L. Morrison, and M. C. Shevky. 1947. "Applied Medical Bacteriology," 340 pp. Lea & Febiger, Philadelphia.

Mueller, J. H., and E. R. Johnson. 1941. Acid hydrolysates of casein to replace peptone in the preparation of bacteriological media. *J. Immunol.*, **40,** 33–38.

———, and P. A. Miller. 1941. Production of diphtheric toxin of high potency (100 Lf) on a reproducible medium. *J. Immunol.*, **40,** 21–32.

———, and ———. 1943. Large scale production of tetanal toxin on a peptone-free medium. *J. Immunol.*, **47,** 15–22.

———, and ———. 1948. Unidentified nutrients in tetanus toxin production. *J. Bacteriol.*, **56,** 219–233.

Paulson, M. 1937. The clinical use of desoxycholate and desoxycholate-citrate agars—new culture media—for the isolation of intestinal pathogens. *Am. J. Med. Sci.*, **193,** 688–690.

Petran, E. 1948. Trypticase tellurite as a primary plating medium for the identification of *C. diphtheriae*. *Public Health Lab.*, **6,** 39–40.

Prescott, S. G., and C. G. Dunn. 1949. "Industrial Microbiology," 2d ed., 923 pp. McGraw-Hill Book Company Inc., New York.

Schaub, I. G., and M. K. Foley. 1952. "Diagnostic Bacteriology: A Textbook for the Isolation and Identification of Pathogenic Bacteria," 4th ed., 356 pp. The C. V. Mosby Company, St. Louis, Mo.

Simmons, J. S., and C. J. Gentzkow. 1955. "Medical and Public Health Laboratory Methods." Lea & Febiger, Philadelphia.

Society for General Microbiology. 1956. Constituents of bacteriological culture media. Special Report, edited by G. Sykes. Cambridge University Press, London.

Spray, R. S. 1936. Semi-solid media for cultivation and identification of the sporulating anaerobes. *J. Bacteriol.*, **32,** 135–155.

Starkey, R. L. 1935. Isolation of some bacteria which oxidize thiosulfate. *Soil Sci.*, **39,** 197–219.

Starr, M. P. 1941. Spirit blue agar: A medium for the detection of lipolytic microorganisms. *Science*, **93,** 333–334.

Sterges, A. J. 1942*a*. Adaptability of silica gel as a culture medium. *J. Bacteriol.*, **43,** 317–327.

———. 1942*b*. Simplified method for the preparation of silica gels. *J. Bacteriol.*, **44,** 138.

Stevenson, J. W., V. A. Helson, and G. B. Reed. 1947. A casein digest medium for toxin production by *Clostridium*. *Can. J. Research*, Sect. E., **25,** 9–13.

Stitt, E. R., P. W. Clough, and S. E. Branham. 1948. "Practical Bacteriology, Hematology and Parasitology," 10th ed., 991 pp. The Blakiston Division, McGraw-Hill Book Company, Inc., New York.

Sulkin, S. E., and J. C. Willett. 1940. A triple sugar-ferrous sulfate medium for use in identification of enteric organisms. *J. Lab. Clin. Med.*, **25,** 649–653.

Temple, K. L. 1949. A new method for the preparation of silica gel plates. *J. Bacteriol.*, **57,** 383.

Tittsler, R. P., and L. A. Sandholzer. 1936. The use of semi-solid agar for the detection of bacterial motility. *J. Bacteriol*, **31,** 575–580.

Tsuchiya, H. M., and H. O. Halvorson. 1947. The preparation of glycolytically active washed cells of lactobacilli. *J. Bacteriol.*, **53,** 719–727.

United States Pharmacopeia XV. 1955.

Untermohlen, W. P., Jr., and C. E. Georgi. 1940. A comparison of cobalt and nickel salts with other agents for the detection of hydrogen sulfide in bacterial cultures. *J. Bacteriol.*, **40,** 449–459.

Van Niel, C. B. 1944. The culture, general physiology, morphology, and classification of the non-sulfur purple and brown bacteria. *Bacteriol. Revs.*, **8,** 1–118.

———. 1949. The "Delft School" and the rise of general microbiology. *Bacteriol. Revs.*, **13,** 161–174.

Vaughn, R. H. 1942. The acetic acid bacteria. *Wallerstein Labs. Communs.*, **5,** (14), 5–26.

———, and M. Levine. 1936. Hydrogen sulfide production as a differential test in the colon group. *J. Bacteriol.*, **32,** 65–73.

Vera, H. D. 1948. A simple medium for identification and maintenance of the gonococcus and other bacteria. *J. Bacteriol.*, **55,** 531–536.

———. 1949. Accuracy and sensitivity of fermentation tests. *Soc. Am. Bacteriologists, Abstr. Papers*, 49th gen. meeting, p. 6.

Wadsworth, A. B. 1947. "Standard Methods of the Division of Laboratories and Research of the New York State Department of Health," 3d ed., 990 pp. The Williams & Wilkins Company, Baltimore.

Whitley, O. R., and S. R. Damon. 1949. Raffinose serum tellurite agar slants as a replacement for Loeffler's medium in diphtheria diagnosis. *Public Health Rept.*, **64,** 457–460.

Wood, A. J., and I. C. Gunsalus. 1942. The production of active resting cells of streptococci. *J. Bacteriol.*, **44,** 333–341.

APPENDIX TO CHAPTER III

Specifications for Bacteriological Grade Agar, Gelatin, and Peptones

BACTERIOLOGICAL GRADE AGAR[1]

Suggested Specifications

Definition. Agar is any phycocolloid derived from *Rhodophyceae* which meets the requirements given below for gelation temperature and gel melting temperature. Bacteriological grade agar meets all the stated requirements.

Forms. Bacteriological grade agar shall be in the form of shreds, flakes, strips, sheets, or granules.

TABLE 4. REQUIREMENTS

Determination	Limits	
	Max	Min
Total solids	.....	78
Solubility, cold, %	2.0	
Solubility, hot, %	.....	99.8
Gelation temperature, °C	39	33
Gel melting temperature, °C	.....	70
Rate of dissolution, min	15	
Sol turbidity, ppm	10	
Threshold gel concentration, %	0.25	
Protein nitrogen, %	0.32	
Reducing substances as galactose, %	10	
Chlorides as NaCl, %	1.5	
Viable spores, per g	3	
Debris count, per g	30	

Analytical Methods

Total solids. Dry accurately weighed duplicate samples of 0.6–1.0 g for 5 hr at 105°C. Cover, cool to ambient temperature in desiccator, and weigh. Caution:

[1] These specifications were prepared by Mr. H. H. Selby of the American Agar and Chemical Co., San Diego 12, California. They were originally presented to the Society of American Bacteriologists Committee on Bacteriological Technic at the 53d general meeting, San Francisco, California, 1953.

Agar itself is a good desiccant when dry. Weighings should be made promptly after cooling and done rapidly. An efficient desiccant is essential.

Solubility, cold. Dissolve 1.5 g of solids (1.87 g of agar of 80 per cent of total solids, for example) in 100 ml of H_2O by autoclaving at 120°C for 20 min in 250-ml tared flask. Remove flask at 99–100°C, swirl thoroughly to mix, and make up to 100 g net. Take 15.2–15.5 ml of solution into Luer-type syringe. Cap with pinched-off needle. Weigh to 0.1 g. Uncap syringe, and eject contents slowly into 50-mm aluminum-foil dish. Recap Luer, and reweigh. Difference is weight of aliquot (A). After 15 min cut gel freehand with knife into 3- to 4-mm squares and place dish in 3- to 5-mm layer of H_2O in ice-cube tray or equivalent. Put tray on 1- to 2-mm cardboard in freezing compartment at −5 to −10°C. Hold overnight. Transfer dish contents to petri plate, break into squares with single-edged razor blade, and retransfer to 30-ml medium-porosity fritted glass crucible. Wash with four 10-ml portions of H_2O at 10–15°C, using suction at end of each wash. Stir and gently press flakes with flattened rod while washing. Evaporate combined washings in porcelain dish tared to 0.1 mg on steam bath. Dry residue 2 hr at 105°C, cool in desiccator, and weigh. Residue is B.

$$\text{Per cent cold solubles} = \frac{6{,}667B}{A}$$

Hot-water solubles. Dissolve 1.5 g of solids in 100 of ml H_2O at 120°C for 20 min. Swirl at 99–100°C. Filter at 80–90°C through fine-porosity 30-ml fritted glass crucible tared to 0.1 mg. Wash flask and crucible three times with 25 ml of 80–90°C H_2O. Dry crucible 2 hr at 105°C, and weigh. Net is C. Run a blank to obtain correction for solubility of sintered glass membrane (D). Correction may approach 1 mg.

$$\text{Per cent hot solubles} = \frac{100[1.5 - (C + D)]}{1.5}$$

Gelation temperature. Dissolve 1.5 g of solids in 100 ml of H_2O at 120°C for 20 min. Swirl at 99–100°C. Cool to 60–70°C, and make to 100 g net. Place 15 ml in 15- by 150-mm test tube, and insert calibrated Weston-type dial thermometer. Insert 2- to 3-mm-OD glass tube with 90° bent tip and 0.5-mm orifice. Run air through tube at ½ to 1 bubble per second. Note temperature at which course of rising bubbles becomes impeded.

Gel melting temperature. Dissolve 2.0 g of solids in 100 ml of H_2O at 120°C for 20 min. Swirl at 99–100°C. Cool to 60–70°C, and make to 100 g net. Place 10 ml in 15- by 150-mm test tube. Stopper, and support tube at 30° angle. Hold 1 hr at 15–25°C. Place tube vertically in 70°C water bath 1 hr. If slant does not collapse, sample passes test.

Rate of dissolution. Arrange a 500-ml spherical three-necked flask with reflux condenser, sealed agitator with circular segment paddle, and stopper. Place 200 ml of hot water in flask, start agitator, and so place and adjust a bunsen burner that the flame covers approximately half the wetted portion of the flask and maintains slow but definite ebullition (3 to 5 drops per second). Place 3 g of solids in length of glass tubing small enough to enter neck of flask. Provide a glass-rod–rubber-stopper plunger for tubing. Force sample from tube into flask with plunger in such a manner that particles do not become attached to flask. Restopper flask. Stop heating and agitation after 10 min. If no undissolved agar remains in liquid, sample passes test.

Sol turbidity. Dissolve 2.0 g of solids in 100 ml of H_2O at 120°C for 20 min. Swirl at 99–100°C. Cool to 60–70°C, and make to 100 g net. Stopper, and hold at 45–47°C for 3 hr. Simultaneously hold the glassware of a Jackson Coleman or Hellige turbidimeter at 45–60°C. Run turbidity estimation as with waters, using standard notation based on turbidities equivalent to SiO_2 content in parts per million.

Threshold gel concentration. Use solution from gel-melting-temperature determination. To duplicate 140-mm lengths of 14.5- to 15.5-mm-ID glass tubing, stoppered at one end, add 17.5-ml portions of 70–90°C H_2O, using Luer-type syringe without needle. Immediately add 2.5-ml portions of sample solution with syringe, making 20 ml per tube. Stopper, and mix by inverting six times. Hold tubes 1 hr at 20–25°C, then 1 hr at 0–5°C, and, finally, 1 hr at 20–25°C. Remove top stoppers, and gently add 3–1 ml of H_2O to tubes. Taking each tube individually, lay on horizontal wooden surface, remove bottom stopper, and induce cylinder of gel to slide from tube by elevating bottom end slightly and moving tube backward as cylinder moves forward relative to tube. Object is to remove tube without motion of gel with respect to wood. If neither cylinder of gel breaks under pull of gravity in 10 sec, sample passes test.

Protein nitrogen. Use Henwood & Garey (Hengar) modification of Kjeldahl method with 0.1 g of solids. Not more than 0.32 per cent of N (approximately 2 per cent protein) shall be found after correcting for reagents by running a blank on sucrose.

Reducing substances as galactose after autoclaving. Dissolve 1 g of solids in 100 ml of H_2O in 500-ml short-neck spherical flask containing one 3-mm glass bead. Use 120°C for 40 min. Swirl at 99–100°C. Transfer immediately to flame or heater preadjusted to maintain a steady boil. Add 0.2 g of NaOH and 4 mg of methylene blue chloride, dry or as freshly made solution. Close flask with two-hole stopper, one hole of which holds 90° vent tube. Titrate through open stopper hole to blue which is stable 0.5 min. Total titration time 2.5 ± 0.25 min. Use Soxhlet's solutions standardized according to *National Bureau of Standards Circular* 0440, page 187, 10 ml of solution A plus 10 ml of solution B diluted to 100 ml on day of use. No more titrating solution must be needed by the sample than the quantity necessary to react with 80 mg of Eastman galactose No. 141 dried to constant weight over boiled H_2SO_4 and carried through the same procedure.

***Chlorides as* NaCl.** Blend 1 g of solids in 200 ml of H_2O in Waring blender or equivalent device provided with rheostat or autotransformer speed control. Start slowly to avoid splash. Blend, covered, at top speed until visually textureless. Cut speed to point of splash-free agitation; rinse top and sides into mixture. Add 2 ml of 10 per cent K_2CrO_4, and titrate in blending jar with 0.0171 *N* $AgNO_3$ to an end point permanent for 1 min. Subtract the value of a blank run on 200 ml of water. Each milliliter of $AgNO_3$ solution represents 0.1 per cent of NaCl.

Viable spores. Transfer 2-g sample as received to 100 ml of sterile tryptone-glucose-beef extract broth in 6-oz screw-capped prescription bottle, using sterilized vegetable parchment for weighing and transferring. Close bottle tightly, and autoclave at 115°C for 5 min. Agitate gently at 90–95°C, cool to 50–60°C, add 1 ml of sterile skim milk, and pour entire charge into three petri dishes, glass-covered. Invert when cool, and incubate at 35°C for 36 hr. Divide total colonies visible in Quebec-type counter by 2 to get spores per gram.

Debris count. Dissolve 1.5 g of solids in 100 ml of H_2O at 120°C for 20 min. Swirl at 99–100°C, and immediately pour entire charge into two 90-mm petri dishes. Cover with porous tops, and allow to congeal. Examine plates on Quebec-type colony counter in dim light at standard magnification of 1.5 diam, adjusting lens position for maximum resolution. Note all objects which a trained microbiologist might mistake for a microbial colony. To avoid the counting of bubbles, examine each object found with a 7.5 X magnifier without moving petri dish or 1.5 X lens. The smaller magnifier is to be used for examination only, not for counting. Count only true debris particles clearly visible at 1.5 X. Run a control with each series of samples, omitting agar. Divide total number of counted particles by 1.5 to obtain count per gram.

GELATIN[1]

"Gelatin is a product obtained by the partial hydrolysis of collagen derived from the skin, white connective tissue, and bones of animals. Gelatin derived from an acid-treated precursor exhibits an isoelectric point between pH 7 and pH 9, known as Type A, while Gelatin derived from an alkali-treated precursor has an isoelectric point between pH 4.7 and pH 5, known as Type B.

"***Description.*** Gelatin occurs in sheets, flakes, or shreds, or as a coarse to fine powder. It is faintly yellow or amber in color, the color varying in depth according to the particle size. It has a slight, characteristic bouillon-like odor. It is stable in air when dry, but is subject to microbic decomposition when moist or in solution.

"***Solubility.*** Gelatin is insoluble in cold water, but swells and softens when immersed in it, gradually absorbing from 5 to 10 times its own weight of water. It is soluble in hot water, in acetic acid, and in a hot mixture of glycerin and water. It is insoluble in alcohol, in chloroform, in ether, and in fixed and volatile oils.

"***Identification A.*** To a solution of Gelatin (1 in 100) add trinitrophenol T.S. or a solution of potassium dichromate (1 in 15) previously mixed with about one-fourth its volume of diluted hydrochloric acid: a yellow precipitate is formed.

"***Identification B.*** To a solution of Gelatin (1 in 5000) add tannic acid T.S.: turbidity is produced.

"***Residue on ignition,*** page 912—Incinerate 5.0 Gm. of Gelatin without the use of sulfuric acid: the weight of the residue does not exceed 100 mg. (2 per cent). Save the residue.

"***Odor and water—insoluble substances***—A hot solution of Gelatin (1 in 40) is free from any disagreeable odor, and when viewed in a layer 2 cm. thick is only slightly opalescent.

"***Sulfite***—Dissolve 20 Gm. of Gelatin in 150 ml. of hot water in a flask having a round bottom and a long neck, add 5 ml. of phosphoric acid and 1 Gm. of sodium bicarbonate, and at once connect the flask with a condenser. Distil 50 ml., receiving the distillate under the surface of 50 ml. of 0.1 *N* iodine. Acidify the distillate with a few drops of hydrochloric acid, add 2 ml. of barium chloride T.S., and heat on a steam bath until the liquid is nearly colorless. The precipitate of barium sulfate, if any, when filtered, washed, and ignited, weighs not more than 3 mg., corresponding to not more than 40 parts per million of sulfur dioxide, correction being made for any sulfate which may be present in 50 ml. of the 0.1 *N* iodine.

"***Arsenic***—Heat 15 Gm. of Gelatin with 60 ml. of dilute, arsenic-free hydrochloric acid (1 in 4) in a covered flask until all insoluble matter is flocculated and the Gelatin dissolved. Add an excess of bromine T.S. (about 15 ml.), and heat until the excess bromine is expelled. Neutralize with ammonia T.S., add 1.5 Gm. of sodium phosphate, and allow to cool. Add a slight excess (about 30 ml.) of magnesia mixture T.S., allow to stand for 1 hour, filter, and wash with five 10-ml. portions of ammonia T.S. diluted with 3 volumes of water. Drain the precipitate well, and dissolve it in dilute hydrochloric acid (1 in 4) to make exactly 50 ml. Subject 5 ml. of this solution to the test for *Arsenic*, page 803: the stain, if any, does not exceed in length or intensity of color that produced in a test made with similar quantities of the same reagents and 1.5 ml. of the standard arsenic test solution (1 part per million).

[1] Quoted (with permission) from U.S.P. XV, pp. 305–306.

"*Heavy metals,* page 898—To the residue obtained in the test for *Residue on ignition* add 2 ml. of hydrochloric acid and 0.5 ml. of nitric acid, and evaporate to dryness on a steam bath. To the residue add 1 ml. of 1 *N* hydrochloric acid and 15 ml. of water, and warm for a few minutes. Filter, and wash with water to make the filtrate measure 50 ml. To 25 ml. of the filtrate add 10 ml. of hydrogen sulfide T.S.: the heavy metals limit for Gelatin is 50 parts per million.

"*Gel strength*—Place 1 Gm. of Gelatin, accurately weighed, and 99 ml. of water in a 200-ml. flask, allow to stand for 15 minutes, place the flask in a water bath at 60°, and swirl occasionally until solution is complete. Transfer 10 ml. of the solution to a test tube having an internal diameter of 12 mm., and place the tube in an ice bath, making certain that the top of the solution is below the level of the ice and water. Place the bath containing the tube in a refrigerator, and maintain it at about 0° for 6 hours. When the tube is removed from the bath and inverted, no movement of the gel is observed.

"*Bacterial content*—When Gelatin is examined as directed under *Gelatin—Bacteriological Test,* page 839, the total bacterial count does not exceed 10,000 per Gm., and coliform bacteria are not present in 10 mg. or less.

Gelatin—Bacteriological Test

"*Preparation of sample*—Employ aseptic conditions throughout.

If the Gelatin is in sheets, flakes, or shreds, reduce it to a powder under aseptic conditions in a sterile grinder or mortar. Mix thoroughly and weigh 1 Gm. of the powdered sample into a sterile dilution bottle containing 99 ml. of sterile water. After the Gelatin is thoroughly wetted, place in a water bath heated to between 40° and 45°. Shake well, and allow not more than 15 minutes for solution.

"*Total count*—Using a sterile, 1-ml. pipet, place 1 ml. of the well-shaken Gelatin solution in each of two sterile Petri dishes 10 cm. in diameter and 15 mm. in depth. Promptly add 10 ml. of liquefied Tryptone Glucose Yeast Agar or Milk Protein Hydrolysate Glucose Agar warmed to 40°. Cover the dishes, and mix thoroughly the gelatin solution with the added medium by tilting and rotating the dishes. Allow the contents to solidify as promptly as possible, invert the dishes, and incubate them at 35° to 37° for 48 hours. Using a lens of 25 diameters magnification and of about 78-mm. focal length, count the colonies: the average of the two plates does not exceed 100 colonies (10,000 organisms per Gm.).

"*Coliform bacteria*—Using a sterile, 1-ml. pipet, place 1 ml. of the well-shaken Gelatin solution in each of two fermentation tubes containing Lactose Broth. Incubate at 35° to 37°, and examine the tubes at the end of 24 and 48 hours: coliform bacteria are absent if no gas is produced. If gas is produced, transfer a culture therefrom as soon as possible after gas formation is observed. Streak the culture upon Eosin-methylene-blue Agar and incubate at 35° to 37° for 24 hours. If typical coliform colonies have appeared, transfer a culture from at least two of them both to agar slants and to fermentation tubes containing Lactose Broth. If typical coliform colonies have not developed in 24 hours, continue incubation for another 24 hours, and select at least two of the colonies considered most likely to be species of the coliform group, and transfer as directed above. Incubate the Lactose Broth at 35° to 37° for 24 to 48 hours. No gas is produced. Incubate the agar at 35° to 37° for 24 hours, and examine the growth microscopically following Gram-staining: no Gram-negative, non-sporulating bacilli are observed.

"Formation of gas in the lactose broth and demonstration of Gram-negative, non-sporulating bacilli confirm the presence of coliform bacteria."

In addition to the U.S.P. XV specifications as given above, gelatin should conform to the following requirements as set forth in the Military Medical Purchase Description, ASMPA 1-212-000, dated 14 September 1953:

"A 12 per cent solution in distilled water, after autoclaving for 15 minutes at 121 C., shall be clear and free of gross suspended particles. After solidifying, it shall melt over a range of from 30 to 36 C.

"It shall be free of fermentable carbohydrate when tested under the following conditions:

"Prepare a medium of 4 per cent Gelatin and 0.3 per cent NaCl with sufficient phenolsulphonthalein added to give a readable color, tube in Durham fermentation tubes and autoclave 15 minutes at 121 C. Inoculate with a loop of a 24-hour culture of *Escherichia coli.* Neither acid nor visible gas shall be produced in 48 hours incubation at 37 C."

PAPAIC DIGEST OF SOYBEAN MEAL

A soluble nutrient material prepared by the action of the enzyme papain on soybean meal followed by suitable purification and concentration. It meets the specifications under *Pancreatic Digest of Casein*, except that it shows substantial amounts of reducing sugars. It contains fermentable carbohydrates and gives positive tests for indole, acetylmethylcarbinol, and sulfide upon inoculation and incubation with the specified organisms.

(These specifications are essentially those contained in U.S.P. XV, page 1026, except for changes in wording.)

PEPTIC DIGEST OF ANIMAL TISSUE (A BACTERIOLOGICAL PEPTONE)[1]

"A tan powder, having a characteristic, but not putrescent, odor. Soluble in water; insoluble in alcohol and in ether. An autoclaved solution (2 in 100) is clear and is neutral or nearly so in its reaction.

"***Degree of digestion.*** Dissolve 1 Gm. in 10 ml. of water, and use this solution for the following tests:

"(*a*) Overlay 1 ml. of the digest solution with 0.5 ml. of a solution of 1 ml. of glacial acetic acid in 10 ml. of diluted alcohol: no ring or precipitate forms at the junction of the two liquids, and on shaking, no turbidity results, indicating the absence of undigested protein.

"(*b*) Mix 1 ml. of the digest solution with 4 ml. of saturated zinc sulfate: a small amount of precipitate is formed, indicating the presence of proteoses. Retain the filtrate.

"(*c*) To 1 ml. of the filtrate from the preceding test add 4 drops of bromine T.S.: the light yellow color changes to a red-brown, indicating the presence of tryptophane.

"***Nitrogen content, loss on drying, residue on ignition, and nitrite.*** Proceed as directed under *Pancreatic Digest of Casein.*

"***Microbial content.*** Dissolve 1 Gm. in 10 ml. of water. Spread 0.01 ml. on one square centimeter of a glass slide. Stain by the Gram method, and examine with an oil-immersion lens: not more than a total of 50 microorganisms, or clumps, are visible in 10 consecutive fields.

"***Bacteriologic test.*** It meets the following tests for bacteria-nutrient properties. Prepare media of the following compositions:

"(*a*) 2% of peptone and sufficient phenol red T.S. to give a perceptible color in water

"(*b*) 0.1% of peptone in water

[1] Quoted from U.S. P. XV, pp. 1024–1027.

"(*c*) 0.1% of peptone and 0.5% of dextrose in water

"(*d*) 1% of peptone in water

"Adjust all media to a final pH of 7.2 to 7.4. Place 5 ml. of (*a*) in Durham fermentation tubes, and 5 ml. each of (*b*), (*c*), and (*d*) in ordinary test tubes. Autoclave the media at 121° for 15 minutes. After autoclaving, and after standing for 24 hours, all media are clear.

"***Presence of fermentable carbohydrate.*** Inoculate medium (*a*) with *Escherichia coli* and with *Streptococcus liquefaciens:* acid is produced by *E. coli* but not by *S. liquefaciens* during incubation for 24 hours.

"***Production of indole.*** Inoculate medium (*b*) with *Escherichia coli* and with *Aerobacter aerogenes*, and incubate for 24 hours. Test by adding about 0.5 ml. of *p*-dimethylaminobenzaldehyde T.S.: the appearance of a pink or red color (soluble in chloroform) indicates the production of indole by *E. coli*. The *A. aerogenes* culture gives a negative test.

"***Production of acetylmethylcarbinol.*** Inoculate medium (*c*) with *Escherichia coli* and with *Aerobacter aerogenes*, and incubate for 24 hours. Test by adding to the culture an equal volume of sodium hydroxide solution (1 in 10), shaking well, and allowing to stand at room temperature for several hours: the appearance of a pink color indicates the production of acetylmethylcarbinol by *A. aerogenes*. The *E. coli* culture gives a negative test.

"***Production of hydrogen sulfide.*** Inoculate medium (*d*) with *Salmonella typhosa*. Hold a strip or loop of lead acetate test paper between the cotton plug and the mouth of the test tube so that it hangs about 5 cm. above the medium. Then incubate for 24 hours: the lower part of the lead acetate test paper shows an appreciable amount of brownish blackening (*lead sulfide*).

PANCREATIC DIGEST OF CASEIN (A BACTERIOLOGICAL PEPTONE)

"A grayish yellow powder, having a characteristic, but not putrescent, odor. Freely soluble in water; insoluble in alcohol and in ether. The casein used in preparation of this digest meets the following specifications:

Residue on ignition	not more than 2.5%
Loss on drying	not more than 8%
Free acid (as lactic acid)	not more than 0.25%
Fat	not more than 0.5%
Reducing sugars	not more than a trace
Fineness	All passes through a 20 mesh sieve

"***Degree of digestion.*** Dissolve 1 Gm. in 10 ml. of water.

"(*a*) Overlay 1 ml. of the digest solution with 0.5 ml. of a solution of 1 ml. of glacial acetic acid in 10 ml. of diluted alcohol: no ring or precipitate forms at the junction of the two liquids, and when shaken no turbidity results (indicating the absence of undigested casein).

"(*b*) Mix 1 ml. of the digest solution with 4 ml. of a saturated solution of zinc sulfate: a moderate amount of precipitate is formed (indicating the presence of proteoses). Filter, and retain the filtrate for the next test.

"(*c*) To 1 ml. of the filtrate from (*b*) add 3 ml. of water, and follow with 1 drop of bromine T.S.: a violet-red color is produced, indicating the presence of tryptophane.

"***Nitrogen content.*** Determine the nitrogen content of the digest, previously dried at 105° to constant weight, by the Kjeldahl method (see page 909): not less than 10% of nitrogen (*N*) is found.

"***Loss on drying.*** Weigh accurately about 1 Gm., and dry at 100° to constant weight: it loses not more than 7% of its weight.

"***Residue on ignition.*** Weight accurately about 500 mg., and heat slowly until thoroughly charred. Cool, add 1 ml. of sulfuric acid, and ignite to constant weight: the weight of the residue corresponds to not more than 15%.

"***Nitrite.*** To 5 ml. of a solution of the digest (1 in 50) add 0.5 ml. of sulfanilic-α-naphthylamine T.S., mix, and allow to stand for 15 minutes: no pink or red color develops.

"***Bacteriological test.*** The digest meets the following tests for bacteria-nutrient properties. Prepare media of the following compositions:

"(*a*) 2% of peptone, in water;
"(*b*) 0.1% of peptone, in water;
"(*c*) 1% of peptone, 0.5% of dextrose, in water;
"(*d*) 1% of peptone, in water;
"(*e*) 2% of peptone, 1.5% of agar, in water.

"Adjust all media to a pH of 7.2 to 7.4.

"***Freedom from fermentable carbohydrate.*** To medium (*a*) add sufficient phenolsulfonphthalein T.S. to give a readable color, tube in Durham fermentation tubes, and autoclave. Inoculate with a loop of 24-hour culture of *Escherichia coli:* no acid, or only a trace in the inner tube, and no gas are produced during incubation for 48 hours.

"***Production of indole.*** Inoculate 5 ml. of medium (*b*) with *Escherichia coli*, incubate for 24 hours, and test by addition of about 0.5 ml. of dimethylaminobenzaldehyde T.S.: it shows a distinct pink or red color which is soluble in chloroform.

"***Production of acetylmethylcarbinol.*** Inoculate 5 ml. of medium (*c*) with *Aerobacter aerogenes*, and incubate for 24 hours. Test by adding to the culture an equal volume of sodium hydroxide solution (1 in 10), shake, and allow to stand at room temperature for several hours: appearance of a pink color indicates the presence of acetylmethylcarbinol.

"***Production of hydrogen sulfide.*** Inoculate 5 ml. of medium (*d*) with *Salmonella typhosa*. Hold a strip or loop of lead acetate test paper between the cotton plug and the mouth of the test tube so that it hangs about 5 cm. above the medium. After incubation for 24 hours, the lower tip of the lead acetate test paper shows little if any darkening. After 48 hours, it shows an appreciable amount of brownish blackening (*lead sulfide*).

"***Growth-supporting properties.*** In the foregoing tests the media support good growth of *Escherichia coli*, *Aerobacter aerogenes*, and *Salmonella typhosa*. Medium (*e*) stab-inoculated with a stock culture of *Brucella abortus* shows good growth in the line of the stab after incubation for 48 hours. Slants of medium (*e*), inoculated with *Escherichia coli*, *Aerobacter aerogenes*, *Salmonella typhosa*, *Pseudomonas aeruginosa*, *Staphylococcus aureus*, and *Staphylococcus albus*, show characteristic growth after incubation for 24 hours. Medium (*e*), to which about 5% of rabbit blood has been added and which has been inoculated and poured into Petri dishes, shows characteristic alpha or beta zones about colonies of *pneumococci* and *beta hemolytic streptococci* (serological groups A and B), recognizable within 24 hours and fully developed after 48 hours' incubation. Medium (*e*), to which about 10% of blood has been added and which then has been heated to 80 to 90° until the blood has turned chocolate-brown, permits the growth of *gonococcus* colonies within 48 hours when incubated in an atmosphere containing about 10% of carbon dioxide."

CHAPTER IV

The Measurement of pH, Titratable Acidity, and Oxidation-reduction Potentials[1]

BARNETT COHEN[2]

THE MEASUREMENT OF pH

Originally, pH was defined as the logarithm of the reciprocal of the hydrogen-ion concentration. However, certain assumptions regarding indeterminate factors enter the theoretical treatment of any method of measuring this quantity. It is now recognized that the pH scale is standardized on a basis that is arbitrary with respect to a small and indeterminate uncertainty, although any pH number closely approximates the logarithm of the reciprocal of the corresponding hydrogen-ion *activity*. The activity of any substance is virtually the product of that substance's molar concentration and a factor called the activity coefficient. This factor expresses the departure from that behavior which would obtain were there no van der Waals and Coulomb (attraction and repulsion) forces operating.

The common methods for the measurement of pH are of two types: (1) potentiometric and (2) colorimetric. The theoretical and practical aspects of the subject are treated extensively in the monograph by Clark (1928).

Potentiometric Methods

The several potentiometric methods to be cited depend upon the fact that the pH of a solution suitably incorporated in a so-called half-cell is proportional to the electric potential difference established between this half-cell and some reference half-cell used as a standard.

[1] This presentation is confined to the brief description of general procedures that may be applied in the bacteriological laboratory. For theoretical discussions and the elaboration of detail, the reader should consult the texts, monographs, and original references cited.

[2] Deceased.

The hydrogen electrode method. This is regarded as the basic experimental method whereby the various other methods are standardized. It consists in the measurement of the potential difference (emf) established under conditions of maximum work between the "hydrogen half-cell," or "hydrogen electrode," and a calomel or other half-cell which is employed as a working standard. The standard reference half-cell is usually a calomel electrode.

The hydrogen half-cell consists of a suitable vessel provided with (1) a platinum foil electrode, coated with platinum-black, which is immersed or intermittently dipped in the solution to be measured, and (2) an inlet and outlet for oxygen-free hydrogen to saturate both solution and electrode at atmospheric pressure.

A convenient reference half-cell is the "saturated calomel electrode" which consists of a vessel containing a layer of purified mercury covered with a paste of calomel (Hg_2Cl_2), mercury, and saturated KCl solution; the calomel paste is layered with crystals of KCl, and the rest of the vessel is filled with saturated KCl solution which has been saturated with calomel. A platinum wire provides the electrical lead to the mercury of the calomel cell, and a siphon containing saturated KCl solution provides liquid junction with the solution to be measured in the hydrogen half-cell.

In the normal hydrogen half-cell, which provides the standard of potential for all measurements of potential in electrochemistry, the hydrogen partial pressure is one normal atmosphere and the hydrogen ions are at unit activity. The potential difference between electrode and solution in the normal hydrogen half-cell is assumed to be zero at all temperatures.

In standardizing the pH scale by means of measurements with a cell composed of a hydrogen half-cell and a saturated KCl calomel half-cell, it is customary to ignore the small and indeterminate liquid junction potential between the saturated solution of KCl and the solution in the hydrogen half-cell.

The combination of the two half-cells to make an electric cell is indicated schematically as follows:

(Pt)H_2; H^+ in solution X	sat. KCl	sat. KCl; Hg_2Cl_2; Hg (Pt)
Hydrogen electrode	KCl bridge	sat. calomel (reference) electrode

For a pH determination, purified hydrogen is bubbled through the test solution to saturate it and the platinized platinum electrode until equilibrium is attained as indicated by constancy of the emf determined potentiometrically between the metal terminals of the hydrogen and the calomel half-cells. The observed emf, in volts,[1] is converted to pH by the following equation, where T is the absolute temperature.

$$\text{pH} = \frac{\text{Observed emf} - \text{emf of calomel cell}}{0.000{,}198{,}322T} = \frac{E_h}{0.000{,}198{,}322T} \qquad (1)$$

[1] The electrical units employed herein are based on the "international" system in which, according to the National Bureau of Standards, 1 international volt (U.S.) equals 1.00033 absolute volts. The Bureau has announced that, as of January 1, 1948, absolute electrical units will supersede international units.

However, the effect of this new convention for potentiometry is to introduce changes which may be regarded as negligibly small in ordinary measurements of pH and oxidation-reduction potentials. For example, in Eq. (2) $-\Delta E_h/\Delta$kH equals 0.05912 international volt and 0.05914 absolute volt, at 25°C (298.1° absolute).

For this equation to be applicable, the temperature must be constant. For precise measurements, a correction must be made for any departure of the hydrogen partial pressure from one atmosphere. The correction seldom exceeds 0.001 volt (0.017 unit of pH) for the ordinary ranges of barometric pressure and vapor pressures of solutions.

As indicated by Eq. (2),

$$\frac{-\Delta E_h}{\Delta \text{pH}} = 0.000{,}198{,}322T \tag{2}$$

the slope of the straight line relating potential to pH is a constant dependent on the absolute temperature. For example, at 25°, the potential of the hydrogen electrode becomes more negative by 0.0591 volt[1] for each unit increase in pH. Values of this constant at certain temperatures are shown as constant A on page 75.

Standardization of the saturated calomel half-cell. For ordinary measurements, the values at different temperatures of the saturated calomel half-cell, referred to the normal hydrogen half-cell, are as follows:

°C	E_{cal}, volts	°C	E_{cal}, volts
20	0.250	35	0.238
25	0.246	38	0.236
30	0.242	40	0.234

The potential of this half-cell after continued use may change as a result of dilution and contamination, and it is advisable to check its value regularly as a routine procedure.

The precise standardization of the calomel half-cell is discussed in detail by Clark (1928). It consists in measuring the potential of this half-cell against the hydrogen electrode in a solution of known hydrogen-ion activity or against other carefully constructed half-cells of reproducible, known potential. For measurements of ordinary precision, the quinhydrone electrode (see below) in 0.1N HCl can serve for standardization of the calomel half-cell.

The quinhydrone electrode. Ignoring refinements and minor details, we may state that the potential of a noble-metal electrode in an acid or neutral solution saturated with quinhydrone varies linearly with the pH of the solution, and this so-called quinhydrone electrode may, therefore, be used to measure the pH of such solutions.

The linear relationship of potential to pH holds *only for acid and neutral solutions* to about pH 8. In more alkaline solutions two effects disturb this regularity. One is the ionization of the reductant, and the other is deterioration of the components of the system.

The quinhydrone electrode, within its range of usefulness, may often be employed in cases where the hydrogen electrode cannot be applied.

[1] See footnote on page 73.

It comes to equilibrium rapidly, and its manipulation is simple and convenient. Consult Clark (1928) for fuller details.

Its utilization may be illustrated in the standardization of the saturated calomel half-cell. The potential E_{cal} of this half-cell is to be determined relative to that of a standard solution of fixed pH and saturated with quinhydrone, e.g., 0.1 M HCl, the pH of which is 1.082 at 38°. This is done with purified quinhydrone and accurately prepared HCl solution as follows. Place about 5 ml of the standard HCl solution in a suitable electrode vessel. Add 50–100 mg of quinhydrone crystals to saturate the solution; some quinhydrone in the solid phase must be present. Insert a clean platinum or gold electrode preferably in contact with the solid phase at the bottom of the vessel. Then join this half-cell with the calomel half-cell by means of a siphon containing saturated KCl solution, bring the system to constant temperature, and measure the potential which should reach a constant value in a few minutes.

The observed potential, E_{obs}, is related to the potential of the calomel cell, E_{cal}, as follows:

$$E_{cal} = E_q - E_{obs} - A \cdot \mathrm{pH} \tag{3}$$

E_q and A are constants at any given temperature, and have the following values:

°C	E_q	A
20	0.7029	0.0581
25	0.6992	0.0591
30	0.6955	0.0601
35	0.6918	0.0611
38	0.6896	0.0617

For example, at 38°, with a quinhydrone electrode in 0.1 M HCl,

$$E_{cal} = 0.6896 - E_{obs} - (0.0617 \times 1.082) \tag{4}$$

from which the value of E_{cal} can be calculated after substitution of the experimentally determined value of E_{obs}.

To determine the pH of an unknown solution, proceed as above except that the unknown solution is substituted for the standard HCl.

The "glass electrode." Under suitable conditions, a properly prepared thin membrane of special glass separating two solutions of different pH exhibits an electric potential that is proportional to the difference in pH of the solutions. Based on this property, a device called the glass electrode is now widely used for the comparative determination of pH.

The glass probably most generally employed is that known as Corning No. 015; Beckman-type E glass has been advocated for alkaline solutions (pH 9–14) because of its low sodium error as compared with that of glass 015.

One of the common forms of the glass electrode consists of a tube of the glass terminating in a thin-walled bulb which contains an electrode of definite potential in a solution of fixed pH. A combination of electrode and buffer solution frequently employed is a platinum wire, silver-plated and then coated with AgCl, in a half-cell containing 0.1 *M* HCl. For the construction, operation, and theory of the glass electrode, consult Dole (1941).

The carefully rinsed bulb of the electrode, after seasoning in water or buffer solution, is immersed in the solution to be tested and coupled through a saturated KCl liquid junction with the saturated calomel half-cell as indicated schematically below,

Ag; AgCl; HCl (0.1 *M*) | glass membrane | solution X | KCl (sat.); Hg_2Cl_2; Hg

all parts of the cell being maintained at a uniform temperature. The potential difference between the terminals of this cell can be related to the pH of solution X if the glass electrode has been standardized in buffer solutions of known pH.

Standardization of the glass electrode. The potential of a properly functioning glass electrode should vary linearly with pH, from about pH 1–9, in solutions of low salt content (up to 0.1 *M*). For this range, therefore, the electrode requires standardization in buffer solution at one point of pH, but preferably at two, within this linear range. Standard buffer solutions convenient for this purpose may be selected from Tables 5 and 7.

TABLE 5. SOME STANDARD BUFFER SOLUTIONS

Solution	pH	
	25°	38°
0.1 *M* HCl	1.085	1.082
0.01 *M* HCl, 0.09 *M* KCl	2.075	2.075
0.05 *M* acid potassium phthalate	4.000	4.015
0.025 *M* KH_2PO_4, 0.025 *M* $Na_2HPO_4 \cdot 2H_2O$	6.855	6.835
0.05 *M* $Na_2B_4O_7 \cdot 10H_2O$	9.180	9.070

Such standardization should be performed at least daily; preferably, it should be done immediately before a measurement. As occasion requires, a series of buffer solutions of known pH should be used to establish more carefully the linearity of response of the electrode. In solutions more alkaline than about pH 9, the 015 glass electrode responds also to cations other than H ions, the potential being influenced by the activity and kind of such cations. Sodium and lithium ions produce the most marked effects; potassium and bivalent cations smaller effects. When working under these conditions, it is advisable to standardize the electrode with known buffer solutions of about the same composition and of pH closely above and below the pH of the sample being tested.

The standardization for linearity of response from pH 1 to 9 is a necessary check on the operation of the glass electrode, since its results are comparative, not absolute. The slope, $-\Delta E_h/\Delta\text{pH}$, should be not merely constant at any temperature but also equal or closely equal to $0.000,198,322T$. Obviously, a "pH-meter" with its pH scale adjusted to the theoretical slope for a given temperature cannot give correct readings at all points from pH 1 to 9 if its glass electrode follows a significantly different slope at the same temperature. For a brief discussion of the effects of temperature, see Clark (1948).

Cleaning of the glass surface, by immersion in a hot mixture of concentrated nitric and sulfuric acids followed by soaking in water, may restore a sluggish or erratic electrode to normal functioning. A somewhat drastic procedure that may be effective is to dip the glass electrode for a second or two in dilute HF or in a 20 per cent solution of ammonium bifluoride and then to wash it thoroughly in water. If the electrode still behaves erratically, it should be discarded. For such an emergency, it is highly advisable to have available a reserve electrode. This may obviate any mistaken tendency to carry on with an electrode of doubtful reliability.

The instructions accompanying the various glass-electrode "pH-meters" now on the market are usually sufficient to aid the user in tracing out sources of trouble and error in operation. A major source of trouble is electrical leakage due to accumulation of films of moisture at critical parts of the circuit, and perhaps the most frequent sites of such accumulation are the electrode support and lead, both of which are apt to be spattered with water or salt solution during careless manipulation.

The glass electrodes now available are fairly rugged and easily adaptable to use under a variety of conditions and on different types of biological material (e.g., liquid and "solid" culture media). Measurements with an accuracy of 0.05 pH may be made rapidly in poorly buffered, colored, or turbid solutions and in blood or serum. The monograph by Dole (1941) discusses many of its uses.

The Colorimetric Method

The colorimetric method of measuring pH makes use of acid-base indicators, which, within certain limits, vary in color with the pH of the solution. Such indicators are compounds capable of existing in solution as conjugate proton (H-ion) donor and proton acceptor, with one of the conjugate pair differing in color from the other. The relation of these two forms to pH is defined by the equation

$$\text{pH} = pK' + \log \frac{[\text{proton acceptor}]}{[\text{proton donor}]} \qquad (5)$$

in which brackets represent concentrations, and pK' (= $-\log K'$) is called the apparent ionization exponent of the indicator's proton donor-acceptor system. Simple calculations, using, for example, 0.8, 0.5, and 0.3 as values for the ratio [proton acceptor]/[proton donor] at each of the pK' values 3, 6, and 9, will show that indicators with different pK' values cover different ranges of pH (see Fig. 1). For a full discussion of the properties and uses of pH indicators, see Clark (1928) and Kolthoff and Rosenblum (1937).

Within a short range on the pH scale on each side of the pK' value, every color gradation of the indicator corresponds to a definite pH number; this zone may be called the sensitive range of the indicator. Throughout its sensitive range, an indicator can be used to determine the pH of a solution by comparing its color in the solution with that produced in standard solutions representing known pH numbers.

The indicators. A selection of indicators is presented in Table 6. All but three of the compounds are sulfonphthaleins which are particularly useful in bacteriological work because of their high tinctorial power, low or moderate salt and protein errors, and relative resistance to bacterial action. Table 6 gives the pK' values of the indicators and their sensitive

TABLE 6. ACID-BASE INDICATORS

Name	pK'	pH range and colors	Recommended conc %[a]	Ml of 0.01 M NaOH per 0.1 g[b]
Thymol blue (acid range)	1.7	Red 1.2–2.8 yellow	0.04	21.5
Methyl orange[c]	3.5	Red 3.1–4.4 yellow	0.05	[d]
Bromphenol blue	4.0	Yellow 3.1–4.7 blue	0.04	14.9
Bromcresol green	4.7	Yellow 3.8–5.4 blue	0.04	14.3
Methyl red	5.0	Red 4.2–6.3 yellow	0.02	[e]
Chlorophenol red	6.0	Yellow 5.1–6.7 red	0.04	23.6
Bromcresol purple	6.2	Yellow 5.4–7.0 purple	0.04	18.5
Bromthymol blue	7.1	Yellow 6.1–7.7 blue	0.04	16.0
Phenol red	7.8	Yellow 6.9–8.5 red	0.02	28.2
Cresol red	8.3	Yellow 7.4–9.0 red	0.02	26.2
Thymol blue (alk range)	8.9	Yellow 8.0–9.6 blue	0.04	21.5
Phenolphthalein	9.7	Colorless 8.3–10.0 red	0.10	[f]

[a] Stock solutions in 95 per cent ethanol for the indicator acids, or in water for the indicator salts, unless otherwise specified.

[b] Grind 100 mg of the pure indicator acid with the amount of NaOH specified, and when solution is complete, dilute with water to a volume that will yield the concentration recommended in column 4.

[c] Do not use with phthalate buffers.

[d] Dissolve 50 mg in 100 ml of water.

[e] Dissolve 20 mg in 60 ml of 95 per cent ethanol, and add 40 ml of water.

[f] Dissolve 100 mg in 65 ml of 95 per cent ethanol, and add 35 ml of water.

SOURCE: See Clark (1948) and Kolthoff and Rosenblum (1937).

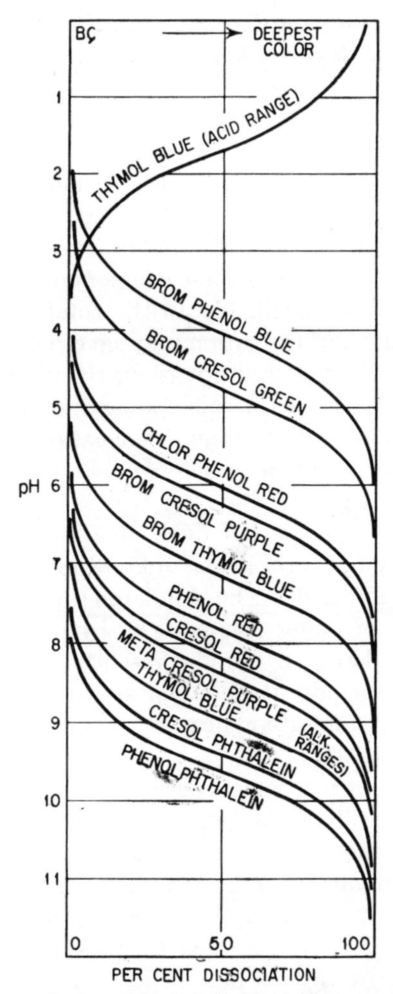

FIG. 1. Ionization curves of some sulfonphthalein indicators, illustrating the general relationships among the acid-base indicators and the applications of Eq. (5). *Note:* In some cases, the positions of the curves on the pH ordinate are approximate. Table 2 should be consulted for accurate values of pK'.

ranges. The last column and footnote *b* of the table give specifications for the preparation of stock solutions of the monosodium salt of each of the sulfonphthaleins.

It will be noted from footnote *a* that ethanolic solutions are ordinarily satisfactory. For precise work, however, aqueous solutions of the indicator salts are preferable to the alcoholic solutions of the free acids. To obviate the labor of preparing the neutralized solutions, some makers now offer the soluble salts of the sulfonphthaleins. They are ammonium, sodium, or possibly other salts of these compounds. In ordinary use, the indicator salts contribute negligibly to the total ions present in a test solution, and the nature of the cation may be of no consequence. However, in some studies of bacterial nutrition, the kind of cation and even the small amounts thus added may be of significance. In such cases, it is advisable to learn from the maker what cations (Na, NH_4, etc.) are present in the indicator salt in order to make due allowance for their possible effects.

The colorimetric method of pH determination depends on matching the color of a suitable indicator in the unknown solution with that of the same indicator in a standard. The standards can be set up in two different ways: by means of buffer standards or by means of "drop ratios." These will be considered in detail presently. In brief outline, the colorimetric method includes these major steps:

1. Selection of the appropriate indicator
2. Preparation of color standards
3. Color comparison for pH determination

Later paragraphs will outline essential specifications that must be observed in each of these steps in order to assure reliable results.

Selection of the appropriate indicator. Test successive small portions (1 ml) of the unknown with a drop of bromthymol blue (BTB). If the color produced is orange or red, then the unknown is probably in the range of pH covered by thymol blue (acid range). If the BTB color is yellow, repeat the test with the indicators of successively lower pK' (see Table 6) until that indicator is found which gives a color within its sensitive or useful range. If the BTB color is blue, proceed in like manner with indicators of higher pK' until the appropriate indicator is found. Of course, if the unknown is more acid than pH 1 or more alkaline than pH 10, none of the indicators listed in Table 6 will serve.

If the unknown solution is unbuffered (e.g., water or saline) or very weakly buffered, the buffering effect of the added indicator may prevail and significantly change the pH of the unknown. In such cases, special methods are required (see Clark, 1928).

It is plain that a rough idea can be obtained as to the pH value of a sufficiently buffered solution by simply finding which indicators give their acid color in it and which give their alkaline color. Indeed, the intelligent employment of indicators with overlapping pH ranges can be made to define the upper and lower limits of a relatively narrow zone of pH within which lies the pH of the solution under study (Small, 1946). Accuracy, however, can be obtained only by actual comparison with the colors produced by the indicators in solutions (buffers) whose pH values are known, or produced by application of Eq. (5) (drop-ratio method, page 85).

Buffer solutions and color standards. A considerable variety of buffer solutions have been proposed; and many of them are discussed and described by Clark (1928). The compositions of the series of buffer standards proposed by Clark and Lubs (1917) are given in Table 7. Preparation of the stock solutions is described by Clark (1928).

After finding the appropriate indicator, prepare or select a series of properly graded standard buffer solutions sufficient in number to bracket the estimated pH of the unknown solution as determined in the preliminary trials. If, for example, the indicator selected is bromcresol green and the estimated pH of the unknown is near 6.0, then not more than five standards, namely, buffers of pH 5.6, 5.8, 6.0, 6.2, and 6.4, should suffice to safely bracket the actual pH of the unknown.

In preparing for the actual measurement, the unknown and the color standards should be contained in clear glass tubes selected for uniform bore, wall thickness, and inherent color. It is essential that the total concentration of indicator in the unknown be exactly the same as that in each of the color standards. This is best accomplished by accurately measuring, with a pipet equal amounts of indicator (e.g., 0.50 ml) into equal amounts (e.g., 10.0 ml) of each of the selected standard buffer solutions. The indicator may be satisfactorily measured in drops provided the dropper tip is properly shaped (not too blunt) and the dropper is held vertically during the measurement. The use of excessive amounts of indicator may introduce difficulties; the minimum quantity necessary to produce recognizable coloration is desirable from the theoretical standpoint. It is essential, of course, that the indicator be uniformly distributed throughout the solutions to which it is added.

Prepared buffer standards can be obtained from supply houses, either as solutions or as powders or tablets to be dissolved as needed. They may also be obtained in sealed glass tubes containing the indicator. Such commercial color standards are convenient and satisfactory. They presuppose the use of comparable concentrations of indicator in the solution under test, and they must be used with the understanding that they are not permanent and may need to be checked or renewed at least once a year. All such indicator standards should be kept in the dark when not in use.

Color comparison. This procedure, commonly miscalled colorimetry, requires intelligent application to yield reliable results. The subject is adequately discussed by Clark (1928, 1948). Accurate color comparison of a standard solution with an unknown requires uniformity of the following conditions: the optical path (i.e., distance through the solutions traversed by the light), transparency, wall thickness and color of the

TABLE 7. COMPOSITION OF MIXTURES GIVING pH VALUES AT 20°C AT INTERVALS OF 0.2
From Clark (1928), pp. 200–201.

KCl, HCl mixtures

pH	0.2 *M* KCl, in ml	0.2 *M* HCL, in ml	Dilute to, in ml
1.2	50	64.5	200
1.4	50	41.5	200
1.6	50	26.3	200
1.8	50	16.6	200
2.0	50	10.6	200
2.2	50	6.7	200

Phthalate, HCl mixtures

pH	0.2 *M* KH phthalate, in ml	0.2 *M* HCL, in ml	Dilute to, in ml
2.2	50	46.70	200
2.4	50	39.50	200
2.6	50	32.95	200
2.8	50	26.42	200
3.0	50	20.32	200
3.2	50	14.70	200
3.4	50	9.90	200
3.6	50	5.97	200
3.8	50	2.63	200

Phthalate, NaOH mixtures

pH	0.2 *M* KH phthalate, in ml	0.2 *M* NaOH, in ml	Dilute to, in ml
4.0	50	0.40	200
4.2	50	3.70	200
4.4	50	7.50	200
4.6	50	12.15	200
4.8	50	17.70	200
5.0	50	23.85	200
5.2	50	29.95	200
5.4	50	35.45	200
5.6	50	39.85	200
5.8	50	43.00	200
6.0	50	45.54	200
6.2	50	47.00	200

TABLE 7. COMPOSITION OF MIXTURES GIVING pH VALUES AT 20°C AT INTERVALS OF 0.2 (*Continued*)

KH_2PO_4, NaOH mixtures

pH	0.2 *M* KH_2PO_4, in ml	0.2 *M* NaOH, in ml	Dilute to, in ml
5.8	50	3.72	200
6.0	50	5.70	200
6.2	50	8.60	200
6.4	50	12.60	200
6.6	50	17.80	200
6.8	50	23.65	200
7.0	50	29.63	200
7.2	50	35.00	200
7.4	50	39.50	200
7.6	50	42.80	200
7.8	50	45.20	200
8.0	50	46.80	200

Boric acid, KCl, NaOH mixtures

pH	0.2 *M* H_3BO_3 0.2 *M* KCl, in ml	0.2 *M* NaOH, in ml	Dilute to, in ml
7.8	50	2.61	200
8.0	50	3.97	200
8.2	50	5.90	200
8.4	50	8.50	200
8.6	50	12.00	200
8.8	50	16.30	200
9.0	50	21.30	200
9.2	50	26.70	200
9.4	50	32.00	200
9.6	50	36.85	200
9.8	50	40.80	200
10.0	50	43.90	200

Notes. Overlapping members of the above series should be checked for consistency; i.e., phthalate "5.8" to "6.2" should match phosphates of the same pH numbers when tested with bromcresol purple; likewise for phosphate and borate "7.8" and "8.0" when tested with cresol red.

According to more recent assumptions used in standardization, the pH values given in the above table are too low about 0.03–0.04 unit of pH.

containers, concentration of indicator in each of the solutions, and radiant power incident upon the systems under comparison. Also, any inherent color in the unknown must be compensated by an equivalent amount in the optical path through the standard. These conditions are met by selecting clear, unscratched tubes of uniform bore, glass thickness, and color, by having the same concentrations of indicator in the unknown and the standard, by dispersing the color uniformly in the solutions, and by employing proper illumination.

The color comparison is conveniently made in a comparator block of the type described by Clark (1928, 1948). Various forms of this are obtainable from supply houses. Two pairs of tubes are arranged in the comparator as follows: (1) a tube containing buffer standard plus indicator behind which is placed a tube containing the unknown solution to compensate for inherent color and (2) a tube containing the unknown solution plus indicator backed by a tube containing distilled water. The two pairs of tubes are viewed against a uniform source of white light so placed that the beams incident upon the two systems are of the same radiant power. The color standards are successively compared with the unknown until a match is obtained, thereby establishing the pH of the unknown. If the color of the unknown falls between those of two adjacent standards, an interpolated pH number may be estimated.

Systems of fixed or "permanent" color standards are also available. These standards consist of colored glasses or other transparent material. Since the spectral absorptions of such standards would hardly be expected to be exactly the same as those of the indicators that they are supposed to match, the applicability and accuracy of these fixed standards must be determined in each case before they are placed in service. Acceptable sets of such standards can be of great convenience in the bacteriological laboratory, especially for approximate determinations.

The drop-ratio standards of Gillespie. If commercial color standards are not available and there are no facilities for making standard buffer solutions, color standards may be prepared by the drop-ratio method as refined by Gillespie (1920). The method of preparing the standards consists in setting up pairs of tubes, containing stepwise proportions, of the full alkaline color and the full acid color of an indicator in such a manner that the resulting color of each pair, when properly viewed, represents a definite pH within the sensitive range of that indicator.

A general notion of the arrangement and composition of the drop-ratio color standards may be obtained from inspection of Table 8. The preparation of the standards is explained in the next two paragraphs and in Table 9.

Although the alcoholic solutions of the indicator acids mentioned in Table 6 may be used, Gillespie recommends for accurate work the use of aqueous solutions of the indicator salts (the preparation of which is specified in Table 6), except in the case of methyl red. Table 9, lower half, gives specifications for the recommended concentrations of seven of the indicator stock solutions. The exact concentration of the indicator solutions is not very significant in much bacteriological work.

Select 18 test tubes of approximately the same bore (between 12 and 15 mm). They can be selected by adding 10.0 ml of water to a large number of test tubes and choosing a lot in which the columns of water come to approximately the same height (i.e., ± 1.5 mm). Place these 18 tubes in two rows in a rack, 9 tubes in each row. To the left-hand tube in the front row add 9 drops of the indicator solution, in the second tube

TABLE 8. DROP-RATIO COLOR STANDARDS FOR pH DETERMINATIONS

Tube pairs	Quantity of indicator solution to be added to each tube later to receive dilute alkali or acid and then brought to final volume of 5 ml	
	Acid tubes	Alkali tubes
1	9 drops*	1 drop
2	8 drops	2 drops
3	7 drops	3 drops
4	6 drops	4 drops
5	5 drops	5 drops
6	4 drops	6 drops
7	3 drops	7 drops
8	2 drops	8 drops
9	1 drop	9 drops

* If a little more accuracy is desired one may use a 1-ml pipet graduated in tenths and use the specified number of tenths of a milliliter instead of drops in preparing these tubes. In that case each tube should be brought up to a total volume of 10 instead of 5 ml.

place 8 drops, and so on to the last tube, which should contain 1 drop. In the back row of tubes place 1 drop in the left hand tube, 2 in the next, etc., up to 9 in the last. Make up approximately $20N$ stock solutions of NaOH and HCl [i.e., 0.2 per cent NaOH and 1 ml of concentrated HCl (sp gr 1.19) diluted to 240 ml]. Then, except in the case of those indicators for which different directions are given in Table 9, add 1 drop of the stock acid solution to each tube in the front row and 1 drop of the stock alkali solution to each tube in the back row; add enough distilled water to each tube to bring its total contents to 5 ml, thoroughly mix the contents of each tube, and return to its place in the rack. It will be seen from Table 9 that two of the indicators, namely thymol blue and bromphenol blue, require more of the alkali or the acid, respectively, than the other standards in order to ensure the appearance of full alkaline or acid color. In the case of thymol blue (alkaline range) and cresol red, the production of the required acid color (yellow) requires not a strong acid but a weaker one such as mono-potassium phosphate or, in the case of thymol blue, distilled water alone.

The arrangement of tube pairs indicated in Table 8 produces progressively different colors corresponding to steps of 10 per cent in the transformation of the indicator from its acid to its alkaline color. That is, each pair of tubes, when aligned between the eye and a source of white light, will show a color mixture corresponding to a definite pH. This pH can be computed by means of Eq. (5), which can be rewritten as

$$\mathrm{pH} = pK' + \log \frac{\text{drops of alkalinized indicator}}{\text{drops of acidified indicator}} \qquad (5a)$$

The fraction on the right side of the above equation is called the drop ratio. The values of the standards for seven of the indicators are given

TABLE 9. DATA FOR DETERMINING pH VALUE BY THE DROP-RATIO METHOD

Pair	No. of drops of indicator		pH value represented by each pair of tubes						
	Alkali tube	Acid tube	Brom-phenol blue	Methyl red	Brom-cresol purple	Brom-thymol blue	Phenol red	Cresol red	Thymol blue
1	1	9	3.0	4.05	5.2	6.15	6.85	7.35	7.95
2	2	8	3.4	4.4	5.6	6.5	7.2	7.7	8.3
3	3	7	3.6	4.6	5.8	6.7	7.4	7.9	8.5
4	4	6	3.8	4.8	6.0	6.9	7.6	8.1	8.7
5	5	5	4.0	5.0	6.2	7.1	7.8	8.3	8.9
6	6	4	4.2	5.2	6.4	7.3	8.0	8.5	9.1
7	7	3	4.4	5.4	6.6	7.5	8.2	8.7	9.3
8	8	2	4.6	5.6	6.8	7.7	8.4	8.9	9.5
9	9	1	4.9	5.95	7.0	8.05	8.75	9.25	9.85

Data as to stock solutions

Per cent concentration of indicator salt or acid	0.008 salt in water	0.008 acid in 95% alcohol	0.012 salt in water	0.008 salt in water	0.004 salt in water	0.008 salt in water	0.008 salt in water
Quantity 0.05*N* NaOH to produce alkaline color*.	1 drop	1 drop	1 drop	1 drop	1 drop	1 drop	2 drops
Quantity of acid† to produce acid color . .	1 ml	1 drop	1 drop	1 drop	1 drop	1 drop†	1 drop†

* If the standards are prepared by the method suggested in the footnote to Table 8 (i. e., measuring the indicator in tenths of 1 ml and diluting to 10 ml) it is well to use 0.1*N* instead of 0.05*N* NaOH to assure proper strength. The exact concentration or the exact number of drops used is of no great importance.

† Use approximately 0.05*N* HCl (or 0.1*N* if the method is modified as indicated in the footnote to Table 8) except in the case of cresol red and thymol blue. In the case of these two indicators a weaker acid must be used. Gillespie recommends 2 per cent KH_2PO_4, or in the case of thymol blue no acid need be used, water alone having a sufficiently high pH value to bring out the full acid color.

in Table 9. They may be computed for the other indicators by using the above equation and the pK' values in Table 6.

For approximate work it is often possible to compare the Gillespie standards with the unknown by merely holding the two tubes of the standard in the hand between the eye and a source of light. For accurate work, however, a comparator block must be used, but one with six holes instead of four, so that a tube of the unknown solution (without indicator) can stand behind the pair of tubes of the standard. The tube of the unknown for comparison with the standard should contain the same amount of indicator as the *sum* of those in the two standard tubes, i.e., 10 drops per 5 ml; and, of course, this tube must be backed by two tubes of water to equalize the optical path through the standard pair.

Indicator papers. Passing mention may be made of these laboratory aids for the approximate measurement of pH. Red and blue litmus papers for the detection of alkalinity and acidity are well known. Papers impregnated with other indicators, singly or in various combinations, can be made or obtained on the market. Those with a single indicator may be of use to detect roughly (about ± 0.3 to 0.4 pH) values within a relatively narrow zone of pH; those with indicator combinations enable one to detect, more roughly, pH values over wider zones of pH. Such papers are more reliable in buffered solutions than in unbuffered ones.

To be emphasized is the fact that the capillary action of the paper and of the sizing materials on the paper fibers may interfere, through selective sorption, with the normal interaction of solution and indicator. Generally speaking, a generous time of soaking of the paper for the establishment of equilibrium seems desirable. On the other hand, a standardization of the procedure may permit a short exposure (30 sec) to yield reproducible results, which are approximate in any case. See Kolthoff and Rosenblum (1937). Indicator papers are not recommended, except when the use of indicator solutions is precluded and a mere approximation is sufficient.

TITRATABLE ACIDITY, BUFFER ACTION, and pH ADJUSTMENT OF CULTURE MEDIA

In the titration of an acid with an alkali, or vice versa, a pH is reached at which the number of equivalents of acid equals the number of those of alkali. This pH is the *equivalence point* ("end point").

If both the acid and the alkali are completely ionized, e.g., HCl and NaOH, it is simple to calculate that this pH is about 7 and that in the case of 0.1*N* reactants, the pH of the HCl solution will sweep precipitously from about pH 4 to 7 upon the addition of the last tenth per cent of NaOH; further, the addition of the first tenth per cent excess of NaOH will cause a shift from pH 7 to about 10. In other words, the titration curve, constructed by plotting pH as ordinates and per cent neutralization as abscissas, is very steep at the equivalence point (pH 7) in this titration.

The ideal indicator for the detection of this equivalence point would be one capable of giving a distinctive color at pH 7, e.g., bromthymol blue. In practice, however, the steepness of the titration curve of the HCl at the equivalence point in the above example will permit this indicator to pass sharply from yellow to blue upon the final addition of a negligibly small excess of NaOH. For this reason, phenolphthalein (pK' 9.7) is frequently used for this purpose because the first appearance of its pink color, at about pH 8.5, is a convenient and usually sufficiently accurate indication of the end point of such a titration.

In fact, except for refinements that may be neglected for ordinary purposes, pH 8.5, detectable by means of phenolphthalein, is a fairly satisfactory end point for the titration of strong acids and of all weak acids with pK' values of less than 6.0. In the case of acids with pK' values greater than 6.0, it is necessary, by application of Eq. (5), to calculate the pH of the equivalence point, and to refine the method of end-point determination. For a discussion of the elementary theory of acid-base titration, see Clark (1928).

Titratable acidity of a culture. The titration of an acid (or a base) to an equivalence point, as discussed above, is a rational application of simple acid-base theory. On the other hand, in the titration of complex mixtures such as milk, tissue extract, or culture media, an equivalence point has no precise meaning. In such a case, the selection of an end-point pH is arbitrary, and fixed by custom (e.g., pH 8.5 with phenolphthalein) or by some special requirement.

In bacteriology, there is frequent need for determining the so-called titratable acidity produced during the growth of a culture in a fluid medium. To do this, it is necessary first to select a base line—that is, a pH number which is to be used as an end point in the titration and for the selection of an appropriate acid-base indicator. In the absence of special criteria, it is reasonable to choose as a base line the pH of the uninoculated medium. The selection of pH 7 as a base line may be acceptable, because many bacteria grow optimally in this region, not necessarily because it represents the pH of theoretical "neutrality." Other base lines may be chosen in accordance with the special requirements for which the titration is to be made.

The titratable acidity of the culture can be measured by titration of a known volume of the fluid with 0.1N NaOH to the predetermined end point as shown by a standardized glass electrode or by the color of a suitable indicator. In the latter case, it is necessary to prepare for comparison an appropriate color standard representing the pH of the chosen end point (see earlier discussion of the essential requirements for adequate color comparison). If the end-point pH is other than that of the uninoculated control, a titration is made of the latter and its titration value is subtracted algebraically as a correction, or "blank," from that of the culture. The result is usually recorded as milliliters of 0.1 normal acid per 100 ml of the culture fluid. If the culture produces an alkaline reaction, the titration is performed with 0.1 NHCl and recorded after correction, if any, in the same way but as a minus quantity of titratable acid. Special precautions are necessary if the titratable acidity is to include all the volatile acids, including CO_2 and bicarbonate, that may be present in the culture that is being titrated.

It should be emphasized that in most cases, the titratable acidity is merely a measure of the buffering capacity (see below) of the medium within the pH range observed. It does not permit further interpretation without additional data on the components of the culture. The titratable acidity is of some importance, along with final pH, in the comparison of high-acid-producing organisms. For such comparisons to be valid, it is necessary that the different organisms be grown in the same medium. Different media which vary in buffering capacity may yield misleading results.

Buffer action. The titration curve of a weak acid has a sigmoid shape, each end of the curve having a large (steep) slope and the main central portion having a small slope. This small slope expresses the buffer action of the system, that is, the ability of the system (comprising the weak acid and its salt) to resist large change in pH on the addition of acid or alkali. The sigmoid shape of the titration curve expresses, therefore, the fact that the buffer action of such a system is maximal at the mid-point and decreases on either side of this point, first gradually and then more extensively as either end of the curve is approached. The limits of the pH zone of effective buffer action may be arbitrarily set at 1.5 pH units greater and less than the pK' of the acid of the buffer system. It is obvious that increasing the concentration of the buffer system will increase its buffer action; therefore buffer action also depends upon the concentration of the buffer system.

The buffer action of a culture medium is dependent on its composition and may vary considerably in different regions of pH. Significant results as to final pH and titratable acidity in cultures depend to a large extent on comparisons made in media having buffer action that is uniform and adjusted in amount to the purpose of the test. A method for estimating such buffer action is as follows:

Assume, for example, that the initial pH of a culture medium is 6.8 and that it is desired to measure the buffering capacity of the medium between the pH limits 5.0 and 8.0. This can be done by titrating an aliquot, e.g., 5 ml, of the medium with $0.05N$ HCl to pH 5.0 and another aliquot with $0.05N$ NaOH to pH 8.0. The sum of these titers gives a simple and useful measure of the buffering capacity of the medium within the pH zone 5.0–8.0. Brown (1921) has described the procedure and some of its practical uses.

The pH-adjustment of a culture medium. This is done with the medium at about 80–90 per cent of its final volume. Prepare approximately normal NaOH and HCl stock solutions and also about 100 ml of each of these solutions diluted with distilled water exactly to one-tenth concentration. Assume, for example, that the adjustment of a colorless medium is to be made to pH 7.0 before sterilization. Test the pH of the medium to establish whether acid or alkali will be required for adjustment to pH 7. To determine the amount required, titrate 5 ml of the medium plus 5 drops of the appropriate indicator (e.g., bromthymol blue) with the diluted acid or alkali until the color almost matches that of 10 ml of standard buffer pH 7.0 plus 5 drops of the same indicator. Next, add water to the tube with the medium to bring the volume to 10 ml, mix well, and make a proper comparison with the standard. If the color difference is small, then small additions of either acid or alkali may be made to bring about a correct match without changing significantly the necessary volume relations. If the color difference is large, the titration should be tried again. (In the case of a medium with inherent color, this should be compensated as previously described.)

From the titration value, a calculation can be made of the amount of the stronger acid or alkali to be added to bring the bulk of the medium to the desired pH. The pH of the medium is checked after the addition, and when the pH is correctly adjusted, the medium is diluted with distilled water to the final volume.

In making a colorimetric pH determination of a well-buffered medium that is already colored, it is permissible to dilute the test sample of the medium 1:5 or 1:10 with distilled water to thin out the inherent color before proceeding with the test. The change in pH due to such dilution of a well-buffered solution is usually negligible. On the other hand, caution must be observed in employing the dilution procedure on poorly buffered solutions, because the results may be misleading should the distilled water, or even the indicator solution, be too far from the desired pH.

THE MEASUREMENT OF OXIDATION-REDUCTION POTENTIALS

Introduction. The oxidation-reduction reaction

$$Cl_2 + 2I^- \rightarrow 2Cl^- + I_2$$

represents an exchange of electrons between the chlorine:chloride system and the iodine:iodide system. These systems may be represented by the hypothetical "half-reactions"

$$Cl_2 + 2_\epsilon \rightleftharpoons 2Cl^-$$
$$I_2 + 2_\epsilon \rightleftharpoons 2I^-$$

to show the participation of electrons. In the interaction, chlorine is the electron-acceptor, and iodide the electron-donor.

The chlorine, the iodine, and a considerable number of other systems can be studied by means of electric cells in which such systems can display their relative oxidation-reduction tendencies in terms of electrode potentials. The latter permit evaluation of the change in Gibbs free energy (see later) in the interaction of any two such oxidation-reduction systems.

Without going into details of derivation or refinements, we may state that the electrode equation for a reversible oxidation-reduction system has the general form:

$$E_h = E_o - \frac{RT}{nF} \ln \frac{[\text{reductant}]}{[\text{oxidant}]} + \left(\frac{\text{a function of pH and}}{\text{dissociation constants}}\right) \tag{6}$$

where E_h is the potential, in volts, referred to that of the normal hydrogen electrode. E_o is a constant characteristic of the system at pH 0; R is the gas constant, 8.315 volt-coulombs per degree per mole; T is the absolute temperature; n is the number of electrons involved in the oxidation-reduction process; F is the faraday (96500 coulombs); ln is the logarithm to the base e; and brackets represent concentrations of the reductant and oxidant. At any fixed pH, the first and last terms on the right side of the above equation may be combined as a constant, E'_o; then

$$E_h = E'_o - \frac{RT}{nF} \ln \frac{[\text{reductant}]}{[\text{oxidant}]} \tag{7}$$

That is, $E_h = E'_o$ at any fixed pH when [reductant] = [oxidant].

It is apparent from Eq. (6) that the potential of such a system may be influenced by the pH of the solution and the potential of one system may vary relative to that of another as the pH is varied. In fact, cases are known where system A can oxidize system B at one pH level, and system B oxidize system A at another. Hence the importance of comparing such potentials at the same pH, as well as the same temperature, and the desirability of specifying pH in connection with a statement of the E_h of a system.

Elaboration of the theory of reversible oxidation-reduction potentials can be found in Clark (1928, 1948), Clark, Cohen, *et al.* (1928), and modern texts on electrochemistry, such as Glasstone (1942).

There are two methods of measuring oxidation-reduction potentials, the potentiometric method and the colorimetric. Each has its advantages and disadvantages, but the potentiometric method is generally preferable for reasons that will appear below. In either case, it is usually necessary to deaerate the container and the solution to be measured by evacuation or by displacing gaseous and dissolved oxygen with an inert gas such as purified nitrogen. Deoxygenation is often accomplished spontaneously in the depths of an actively growing culture of facultative bacteria.

The Potentiometric Method

Electrode vessel. This may be a test tube with a constriction and bulb at its lower end or a more elaborate container depending on the requirements of the experiment. Such vessels are described by Clark, Cohen, *et al.* (1928), Borsook and Schott (1931), Allyn and Baldwin (1932), and Hewitt (1936).

Electrodes. A "noble," or "unattackable," metal is the electrode of choice. A coil of bright platinum wire has been frequently employed, but this is difficult to clean thoroughly, and there is danger of entrapment of particulate material during a measurement. Platinum sheet, about 5 mm square or larger, is preferable.

Gold-plated platinum electrodes seem to have certain advantages. They can be readily replated to provide a clean, new surface and thereby obviate erratic electrode behavior. Second, gold, being relatively impervious to hydrogen, should have less tendency to act as a hydrogen electrode in a culture producing appreciable quantities of molecular hydrogen. However, some observers do not consider this of much practical importance.

Electrodes should be checked for reliability by measuring the potential of a known oxidation-reduction system (*e.g.*, quinhydrone in 0.1 M HCl, $E_h = 0.6351$ at 25°, see page 74). Where possible, duplicate or multiple electrodes should be employed, and one that exhibits persistent erratic behavior should be discarded. Unless the solution or culture under examination is well stirred, the electrode reading may record merely a local oxidation-reduction potential rather than one representative of the solution as a whole. In a heavily growing culture, electrodes may become coated with adherent cell masses, and duplicate electrodes may show widely divergent potentials even when the culture is well stirred.

The common method of cleaning a platinum electrode involves cautious treatment with aqua regia, hot concentrated nitric acid, or hot bichromate cleaning mixture, followed by thorough washing in water. For careful oxidation-reduction work, this procedure may not leave the metal surface altogether "inert." A more suitable procedure is to electrolyze a 1:1 solution of concentrated HCl with the electrode to be cleaned as the anode (gold-plated platinum may be deplated in the same way). The well-washed electrode may be stored in distilled water. If the metal surface remains dry for any length of time, the electrode may be sluggish in reaching an equilibrium potential.

Calomel half-cell (see also page 74). The "saturated" type of any convenient form is generally suitable, preferably one that permits flushing of the siphon outlet with saturated KCl solution in order to wash away contaminations from liquid junction contacts. Liquid junction between the calomel half-cell and culture should be of a kind which can be made aseptically when desired. For ordinary purposes, this is conveniently accomplished by preparing a glass tube partly sealed at one end over a piece of acid-washed asbestos fiber. This tube is filled by means of a capillary-tipped pipet with melted KCl-agar (40 g of KCl per 100 ml of 3 per cent agar in water) and autoclaved. The partly sealed end of the tube is inserted into the culture to provide the "liquid" junction, and the open end is placed in bubble-free contact with saturated KCl solution leading to the calomel half-cell.

Potentiometer and galvanometer. Generally speaking, cell suspensions and bacterial cultures are poorly stabilized with respect to oxidation-reduction potential. Consequently, disturbing polarization may occur if even the small amount of current necessary to operate the usual potentiometer and galvanometer circuit is allowed to pass through the half-cell containing the biological system under measurement, and the observed potential may be of uncertain accuracy and reliability. This difficulty can be minimized by the employment of a vacuum-tube potentiometer-electrometer of the kind now in common use for glass electrode measurements and provided with a scale graduated in volts.

The oxidation-reduction cell is set up by joining the saturated calomel half-cell with the half-cell containing the solution or culture to be measured as indicated in the following scheme:

Pt; solution X or or PtAu; culture X	KCl or KCl-agar bridge	saturated calomel (Pt) half-cell

Ordinarily, the potential of a culture is negative (reducing) to that of the calomel half-cell, and the metal terminals of the above oxidation-reduction cell are connected accordingly to the terminals of the potentiometer. The reading of potential thus obtained will be that referred to the calomel half-cell, and this observed potential, E_{obs}, can be converted to E_h, the potential referred to the standard normal hydrogen half-cell, by adding E_{obs} and E_{cal} algebraically. That is, $E_h = E_{obs} + E_{cal}$. Thus, if $E_{cal} = +0.250$ volt (see page 75) and $E_{obs} = -0.150$ volt, then $E_h = +0.100$ volt.

Significance of E_h measurements. The potentiometric method is direct and relatively simple. The interpretation of the results is, however, another matter. Discounting subsidiary, but sometimes important, instrumental effects such as potentials due to liquid junctions and temperature differences within the oxidation-reduction cell, all of which can be eliminated or minimized (see Clark, 1928), an observed E_h of a system such as ferric:ferrous iron, *under conditions of equilibrium and maximum work*, is a measure of the Gibbs free energy change, $nFE_h = -\Delta G$, in the reaction between the components of the two halves of the oxidation-reduction cell. This is the case for a considerable number of oxidation reductions which, alone or in the presence of catalysts and mediators, can take place more or less rapidly and reversibly as if a transfer of electrons, with or without accompanying protons, were direct and complete. These are reactions between so-called electromotively active systems, the E_h of which is fixed, at constant pH, by a characteristic constant and by the relative concentrations (more accurately, activities) of the components of each such system. For example, a potential of the ferric:ferrous system in acid solution can be defined by the relation which implies the

$$E_h = E'_o - \frac{RT}{F} \ln \frac{[Fe^{++}]}{[Fe^{+++}]} \quad (8)$$

limitation that definite and significant potentials are possible only in the presence of *finite* ratios of oxidant to reductant. In addition, the total concentration of the reversible system may be decreased to and beyond a level at which traces of electromotively active contaminants attain dominance and an observed potential becomes unstable and difficult to interpret.

In contrast to the above-mentioned reversible processes which are readily amenable to E_h measurement, there are a great many oxidation reductions that proceed by a variety of mechanisms that do not permit formulation and precise measurement in terms of equilibrium states. Electrode potentials in such cases are difficult to interpret and of uncertain significance.

In cell suspensions and bacterial cultures, especially when deprived of free access to oxygen, there develops with time a progressively more negative potential which traverses the zones characteristic of reversible oxidation-reduction indicators (see next section). Polarization of the electrode or a small dose of an oxidant may reverse the trend of reduction potential temporarily, but the trend is resumed after a while to levels of potential that may sometimes be associated with the type of cell and the various metabolites in the suspension or culture. Duplicate electrodes in such systems may not be in good agreement at the start, but they will reach about the same limiting value in time. For examples, see Clark, Cohen, *et al.* (1928), Allyn and Baldwin (1932), and Hewitt (1936).

The Colorimetric Method

Introduction. The empirical use of substances such as litmus or methylene blue as indicators of reduction in bacterial cultures is well known. For the determination of various degrees of reduction intensity an appropriate series of indicators is necessary. Among those available are reversible oxidation-reduction systems, the oxidants of which are usually colored and the reductants practically colorless. A number of such indicator systems have been characterized and may be employed, with due precautions, in determining an oxidation-reduction potential colorimetrically.

A selection of such indicators[1] is listed in Tables 10 and 11. Similar

[1] A special comment is necessary in regard to neutral red (compound *t* in Tables 10 and 11). It undergoes reversible reduction in the usual manner, and the colorless solution of reductant formed upon *rapid* reduction reoxidizes very rapidly when exposed to air. However, the reductant on standing in solution at pH 4–6 for a little time undergoes transformation to a fluorescent substance which is stable for days in the presence of air but reoxidizes rapidly upon acidification. As an oxidation-reduction indicator, therefore, neutral red must be employed with due caution and can be used only for rough comparisons.

tabulations are given by Hewitt (1936). Fuller details can be found in Clark, Cohen, *et al.* (1928) and Cohen (1933, 1935). Table 10 gives the names of the indicators, listed in the order of their E'_o values at pH 7.0,

TABLE 10. A SELECTION OF OXIDATION-REDUCTION INDICATORS E'_0 AT pH 7 (30°)

(Values of E'_0 between pH 5 and 9 will be found in Table 11)

	Compound	E'_0, *volts*
a.	Phenol-*m*-sulfonate-indo-2,6-dibromophenol	0.273
b.	*m*-Chlorophenol-indo-2,6-dichlorophenol	0.254
c.	*o*-Chlorophenol-indophenol	0.233
d.	2,6-Dichlorophenol-indophenol	0.217
e.	2,6-Dichlorophenol-indo-*o*-cresol	0.181
f.	1-Naphthol-2-sulfonate-indo-2,6-dichlorophenol	0.119
g.	Lauth's violet (Thionin)	0.062
h.	Cresyl blue	0.047
i.	Methylene blue	+0.011
j.	Indigo tetrasulfonate	−0.046
k.	Methyl Capri blue	−0.061
l.	Indigo trisulfonate	−0.081
m.	Indigo disulfonate	−0.125
n.	Gallophenine	−0.142*
o.	Brilliant alizarine blue	−0.173*
p.	Phenosafranine	−0.252
q.	Tetramethyl-phenosafranine	−0.273
r.	Safranin T	−0.285
s.	Induline scarlet	−0.299
t.	Neutral red†	−0.324
u.	Rosindone sulfonate No. 6	−0.385
	(hydrogen at 1 atm)	(−0.421)

* At 25°.

† See the footnote on p. 94 in text.

and Table 11 gives the corresponding E'_o values at successive levels between pH 5.0 and 9.0. The magnitude of the salt and protein errors of these compounds has not been established.

Each indicator system listed in Tables 10 and 11 involves a two-electron transfer, and the relation of E'_o to other factors at *fixed pH* is given by Eq. (9).

$$E_h = E'_o - \frac{RT}{2F} ln \frac{[\text{reductant}]}{[\text{oxidant}]} \tag{9}$$

Converted to ordinary logarithms after insertion of numerical values, this equation becomes, at 30°C,

$$E_h = E'_o - 0.030 \log \frac{[\text{reductant}]}{[\text{oxidant}]} \tag{10}$$

TABLE 11. SELECTED OXIDATION-REDUCTION INDICATORS, RELATION AT E'_o TO pH (30°)
(Letters refer to compounds listed in Table 10; the values listed are E'_o in volts)

pH	a	b	c	d	e	f	g	h	i	j	k	l	m	n*	o*	p	q	r	s	t†	u
5.0	+0.390	+0.391	....	+0.366	+0.335	+0.262	+0.138	+0.149	+0.101	+0.065	+0.038	+0.032	−0.010	−0.003	−0.040	−0.158	−0.157	−0.197	−0.235	−0.205	
5.5	0.360	0.359	....	0.332	0.300	0.230	0.109	0.117	0.072	0.035	+0.006	−0.002	−0.040	−0.042	−0.080	−0.188	−0.194	−0.227	−0.253	−0.236	
6.0	0.330	0.326	+0.301	0.295	0.261	0.196	0.094	0.189	0.047	+0.006	−0.021	−0.028	−0.069	−0.077	−0.112	−0.215	−0.225	−0.251	−0.268	−0.265	−0.298
6.5	0.301	0.290	0.269	0.255	0.220	0.158	0.077	0.066	0.028	−0.022	−0.043	−0.056	−0.098	−0.110	−0.142	−0.234	−0.252	−0.270	−0.284	−0.294	−0.349
7.0	0.273	0.254	0.233	0.217	0.181	0.119	0.062	0.047	+0.011	−0,046	−0.061	−0.081	−0.125	−0.142	−0.173	−0.252	−0.273	−0.285	−0.299	−0.324	−0.385
7.5	0.246	0.220	0.195	0.182	0.145	0.080	0.047	0.030	−0.005	−0.066	−0.077	−0.103	−0.148	−0.172	−0.203	−0.269	−0.288	−0.300	−0.314	−0.352	−0.425
8.0	0.218	0.188	0.155	0.150	0.112	0.046	0.030	+0.015	−0.020	−0.083	−0.093	−0.121	−0.167	−0.202	−0.226	−0.284	−0.303	−0.316	−0.329	−0.382	−0.460
8.5	0.192	0.159	0.117	0.119	0.081	+0.016	+0.017	−0.001	−0.035	−0.099	−0.108	−0.137	−0.184	−0.232	−0.251	−0.299	−0.319	−0.331	−0.344	−0.410	−0.491
9.0	0.168	0.133	0.082	0.089	0.051	−0.012	−0.001	−0.016	−0.050	−0.114	−0.123	−0.152	−0.199	−0.262	−0.279	−0.314	−0.334	−0.347	−0.359	−0.438	−0.520

* At 25°.
† See the footnote on p. 94 in text.

The relation of percentage reduction to potential as defined by the last term in Eq. (10) is given in Table 12. For example, if methylene blue is observed to be 80 per cent reduced at pH 7,

$$E_h = 0.011\text{–}0.018 = -0.007 \text{ volt.}$$

Color standards. Since the compounds listed in Tables 10 and 11 are practically one-color oxidation-reduction indicators, color standards of sufficient approximation can be prepared simply by graded dilutions of the colored component, the oxidant. It should be borne in mind that some of the compounds are also acid-base indicators; therefore it may be necessary to set up the color standards in a buffer at the same pH as the solution or culture under test.

Colorimetric measurement. The general principles of color comparison as outlined for the indicator method of pH determination are applicable here. In addition, special precautions are required to make certain that the measurement is a valid one. An indicator may fade in a test solution for reasons other than simple reduction. The compound may precipitate or adsorb on suspended particles, or it may be decomposed; in such cases judicious treatment with a suitable oxidizing agent (e.g., ferricyanide or air) will not immediately restore the initial color of the oxidant. Moreover, many reversible oxidation-reduction systems are so sensitive to oxygen as to require extreme precaution for its exclusion. This applies to the electrometric method as well as to the colorimetric.

TABLE 12

Reduction, %	−0.03 log ratio, volts	Reduction, %	−0.03 log ratio, volts
1	+0.060	60	−0.005
10	0.029	70	−0.011
20	0.018	80	−0.018
30	0.011	90	−0.029
40	0.005	99	−0.060
50	0.000	100	(− ∞)

It is a fact that many biological systems act as if they contain, at any moment, only minute amounts of electromotively active oxidation-reduction substances; therefore the addition to such a system of even a small amount of indicator oxidant may suffice to oxidize the system at once without appreciable reduction of the indicator. This drawback cannot be overcome except by allowing sufficient time for the biological system to overcome the poising[1] effect of the added indicator. However, the

[1] Poising action of an oxidation-reduction system is analogous to buffer action of an acid-base system. (Compare paragraph on buffer action, p. 89.)

time required may be very long (especially in relation to the most active period of a growing bacterial culture) so that it may be difficult or impossible to determine successive E_h values colorimetrically at brief intervals.

Furthermore, the indicator may not merely come into simple oxidation-reduction equilibrium with the components of the system under test. It may act catalytically to displace the oxidation-reduction equilibrium that it is supposed to measure, or it may be toxic toward living cells or combine chemically with components of the system under test.

In summary, the indicator method, often applicable where it is impossible to employ an electrode, may give results that require considerable caution in interpretation, especially the results obtained on unstable oxidation-reduction systems or on biological material containing them.

REFERENCES

Allyn, W. P., and I. L. Baldwin. 1932. Oxidation-reduction potentials in relation to the growth of an aerobic form of bacteria. *J. Bacteriol.*, **23,** 369–398.

Borsook, H., and H. F. Schott. 1931. The rôle of the coenzyme in the succinate-enzyme-fumarate equilibrium. *J. Biol. Chem.*, **92,** 535–557.

Brown, J. H. 1921. Hydrogen ions, titration and the buffer index of bacteriological media. *J. Bacteriol.*, **6,** 555–568.

Clark, W. M. 1928. "The Determination of Hydrogen-Ions," 3rd ed. The Williams & Wilkins Company, Baltimore.

———. 1948. "Topics in Physical Chemistry." The Williams & Wilkins Company, Baltimore.

Clark, W. M., Barnett Cohen, *et al.* 1928. Studies on Oxidation-Reduction, I-X. *U. S. Public Health Service, Hyg. Lab. Bull.* 151.

Clark, W. M., and H. A. Lubs. 1917. The colorimetric determination of hydrogen-ion concentration. *J. Bacteriol.*, **2,** 1–34, 109–136, 191–236.

Cohen, Barnett. 1926. Indicator properties of some new sulfonphthaleins. *Public Health Rpts.*, **41,** 3051–3074.

———. 1933. Reversible oxidation-reduction potentials in dye systems; (*also*) Reactions of oxidation-reduction indicators in biological material, and their interpretation. *Cold Spring Harbor Symposia Quant. Biol.*, **1,** 195–204, 214–223.

———. 1935. Oxidations and Reductions, chap. 19 in "A Textbook of Biochemistry," by B. Harrow and C. P. Sherwin. W. B. Saunders Company, Philadelphia.

Dole, M. 1941. "The Glass Electrode." John Wiley & Sons, Inc., New York.

Gillespie, L. J. 1920. Colorimetric determination of hydrogen-ion concentration without buffer mixtures, with especial reference to soils. *Soil Sci.*, **9,** 115–136.

Glasstone, Samuel. 1942. "An Introduction to Electrochemistry." D. Van Nostrand Company, Inc., Princeton, N.J. (See Chap. 8.)

Hewitt, L. F. 1936. "Oxidation-Reduction Potentials in Bacteriology and Biochemistry," 6th ed. The Williams & Wilkins Company, Baltimore.

Kolthoff, I. M., and Charles Rosenblum. 1937. "Acid-base Indicators." The Macmillan Company, New York.

Small, James. 1946. "pH and Plants." D. Van Nostrand Company, Inc., Princeton, N.J.

CHAPTER V

Maintenance and Preservation of Cultures

Freeman A. Weiss

INTRODUCTORY

The title of this chapter suggests that there are two somewhat distinct aims in keeping cultures of microorganisms—*maintenance* and *preservation,* or perhaps conservation would be a better term. Sometimes the preservation of cultures is only a continuation of the process of maintaining them, but often it implies something more. Maintenance means essentially supporting the culture and keeping it alive, pure, and in recognizably typical growth. Preservation has the connotation of long-term maintenance, but in addition, in view of the propensity of all organisms to vary—especially microorganisms because of the high frequency of generations of which they are capable—it also means maintaining them at essentially constant biological potentials. A preserved culture, for example one lyophilized (from *lyo,* loose, + *philos,* loving) or in dry soil, may appear macroscopically anything but typical, but it can be restored to its typical morphology and usually to its initial physiological and biochemical characteristics by suitable manipulation. On the other hand, cultures that are merely maintained may show typical form, may grow luxuriantly, but if they have lost some biological characteristic for which they were primarily selected, they have not been satisfactorily preserved.

The basic function of a culture collection is to conserve cultures, not merely maintain them. This distinction does not always appear, but it exists, and it is with this in view that the methods and materials recommended in this chapter were selected. They are essentially the procedures adopted by the American Type Culture Collection after long experience in both maintaining and preserving cultures. These methods also are subject to constant change as better techniques or materials are discovered, and no claim to universal application or general superiority is or can be made. In fact the procedures outlined here should always be supplemented, at least for routine maintenance of cultures, by technics recommended in the manuals published by the manufacturers of dehydrated and other ready-prepared culture media (Difco Laboratories, Baltimore Biological Laboratory, and others).

SELECTION OF MATERIALS

As the maintenance of cultures is largely a mechanical task (for which, however, no suitable machine has been devised), it is desirable to reduce all mechanical factors to a minimum consistent with efficient manipulation of the cultures. The size and number of culture vessels, their convenience in handling, the volume of culture media required, the storage space for cultures, especially refrigerated storage, all must be taken into consideration. For the conditions under which the American Type Culture Collection operates, the following specifications for materials have proved very satisfactory and are recommended to others, though recognizing that other laboratories may have varied preferences of their own.

Culture tubes. These are 125 by 16 mm with straight sides (without flare) for cotton plugs or screw-cap tubes with rubber liners (plastic liners are prone to drop out, at times they leak, and they may give off toxic emanations that suppress growth of some cultures). These tubes hold 5–6 ml of culture media. For slow growers like *Mycobacterium tuberculosis*, and those requiring special media, as marble chips for *Nitrobacter* or a large surface of floating sulfur for *Thiobacillus*, larger culture tubes or flasks are needed.

Shell vials for lyophilizing. Outer vial is 85 by 14 mm; inner vial 35 by 9.5–10.00 mm (outside dimensions).

Culture tube racks. For the size of tube specified, a standard zinc-coated wire rack about 7¾ by 3¼ by 3½ in. will hold 40 tubes; two of these joined end to end will fit the shelf depth of average-sized household or laboratory refrigerators. A storage space 40 by 16 by 16 in. will accommodate 800 cultures. Special racks or trays for holding smaller culture tubes or lyophilized specimens that are even more economical of space are commercially available.

Milk solution for lyophilizing. Although various fluids, including serum (usually from beef blood), corn-steep liquor, and solutions of lactose or glucose, have been proposed for obtaining cell suspensions for lyophilizing, skim-milk powder, reconstituted at twice the original solid content of milk, possesses general advantages. It is readily obtainable and of uniform quality, can be stored without deterioration for many months (refrigerated), and has no adverse chemical or physical effects on most organisms. It is not suitable for all bacteria, however; for example, *Vibrio comma* must be lyophilized from serum. Milk must be carefully prepared for use in lyophilizing, as an excess of heat will cause clotting or partial caramelization, making it difficult to obtain a uniform suspension of cells and perhaps interfering with dehydration. The following specifications have given consistent satisfaction.

Warm 200 ml of distilled water, and add 40 g of powdered skim milk. Stir thoroughly, preferably with an electric mixer. Filter through cheesecloth and tube, 5–6 ml per tube. Autoclave exactly 13 min at 115°C (approximately 11 lb). Ample space for the circulation of heat around the tubes must be provided to ensure sterilization at this exposure.

Vaspar, a mixture of equal parts of 45° mp paraffin and white petrolatum, makes an excellent seal for cultures of anaerobes and for liquid cultures that are transported or mailed. Sometimes a layer of melted agar is added above the culture, for example of *Lactobacillus* in milk, to accomplish the same purpose. This effects a tighter seal, since wax may shrink away from the glass and permit leakage of fluid, but a wax seal has the advantage, on the other hand, that a pipet can be pushed through it without disturbing the seal or clogging, thus permitting easy transfer of the culture.

White mineral oil or liquid petrolatum, of the medicinal grade, is commonly used to seal cultures of various kinds of microorganisms, on agar slants or stabs and also broth cultures, against desiccation and oxidation, thus preserving them for months or sometimes several years beyond the survival of unsealed cultures. Both vaspar and mineral oil can be sterilized by autoclaving at 15 lb pressure for 60 min and then driving off any entrapped moisture by heating in a drying oven at 110°C for about 1 hr.

DORMANT CONSERVATION

Lyophilization. The essentials of the freeze-drying method of preserving cultures may be stated as follows: (1) Obtain as dense as possible a suspension of cells of the organism to be preserved by washing down a culture that is grown to the proper stage of maturity with a suitable fluid, commonly serum, milk, or 3 per cent lactose; (2) transfer 0.1–0.2 ml of the suspension to vials; (3) quick-freeze these in a mixture of dry ice and 95 per cent alcohol; (4) evacuate while frozen until completely desiccated; (5) hermetically seal the vials while evacuated. There are many variations of the method, and some elaborate types of equipment have been devised to accommodate a large number of vials at one time and to control environmental factors.

In commercial establishments and some institutions where lyophilized preparations are produced in great quantities, it is customary to use glass tubing, cut in short lengths and sealed at one end, or very small culture tubes, which are handled en bloc through the freezing, dehydrating, and sealing processes, by attaching them to a multioutlet manifold. The tubes or vials may be as small as 3–4 mm in outside diameter, 1.5–2 mm inside. To fill such small vials or reconstitute the culture requires a capillary pipet or hypodermic syringe, and the vials, especially if hard glass is used, are difficult to open and are subject to considerable risk of contamination by inrush of air when the vacuum is broken.

The procedure which has long been in use (15 years) at the American Type Culture Collection eliminates most of these difficulties, and especially the hazard of contamination when the culture is opened, by using a double-vial arrangement, one within the other, the inner one plugged with cotton like an ordinary culture tube.

Specifications for the vials have been given. The small vials are cleaned, cotton-plugged, and autoclaved in advance. Prior to use they are labeled with glass-marking ink. Filling is done with an ordinary measuring pipet, 1 ml size, graduated in 0.1 ml to a basal line (not to the tip, as it is undesirable to blow the last drop of fluid into the small vial). For most cultures, the use of slightly over 1 ml of fluid to wash down the slope will provide a cell suspension of sufficient density so that 0.15–0.2 ml dispensed in each vial yields satisfactory lyophilized preparations. Thus from one growing culture five preserved ones can be obtained. The last drop or two from the pipet is customarily placed on an agar slant and spread evenly over the surface or transferred to other suitable media as a check on the purity of the culture during these manipulations.

After filling, the cotton plugs are trimmed just above the rim of the vials, and the vials are quickly transferred to a freezing mixture of roughly equal parts of finely crushed dry ice and alcohol. A convenient receptacle for this is a large petri dish, covered with a coarse wire screen through which the vials will pass and which holds them upright. The freezing mixture need not be more than about 1 cm deep. As soon as the culture has solidified, the vials are transferred by means of long forceps to a stout bottle having a ground-glass top, to which is connected a U tube terminating in a moisture trap and having a side outlet near the top which is connected to a vacuum pump. The vessel holding the vials is enclosed in a suitable container which is packed with coarsely crushed dry ice, serving as a cold jacket to keep the material frozen through the initial stage of dehydration. This will ordinarily be within an hour or two, and this time should not be greatly exceeded, as prolonged freezing at this temperature is inimical to survival. Throughout the process of dehydration the distal end of the U tube, with the moisture trap, is kept in a thermos vessel surrounded with dry ice and alcohol, hence serving as the cold element of the distilling apparatus. The moisture trap serves to protect the oil in the vacuum pump from absorption of water. The pump must run until desiccation is complete, which will ordinarily require about 6 hr, but overnight operation is a convenient way to ensure enough time, and the sealing is completed the next morning.

In the meantime a set of the large vials is prepared to receive the small ones by placing a pad of cotton in the base; then the small vial is inserted and in turn is covered with a wad of shredded asbestos which is pressed rather firmly against its top. The asbestos plug protects the cotton from scorching and the inner vial from excessive heat while the upper part of the large vial is drawn out to a thin neck. This step is carried out by heating the vial, about 4 cm from its open end, in a gas-air torch set to give a narrow pointed flame, rotating it so as to soften and constrict uniformly, until an isthmus about 5 cm long and 2–3 mm in outside diameter can be drawn out. Great care must be taken that the bore of the isthmus is not closed in this process. The open end of the vial, which still has its original diameter, serves for attachment to a manifold for the final exhaustion and sealing.

A six-outlet manifold is convenient for this purpose, and for expeditious progress if many vials are to be sealed, two such manifolds, each provided with a shutoff valve, are connected by a Y tube to a manometer and then to a vacuum pump. The connection between vial and manifold outlet is made through a snug-fitting rubber cork, which must be kept well lubricated with stopcock grease. The vials remain on the manifold under as high a vacuum as the apparatus will pull (it should be at least 100 to 150 μ, and 50 is preferable) for about 10 min; then each is separated in succession by burning off the neck in a micro bunsen burner. Care must be taken not to overheat the rounded end of the vial back of the tip; otherwise it may collapse because of the vacuum within and thus form a precarious seal.

Subsequently, usually within a day or two and again for about 10 days, before the preparations are filed away as reliable, the vials are tested with a high-frequency vacuum tester. A glow within the vial when the tip of the apparatus is brought near it is indicative of a suitable vacuum, but only momentary exposure to the discharge is advisable, as microscopic perforations of the glass may result, with loss of vacuum.

If the survival capacity of the organism under lyophilization is not known, a check for viability should be made after one month and again after six. If viability is demonstrated at these intervals, it may be assumed that the preparation will remain viable for the average survival period of the species. This may vary from 2 to 10 or more years. The over-all viability gradually declines, the rate depending largely on the number of viable cells in the original preparation.

Most lyophilized preparations survive well if stored at room temperature, but some species of *Hemophilus*, *Lactobacillus*, *Neisseria*, *Pasteurella*, and doubtless others are short lived unless refrigerated.

For reconstituting the culture the tip of the vial above the asbestos wad is heated moderately in a gas flame, then a drop of water is placed on it. The resultant cracking breaks the vacuum, and the inner vial, protected by its cotton plug from contamination, is withdrawn. Reviving the culture is the reverse of the filling process. Enough nutrient broth to liquefy the pellet is added by means of a pipet, then withdrawn and transferred to a tube of broth or an agar slant. It is often desirable to inoculate both a broth culture and a slant. The latter permits spreading the inoculum out so that often discrete colonies develop from which subsequent transfers can be made. The broth, on the other hand, permits enfeebled cultures to start development when direct inoculation of a slant might fail.

Oil Sealing

Cultures of many kinds of bacteria can be preserved for periods greatly exceeding the viability of ordinary agar or broth cultures by sealing them with sterile paraffin oil. This is essentially a process of preventing moisture loss from the medium, although suppression of growth by excluding oxygen may also contribute to longevity. Among the bacteria, as a rule only species that produce a copious surface growth and can therefore be easily transferred are well adapted to this technique. Some kinds of *Rhizobium* (but not all species) have survived for 4 or 5 years by this method. Other genera that survive well in oil-sealed cultures are *Achromobacter*, *Bacillus*, most of the *Enterobacteriaceae*, *Flavobacterium*, *Micrococcus* (some species), *Proteus*, *Pseudomonas*, *Sarcina*, *Serratia*, *Streptococcus*, *Vibrio* (saprophytic forms). On the other hand, cultures of the following have not been successfully preserved under oil seals in our experience: *Azotobacter*, *Lactobacillus*, *Leuconostoc*, *Mycobacterium*, *Rhodospirillum*, *Salmonella*.

The procedure of oil sealing is simplicity itself. Screw-cap tubes are preferable to cotton-plugged ones, as there is always some risk of getting oil on cotton plugs. The preparation of the oil has been described. For treating only a few cultures the oil is simply poured over the culture until all the agar is covered. The use of a separatory funnel as a reservoir for the oil and of a glass sleeve enclosing the tip of the funnel, into which the culture tube is inserted as a shield against atmospheric contamination, may be advisable when a considerable number of cultures are to be oiled. It is important not to let the oil get too warm from repeated flaming of the reservoir tube or from contact with the flamed lip of the culture tube. No portion of the agar must remain uncovered; otherwise loss of water will continue through any exposed surface. It is advisable to store oiled cultures at the same temperature at which unsealed ones are kept for preservation.

Soil Cultures

Soil is sometimes used directly as a growing medium (though more often for molds and Streptomycetes than for bacteria), and it is also used as an absorbent or desiccating medium for preserving cultures. In the first case soil that is suitably pulverized and screened is placed in culture tubes, and enough water or very dilute culture medium is added to raise it to about 60 per cent of its maximum water-holding capacity. It must be autoclaved at least 1 hour[1] and tested for sterility before use. After the culture has sufficiently grown, it is simply allowed to become air dry and is then suitably sealed and stored at room temperature or refrigerated.

If soil is to be used merely as an absorbent, it is first air dried, then tubed and autoclaved, and finally heated in a drying oven until moisture-free, after which the tubes are stored in a desiccator until needed. They are then inoculated with a broth culture or with a cell suspension washed from an agar slope, as described under "Lyophilization." For preparing a cell suspension a 2 per cent solution of peptone is useful. A 0.05- to 0.1-ml portion of this suspension is placed with a pipet on 1 ml of soil, and the tubes are returned to a desiccator until dry or may be tightly stoppered if the soil already appears dry.

A variation using silica gel granules as an absorbent has been proposed by Dr. J. Lederberg, of the University of Wisconsin. Silica gel of the grade known as 40, 6-16 mesh is used, and 1–1.5 g is placed in culture tubes about 75 by 8 mm, which are then plugged and baked in an oven at 170°C for 2 hr. They are cooled and stored in a desiccator until needed. A cell suspension in 2 per cent peptone is prepared and pipetted on to the granules as described above. The tubes may then be hermetically sealed, without evacuating, or stoppered and stored in a desiccator. The method is still in the trial stage but has given promise of being successfully adaptable.

Other Desiccants

Pieces of thread and strips of filter paper have long been used as absorbents of spore suspensions of both aerobes and anaerobes. After immersion in a spore suspension, followed by air drying, these materials can be stored in glassine envelopes or culture tubes for many months with satisfactory retention of viability.

Deep Freezing

The storage temperatures indicated in the following section for holding cultures of various bacteria between routine transfers are only low enough to retard or suppress normal growth; they are not necessarily the minimum temperatures at which the cultures will survive. Temperatures

[1] Some technicians advise much more drastic sterilization, up to 2 hr at a time and repeated three times at 1-day intervals. Excessive heating of soil should be avoided, however, because it breaks down the organic components and may release toxic products. We consider 1 hr at 15 lb as generally enough.

TABLE 13. MAINTENANCE METHODS (AMERICAN TYPE CULTURE COLLECTION)

Genus	Species	Medium formula No.	Transfer time	Incubation temp, °C	Storage temp, °C	Lyophilization*
Acetobacter	Glucose positive	2,3	1 mo	28	10	+
Acetobacter	Mannitol positive	4	1 mo	28	10	+
Achromobacter	Various	23†	1 to 2 mo	Room	10	+
Achromobacter	Fischeri	23*a*	1 mo	Room	Room	+
Actinomyces	Bovis	10 (stab culture under seal)	1 mo	34–37	Room	−
Aerobacter	Various	23†	2 mo	30	10	+
Agrobacterium	Various	23†	2 mo	Room	10	+
Alcaligenes	Various	23†	3 mo	34–37	10	+
Arthrobacter	Various	23†	3 mo	Room	10	+
Azotobacter	Glucose positive	5	4 mo	Room	10	+
Azotobacter	Mannitol positive	6	4 mo	Room	10	+
Bacillus	Various	23†	12 mo-indefinite	28‡	10	+
Bacillus	Starch-hydrolyzing	27	12 mo-indefinite	28	10	+
Bacillus	Urea-decomposing	36	12 mo-indefinite	28	10	+
Bacteroides	Various	8,10	1 mo	37	28	+
Brucella	Various	26	6 wk	34–37	10	+
Cellulomonas	Various	23†	3 mo	30	10	+
Cellvibrio	Vulgaris	12	1 mo	Room	10	−
Chromobacterium	Various	23†	6 wk	Room	18	+
Clostridium	Various	7,17	6 mo-indefinite	28	Room	+
Corynebacterium	Acnes	10	1 mo	34–37	18	+
Corynebacterium	Diphtheriae	10,20	1 mo	35–37	18	+
Corynebacterium	Animal sources	8,10	1 mo	30	18	+
Corynebacterium	Plant pathogens	23,25	1–2 mo	Room—28	18	+
Desulfovibrio	....	30	6 mo	34–37	10	−
Diplococcus	Pneumoniae	8	....	35–37	......	++
Erwinia	Various	23†	3 mo	Room	10	+
Erwinia	Tracheiphila	20	1 mo	Room	Room	−
Erysipelothrix	Insidiosa	8	1 mo	34–37	10	+
Escherichia	Various	23†	3 mo	34–37	10	+
Flavobacterium	Various §	1	3 mo	30	10	+
Gaffkya	Tetragena	20	3 mo	35–37	10	+
Hemophilus	Influenzae	8,24	....	35–37	......	++
Hemophilus	Pertussis	9 with blood	1 mo	35–37	10	+
Klebsiella	Various	20	2 mo	34–37	10	+
Lactobacillus	Various	17,33,34	2–4 wk	34–37†	10	+
Leuconostoc	Various	17,33	6 wk	Room	10	+
Listeria	Various	10	4 mo	34–37	10	+
Malleomyces	Mallei	16	1 mo	34–37	Room	+
Micrococcus (including *Staphylococcus*)	Various	23†	3 mo	30‡	10	+
Mycobacterium	Saprophytic	16	4 mo	30‡	10	−
Mycobacterium	Tuberculosis	14,18	3 mo	34–37	10	−
Moraxella	Various	8	....	34–37	......	++
Neisseria	Gonorrhoeae	11 with CO_2	10 days	35–37	35–37	++
Neisseria	Meningitidis	8,20	10 days	35–37	35–37	++
Neisseria	Saprophytic	10,13,20	1 mo	34–37	Room	+
Nitrobacter	Agile	21	1 mo	Room	Room	−
Nitrosomonas	Europeae	22	1 mo	Room	Room	−

TABLE 13. MAINTENANCE METHODS (AMERICAN TYPE CULTURE COLLECTION) (*Continued*)

Genus	Species	Medium formula No.	Incubation time	Storage temp, °C	Storage temp, °C	Lyophilization*
Pasteurella	Various	10,13,20 with blood	1 mo	34–37	18	++
Propionibacterium	Various	33	2 mo	30	10	+
Proteus	Various	23†	3 mo	34–37	10	+
Pseudomonas	Aeruginosa	23†	3 mo	34–37	10	+
Pseudomonas	Plant pathogens	23,25,37	3 mo	Room—28	10	+
Rhizobium	Various	28‖	1 mo	Room	18	?
Rhodospirillum	Rubrum	23	6 wk	Room	18	−
Salmonella	Various	23†	3 mo	34–37	10	++
Sarcina	Various	23†	3 mo	30	10	+
Serratia	Various	23†	3 mo	30	10	+
Shigella	Various	23†	....	34–37	......	++
Spirillum	Various	23,23*a*	1 mo	Room	Room	−
Streptococcus	A and B groups	19 with blood under oil	1 mo	34–37	10	+
Streptococcus	D group	33,34	1 mo	30‡	10	+
Streptomyces	Various	15	3 mo	Room‡	10	+
Thiobacillus	....	31,32	7–10 days	26	26	−
Vibrio	Comma	20	2 mo	34–37	10	−
Xanthomonas	Various	25,38	2 mo	Room—28	10	+

* Plus sign indicates successful preservation by this method, ++ that the organism in question should preferably be kept lyophilized; minus sign that lyophilization is unsuccessful or not tried at ATCC.

† Tryptone-glucose-yeast medium (No. 35) as used by the Northern Regional Research Laboratory, U.S. Department of Agriculture, may be substituted.

‡ The optimum temperature for some species is different; consult Bergey's Manual.

§ A special medium is required by certain species.

‖ Some strains are favored by the addition of $CaCO_3$ to this medium.

below 0°C are sometimes employed for long-term preservation of bacterial cultures, including some that are not amenable to dry methods of storage. The principle involved is similar to that applied in quick freezing of food products, red blood cells, spermatozoa, and vital tissues. Relatively little is known as yet about the effect of subfreezing temperatures on the viability of bacterial cultures, including such factors as minimum temperature that is tolerated and duration and rate of freezing. The available information is reviewed in references cited at the end of this chapter.

Since essentially aqueous solutions or emulsions such as broth and agar become solid at these temperatures, often with rupture of glass containers even if the organisms survive, some means of counteracting the physical effect of ice formation is desirable. Hollander and Nell (1954) have suggested the use of 15 per cent glycerol to accomplish this purpose. In general, it appears that moderately low temperatures, as −10 or −20°C, have been more favorably regarded for the preservation of bacterial cultures than lower or higher subfreezing temperatures. There are few conclusive data, however.

MEDIA

Formulas employed by the American Type Culture Collection:

1. *Ayers & Johnson Agar*

Stock culture agar (Difco)	50 g
Distilled water	1,000 ml

pH 7.4

2. *Acetobacter Agar* (*Glucose*)

Autolyzed yeast	10 g
$CaCO_3$	10 g
Agar	15 g
Distilled water	1,000 ml
Heat to 100°C, and add	
Glucose	3 g

In tubing, the $CaCO_3$ should be distributed evenly between tubes. After autoclaving, the tubes should be shaken, then cooled quickly and slanted so as to keep the $CaCO_3$ in suspension.

3. *Acetobacter Agar with Liver Extract*

Liver extract	100 ml
Tryptone	5 g
Agar	20 g
Distilled water	900 ml
Heat to 100°C, and add	
Glucose	20 g
$CaCO_3$	10 g

Observe precautions as in formula 2 to keep $CaCO_3$ evenly suspended.

4. *Acetobacter Agar* (*Mannitol*)

Yeast extract	5 g
Peptone	3 g
Agar	15 g
Distilled water	1,000 ml
Heat to 100°C, adjust pH to 7.4, and add	
Mannitol	25 g

5. *Azotobacter Agar* (*Glucose*)

K_2HPO_4	1.0 g
$MgSO_4$	0.2 g
NaCl	0.2 g
$FeSO_4$	Trace
Agar	15 g
Soil extract	100 ml
Tap water	900 ml

Dissolve, adjust pH to 7.6, and autoclave. Add 1 ml of sterile 10 per cent glucose to each tube for use. To prepare soil extract place 500 g of air-dry field soil in 1,300 ml of 0.1 per cent Na_2CO_3, autoclave for 1 hr, filter through paper, and make up volume to 1,000 ml.

6. *Azotobacter Agar (Mannitol)*

K_2HPO_4	1.0 g
$MgSO_4$	0.2 g
NaCl	0.2 g
$FeSO_4$	Trace
Agar	15 g
Soil extract	100 ml
Tap water	900 ml
Dissolve, add mannitol	20 g

Adjust pH to 8.3, autoclave.

7. *Beef Liver for Anaerobes*

Ground beef liver	500 g
Tap water	1,000 ml
Peptone	10 g
K_2HPO_4	1 g

Soak liver in 1 liter of water overnight in refrigerator; skim off fat. Autoclave 10 min at 15 lb. Strain through cheesecloth; save the meat. Add peptone and K_2HPO_4 to broth, and heat to 100°C. Adjust pH to 9.0, filter through paper, and make up to 1 liter with tap water.

Place a small amount of $CaCO_3$ in each tube; add meat to depth of ½ in.; cover with broth to total depth of 2 in. Autoclave.

8. *Blood Agar*

Heart infusion agar (Difco)	40 g
Distilled water	1,000 ml

Heat to 100°C; adjust pH to 7.4. Tube, and autoclave, but do not slant. Melt blood base agar, cool to 45°C, add 0.5 ml of sterile blood aseptically, mix, and slant.

8*a*. *Blood Glucose Agar*

Same as formula 8 with addition of 0.5–0.75 ml of sterile 10 per cent glucose solution per tube, besides blood.

9. *Bordet-Gengou Medium*

Potato infusion	100 ml

Prepare by dicing 12.5 g of potato in 100 ml of distilled water, let stand overnight at 60°C, filter through cheesecloth, and restore volume.

Agar	1.5 g
NaCl	0.55 g
Proteose peptone	1.0 g
Heat to 100 C, adjust pH to 7.4, and add	
Glycerol	1.0 ml

10. *Brain-Heart Infusion Agar*

Brain-heart infusion agar (Difco)	37 g
Agar	15 g
(for semisolid)	2 g
Distilled water	1,000 ml

pH 7.4

11. *Chocolate Agar*

Heat blood agar (formula 8) to 80°C; reslant.

12. *Cellvibrio Medium*

$NaNO_3$	2.0 g
$MgSO_4$	0.5 g
KCl	0.5 g
$Fe_2(SO_4)_3$	0.01 g
KH_2PO_4	0.14 g
K_2HPO_4	1.2 g
Yeast extract	0.02 g
Agar	7.5 g
Distilled water	1,000 ml

Place a strip of filter paper ½- by 4–5 cm on the agar slope, or suspend in liquid medium.

13. *Cystine-trypticase Medium*

Cystine-trypticase agar (BBL)	29.5 g
Carbohydrate (optional)	5–10 g
Distilled water	1,000 ml

Mix well; heat gently with agitation; boil for 1 min. Tube, and autoclave 15 min at 12 lb. Store at room temperature.

14. *Dorsett Egg Medium*

Commercial preparations are advised.

15. *Emerson Medium*

NaCl	2.5 g
Peptone	4.0 g
Yeast extract	1.0 g
Beef extract	4.0 g
Distilled water	1,000 ml
Adjust to pH 7.0 with KOH; add	
Glucose	10 g
Agar	30 g

16. *Glycerol Agar*

Blood base agar (Difco)	20 g
Nutrient agar	15.5 g
Distilled water	1,000 ml
Heat to 100°C; adjust pH to 7.3; add	
Glycerol	60 ml

17. *Liver Medium (Northern Regional Research Laboratory)*

Liver extract (commercial)	10 g
Yeast extract	5 g
Tryptone	10 g
K_2HPO_4	2 g
Glucose	5 g
Distilled water	1,000 ml

pH 7.4

If stabs are desired, add 10 g of agar. Liver extract can be prepared by placing 1 lb of ground beef liver in 2 liters of water and heating in flowing steam for about 3 hr until the liquid assumes a yellow fluorescence. Filter through cheesecloth, place in flasks, and autoclave. Keep aseptically, and use 100 ml in above formula; reduce water to 900 ml. Save the solid meat, and place a few particles in each tube.

18. *Lowenstein-Jensen Medium*

Commercial preparations are advised.

19. *Meat-infusion Broth*

Lean ground beef	500 g
Distilled water	1,000 ml

Mix thoroughly; store in refrigerator overnight. Skim off fat, strain the infusion through cheesecloth, and add

Peptone	10 g
NaCl	5 g

Dissolve by heating, strain through cheesecloth, and make up to volume. Adjust pH to 7.4. Heat in autoclave 10 min at 15 lb, filter through paper, and adjust pH. Dispense in tubes, and autoclave 20 min at 15 lb.

20. *NIH Semisolid Medium*

(National Institutes of Health)

Meat infusion broth	675 ml
Nutrient agar	1.8 g
Distilled water	75 ml
KCl (1.5 g in 10 ml of water)	1 ml
$CaCl_2$ (1.5 g in 10 ml of water)	0.5 ml

Heat to 100°C; adjust pH to 7.0, tube for stabs.

21. *Nitrobacter Medium*

(*a*) $NaNO_2$	10 g
K_2HPO_4	0.5 g
NaCl	0.3 g
$MgSO_4$	0.5 g
$MnSO_4$	Trace
$Fe_2(SO_4)_3$	Trace
Distilled water	1,000 ml

pH 7.5

Put 100-ml portions in flasks. Autoclave.

(*b*) Wash marble chips thoroughly in distilled water; place in large test tubes about one-third full; autoclave 1 hr at 15 lb. Add (*a*) to (*b*) aseptically.

22. *Nitrosomonas Medium*

$(NH_4)_2SO_4$	2.0 g
K_2HPO_4	1.0 g
NaCl	0.5 g
$MgSO_4$	0.5 g
$MnSO_4$	Trace
$Fe_2(SO_4)_3$	Trace
Distilled water	1,000 ml

pH 8.5

Dispense 100-ml portions in flasks, autoclave, and add to tubes of marble chips as in formula 21.

23. *Nutrient Agar*

Blood base agar (Difco)	20 g
Peptone	2.5 g
Beef extract	1.5 g
Agar	15.5 g
Distilled water	1,000 ml

pH 7.4

23*a*. *Alternate Formula with Salt*

Peptone	5 g
Beef extract	3 g
NaCl	30 g
Agar	14 g
Distilled water	1,000 ml

24. *Peptic Digest of Blood*

Autoclave 150 ml of 0.9 per cent NaCl in a bottle having a ground-glass stopper but with the stopper separately wrapped in paper and the bottle cotton-plugged. Place in 55°C water bath, and add

HCl (conc)	6 ml
Sterile sheep blood, defibrinated	50 ml
Pepsin	1 ml

Keep in water bath at 55°C overnight, shaking occasionally during first 2 hr. Add 12 ml of 20 per cent NaOH, then slowly with intermittent shaking enough more to raise the pH to 7.6. Then restore to pH 7.0–7.2 with HCl. Add 0.25 per cent chloroform; store glass-stoppered in refrigerator. For use add 0.1 ml to brain-heart infusion agar or semisolid medium.

25. *Potato Dextrose Agar*

Potato, peeled and diced	300 g
Distilled water	1,000 ml

The potatoes should be handled with minimum exposure to air. Boil in 500 ml of water until thoroughly cooked. Filter through cheesecloth, make up the volume to 1,000 ml, and add

Agar	15 g
Glucose	20 g

Dissolve by heating. Autoclave.

26. *Potato Agar*

Potato infusion	500 ml
Water	500 ml
Peptone	10 g
Beef extract	5 g
NaCl	5 g
Agar	30 g

Heat to 100°C, adjust pH to 7.3, autoclave 5 min at 15 lb, filter through cotton, and add

Glucose	10 g

Tube, autoclave, and slant.

The potato infusion is made by slicing 250 g of peeled potatoes into 500 ml of distilled water. Cover, and hold in 60°C incubator overnight. Filter through cheesecloth, and make up to 1,000 ml.

27. *Potato-starch Agar*

(*a*) Nutrient agar	3.1 g
Distilled water	80 ml
pH 7.4	
(*b*) Potato starch	1 g
Distilled water (cold)	20 ml

Mix thoroughly, then heat with constant stirring until a smooth paste is formed. Combine with (*a*), autoclave, and slant.

28. *Rhizobium Medium*

Yeast extract	1 g
Agar	15 g
Soil extract (see formula 5 for preparation)	200 ml
Tap water	800 ml
Heat to 100°C, adjust pH to 7.4 and add	
Mannitol	10 g

29. *Spirillum Medium*

a. Fresh-water species

Peptone	5 g
Beef extract	3 g
Yeast autolysate	3 g
Calcium lactate	1 g
Agar	2 g
Tap water	1,000 ml

b. Marine species

Use sea water or sea-salt solution of equivalent concentration. An artificial sea salt can be compounded of

NaCl	2.75 g
$MgCl_2$	0.50 g
$MgSO_4$	0.20 g
$CaCl_2$	0.05 g
KCl	0.10 g
$FeSO_4$	Trace
Distilled water	100 ml

30. *Sporovibrio Medium* (*Starkey*)

Peptone	5.0 g
Beef extract	3.0 g
Yeast extract	0.2 g
$MgSO_4$	1.5 g
Na_2SO_4	1.5 g
Ferrous ammonium sulfate	0.1 g
(increase to 0.2 g if medium does not blacken after growth)	
Agar	15 g
Tap water	1,000 ml
Heat to 100°C, adjust pH to 7.4, add	
Glucose	5 g

Tube for stabs, using long (150-mm) tubes. Boil tubes to be inoculated for 10 min; cool quickly. After inoculating, push cotton plug down about half way, place two pieces (0.5 mm) of pyrogallic acid on it, add 8 to 10 drops of strong NaOH. Cork tightly with rubber stopper, invert, and incubate at 37°C.

31. *Thiobacillus Thioparus Medium*

(*a*) K_2HPO_4	2.0 g
$CaCl_2$	0.1 g
$MgSO_4$	0.1 g
$MnSO_4$	Trace
$FeSO_4$	Trace
Tap water	900 ml

pH 7.8

Dispense 90 ml in 250 ml-flasks; autoclave.

(*b*) $Na_2S_2O_3$	10 g
Tap water	50 ml
Autoclave in flasks.	
(*c*) $(NH_4)_2SO_4$	0.1 g
Tap water	50 ml

Autoclave in flasks.
At time of inoculating add 5 ml aseptically of each of (*b*) and (*c*) to (*a*)

32. *Thiobacillus Thiooxidans Medium*

$(NH_4)_2SO_4$	0.2 g
$MgSO_4$	0.5 g
$CaCl_2$	0.25 g
$FeSO_4$	Trace
KH_2PO_4	3.0 g
Precipitated sulfur	10 g
Tap water	1,000 ml

One-gram portions of sulfur are placed in 10 dry 250-ml flasks, and 100 ml of a solution of the other components is carefully poured into each flask so that the sulfur floats. Sterilize in flowing steam for ½ hr on 3 consecutive days.

33. *Tomato-juice Agar*

Tryptone	10 g
Yeast extract	10 g
Agar	12 g
Distilled water	800 ml
Heat to 100°C; adjust pH to 7.2; add	
Tomato juice	250 ml

Tube for stabs; autoclave. The tomato juice must be taken from canned unseasoned tomatoes, *not from commercial juice preparations.* It is prepared by filtering the juice from a No. 2 can, keeping refrigerated overnight, then adjusting pH to 7.0.

34. *Tomato, Yeast, Milk Medium*

Dehydrated skim milk	100 g
Yeast extract	5 g
Distilled water	1,000 ml
Mix well and add	
Tomato juice, pH 7.0, as in formula 33	100 ml
Methylene blue 1.5% solution in alcoholic KOH	2 ml

Filter through cheesecloth, tube, and autoclave.

35. *Tryptone Glucose Yeast Agar*

(Northern Regional Research Laboratory)

Tryptone	5 g
Yeast extract	5 g
Glucose	1 g
K_2HPO_4	1 g
Agar	20 g
Tap water	1,000 ml

pH 7.0

36. *Urea Agar*

Urea	10 g
Distilled water	100 ml

Sterilize by filtration. Add ½ ml per tube aseptically to nutrient agar (No. 23) after melting and cooling to 45–50°C. Mix well; cool for slants.

37. *V-8 Juice Agar*

V-8 juice (Campbell Soup Co.)	200 ml
$CaCO_3$	3 g
Agar	15 g
Tap water	800 ml

pH 7.2

The V-8 juice can be filtered or not depending on the clarity desired, and the amount of $CaCO_3$ may be varied to give a different pH.

TABLE 14. COMMERCIALLY AVAILABLE MEDIA RECOMMENDED FOR CULTURE MAINTENANCE

Organism	Recommendations of					
	Baltimore Biological Laboratory, Inc.			Difco Laboratories, Inc.*		
	Media	Storage temp, °C	Suggested transfer interval	Media	Storage temp, °C	Suggested transfer interval
Actimomyces	CTA medium (cystine-trypticase‡ agar) Thioglycollate medium 135C with $CaCO_3$	Room	3–6 mo	Brain-heart infusion, or same with 0.2 or 1.5% agar Bacto-tryptose agar	Room (less than 25)	1–12 mo
Algae†						
Euglena	Trypticase agar slants	Above 6				
Chlorella	Trypticase salts medium					
Brucella	CTA medium	Room	3–6 mo	Tryptose agar Brain-heart infusion agar	Room	3 mo
Clostridia	Cooked-meat phytone§ medium Thioglycollate medium 135C with $CaCO_3$ Trypticase agar base	Room	Yearly	Cooked-meat medium Egg meat medium	Room	10 years plus
Coliform group: Salmonella and other enteric bacilli	Trypticase agar base Trypticase soy agar with oil	Room or refrigerator	6–12 mo	Cooked-meat medium Nutrient agar	Room	1 year plus
Corynebacteria	CTA medium Trypticase soy agar under oil	Room	3–6 mo	Loeffler blood serum Tryptose agar Heart-infusion agar	Room	1 mo
Hemophilus: blood-requiring	CTA medium Trypticase soy agar with blood	Room	3 mo	Chocolate agar prepared with proteose No. 3 agar or GC Medium base enriched with hemoglobin and bacto-supplement B or C	37	Weekly
Lactobacilli and Leuconostoc	Lactobacillus agar with oil Thioglycollate medium 135C with $CaCO_3$	Room	Monthly to 6–8 wk	Micro assay culture agar Tomato-juice agar Litmus milk	Room	1–12 mo

* Trade names of all products of this company include the prefix "Bacto-." It is omitted in this table to economize on space.

† Media and storage times and temperatures vary with organisms. Periods of transfer may vary with types of closure of container and concentration of media. Dehydration occurring in cotton-plugged tubes may impair longevity of cultures.

‡ U.S.P. pancreatic digest of casein.

§ U.S.P. papaic digest of soybean meal.

TABLE 14. COMMERCIALLY AVAILABLE MEDIA RECOMMENDED FOR CULTURE MAINTENANCE (*Continued*)

Organism	Recommendations of: Baltimore Biological Laboratory, Inc.			Difco Laboratories, Inc.*		
	Media	Storage temp, C	Suggested transfer interval	Media	Storage temp, °C	Suggested transfer interval
Micrococci	CTA medium Trypticase soy agar with oil	Room	3–6 mo	Cooked-meat medium Brain-heart infusion, or same with 0.2 or 1.5 per cent bacto agar Blood agar base	Room	3 mo
Mycobacteria	Dorset medium ATS medium Lowenstein-Jensen medium	Room or refrigerator	6–12 mo	Dorset medium Petragnani medium Lowenstein medium Glycerol agar	Room	1 year plus
Nonpathogens	Same, or CTA medium or trypticase soy agar with oil					
Neisseria	CTA medium	Room	3 mo	Dextrose starch agar 37 ½ strength + supplement C		4 wk
Gonococcus	CTA medium	35	10 days	GC medium base with hemoglobin + supplement C		
Pasteurella-Listeria				Cooked-meat medium	Room	1 year
P. multocida	CTA medium	Room	3–6 wk	Tryptose agar		
P. tularensis	Cystine heart agar with blood	Room	1 mo.	Cystine heart agar + bacto-hemoglobin		
Pleuropneumonia	CTA medium with inactivated sheep serum PPLO agar with serum	35	10–21 days	PPLO agar with ascitic fluid or PPLO serum fraction	37	1 wk
Pneumococci	CTA medium	Room	3–6 mo	Cooked-meat medium Brain-heart infusion agar Tryptose blood agar base + blood Stock culture agar	Room	1 mo

* Trade names of all products of this company include the prefix "Bacto-." It is omitted in this table to economize on space.

† Media and storage times and temperatures vary with organisms. Periods of transfer may vary with type of closure of container and concentration of media. Dehydration occurring in cotton-plugged tubes may impair longevity of cultures.

TABLE 14. COMMERCIALLY AVAILABLE MEDIA RECOMMENDED FOR CULTURE MAINTENANCE (*Continued*)

Organism	Recommendations of: Baltimore Biological Laboratory, Inc.			Difco Laboratories, Inc.*		
	Media	Storage temp, °C	Suggested transfer interval	Media	Storage temp, °C	Suggested transfer interval
Protozoa†	Simplified trypticase	Room	2 weeks	Endamoeba medium overlayed with horse serum saline with rice flour	37	14 days overlaid
Trichomonas vaginalis	Serum medium	35	3 days	Lash serum medium		
Salmonella-Shigella (see also Coliform group)	Trypticase soy agar under oil	Room or refrigerator	6–12 mo	Cooked-meat medium Nutrient broth Nutrient agar	Room	2 years plus
Streptococci	CTA medium Thioglycollate medium 135C Trypticase soy agar under oil	Room	3–6 mo	Cooked-meat medium Brain-heart infusion agar Tryptose blood agar base + blood Stock culture agar	Room	1 mo
Streptomyces	CTA medium Trypticase agar base Trypticase soy agar or mycophil agar under oil	Room or refrigerator	6 mo or more	Brain-heart infusion agar Potato dextrose agar	Room	3 mo†
Yeasts	Mycophil agar under oil	Room or refrigerator	6 mo or more	Dextrose starch agar Potato dextrose agar Mycological agar	Room	3 mo
Molds	Mycophil agar under oil	Room or refrigerator	6 mo or more	Sabouraud maltose agar Mycological agar Littman oxgall agar	Room	12 mo

* Trade names of all products of this company include the prefix "Bacto-." It is omitted in this table to economize on space.

† Media and storage times and temperatures vary with organisms. Periods of transfer may vary with type of closure of container and concentration of media. Dehydration occurring in cotton-plugged tubes may impair longevity of cultures.

REFERENCES

British Commonwealth Collections of Microorganisms. 1954. "A Discussion on the Maintenance of Cultures by Freeze Drying," 48 pp. Her Majesty's Stationery Office, London.

Fennell, Dorothy I., K. B. Raper, and May H. Flickinger. 1950. Further investigations on the preservation of mold cultures. *Mycologia*, **42**, 135–147.

Flosdorf, Earl W. 1949. "Freeze-Drying. Drying by Sublimation," 280 pp. Reinhold Publishing Corporation, New York.

Gordon, Ruth E., and N. R. Smith. 1947. Preservation of certain microorganisms under paraffin oil. *J. Bacteriol.*, **53**, 669.

Harris, R. J. C. 1954. "Biological Applications of Freezing and Drying," 415 pp. Academic Press, Inc., New York.

Hartsell, S. E. 1947. The longevity of bacterial cultures under paraffin oil. *J. Bacteriol.*, **53**, 801.

———. 1953. The preservation of bacterial cultures under paraffin oil. *Appl. Microbiol.*, **1**, 36.

Hauduroy, Paul. 1951. "Techniques bactériologiques," 167 pp. Masson et Cie, Paris.

Haynes, W. C., L. J. Wickerham, and C. W. Hesseltine. 1955. Experience in maintaining industrially important microorganisms. *Appl. Microbiol.*, **3**, 361–368.

Hollander, David H., and E. Ellen Nell. 1954. Improved preservation of *Treponema pallidum* and other bacteria by freezing with glycerol. *Appl. Microbiol.*, **2**, 164–170.

Proom, H., and Louis M. Hemmons. 1949. The drying and preservation of bacterial cultures. *J. Gen. Microbiol.*, **3**, 7–18.

Raper, K. B., and D. F. Alexander. 1945. Preservation of molds by the lyophil process. *Mycologia*, **37**, 499–525.

Weiser, R. S., and C. M. Osterud. 1945. Studies on the death of bacteria at low temperatures. I. The influence of the intensity of the freezing temperature, repeated fluctuations of temperature, and the period of exposure to freezing temperatures on the mortality of *Escherichia coli*. *J. Bacteriol.*, **50**, 413–439.

CHAPTER VI

The Study of Obligately Anaerobic Bacteria[1]

L. S. McClung and Robert B. Lindberg

It is impossible to list here all the methods which have been proposed for the study of anaerobic bacteria; an attempt is made, however, to outline a number of technics which have been used widely and which should ordinarily be suitable for routine studies of anaerobic species. Those interested in other technics are advised to consult Sec. B of the subject index bibliography relating to the anaerobic bacteria (McCoy and McClung, 1939; McClung and McCoy, 1941). The worker who has had no experience with anaerobic bacteria should study some of the articles which deal with principles of anaerobic culture or which record the results of a study of a considerable number of strains: Fildes, 1931; Hall, 1922, 1928, 1929; Heller, 1921; Knorr, 1923, 1924; McCoy *et al.*, 1926, 1930; McIntosh, 1917; Meyer, 1928; Reed and Orr, 1941; Spray, 1936; Zeissler, 1930; Zeissler and Rassfeld, 1928. Valuable suggestions in English will be found in Smith (1955), and in French in Lehert and Tardieux (1952). The recent literature has been reviewed by McClung (1956).

The organisms which we call obligate anaerobes are those that require a low (reduced) oxidation-reduction potential, which can be brought about by exclusion or removal of atmospheric oxygen, alone or in combination with chemical reducing agents added to the medium. It is not easy to answer the question of the best method of determining whether or not a given organism is an obligate anaerobe. The catalase reaction, when applied to pure culture, gives presumptive evidence, for obligate

[1] The methods and technics suggested herein are those recommended for use with the more common sporeforming anaerobic species. Many of these methods are suitable also for the study of the nonsporeforming types, but for the present no attempt will be made to outline particular methods of study for these. If the technics herein outlined do not prove satisfactory, the worker interested in the pathogenic nonsporeformers should consult the review of Dack (1940) and the books by Prévot (1948, 1955) and Smith (1955). Nonpathogenic types exist, as for example, the methane organisms discussed by Barker (1936). For the complete literature on all types refer to Sec. Id (nonsporeformers) in the bibliography of McCoy and McClung (1939) and McClung and McCoy (1941).

anaerobes usually are catalase-negative. For this reaction a plate culture of the organism in question is flooded with a 10 per cent solution of H_2O_2. The evolution of gas bubbles from the colonies denotes the presence of catalase.

If the proper material for the catalase reaction is not available or in doubt, the following technic will usually suffice to characterize an anaerobic strain and to differentiate it from the aerobes: Inoculate, while the agar is molten, several deep tubes (8- to 9-cm columns of medium) of a suitable nutrient agar medium containing 1.0 per cent glucose; allow these to solidify in an upright position and incubate the tubes at several temperatures or at the optimum temperature for the organism in question; adjust the seeding so that relatively few (e.g., 25–50) colonies per tube will result. With an obligate anaerobe, all the colonies should be localized in the bottom of the tube and none should appear on the surface or in the upper 1-cm layer. Likewise, with pathogenic organisms cultured in fluid thioglycollate medium, the growth should be confined to the lower section of the medium and no growth should result in the upper layer wherein the methylene blue is recolorized. If growth does occur in the upper layer of either medium, the culture is not an obligate anaerobe or is contaminated with an aerobic or a facultative species. In the case of clinical cultures, in which speed is important, two blood agar or egg-yolk plates (McClung and Toabe, 1947) may be inoculated and incubated in parallel, one anaerobically, the other aerobically. The appearance of growth in only the anaerobic environment is evidence of presence of an obligate anaerobe.

ANAEROBIC CULTURE METHODS AND EQUIPMENT

All the procedures which have been devised for the cultivation of anaerobic bacteria have the single purpose of excluding *atmospheric* oxygen from the environment in which the growth is to take place. With certain tubed media the oxygen potential may be reduced sufficiently by constituents of the medium to permit anaerobic growth (Brewer, 1940; Hewitt, 1950; Knight, 1931; Reed and Orr, 1943; and Molland, 1944). Since this is rarely possible for surface cultures on a solid medium, usually plate and slant cultures are incubated within a closed container from which the oxygen is removed by one or another means. A study of the various methods shows that no single procedure may be proposed as the best technic but the method of choice will depend upon the prevailing circumstances. A procedure which is ideal for one situation may be impractical or impossible to apply with other conditions. Each of the technics outlined below is recommended within the limits proposed in the discussions.

Use of methylene blue as indicator of anaerobiosis. For all types of anaerobic jars and containers, except individual-plating or tube-culture systems, it is convenient to include an indicator tube which will serve as a check on the development of anaerobiosis. The most commonly used system utilizes the change of methylene blue from the colored (oxidized state) to the leuco form (reduced state). Using the solution prepared as given below, any system which gives sufficient degree of removal of oxygen from the atmosphere for anaerobic growth to develop will cause the blue color of the solution to disappear or will maintain the colorless condition if the solution is boiled (heat reduction) immediately prior to its being placed in the container. A somewhat less sensitive system can, in an emergency, be prepared by adding a tinge of color from Loeffler's alkaline methylene blue to a tube of glucose broth.

The procedure recommended (Fildes, 1931) is: Prepare three stock solutions, (1) 6.0 ml of 0.1*N* NaOH diluted to 100 ml with distilled water, (2) 3.0 ml of 0.5 per cent aqueous methylene blue diluted to 100 ml with distilled water, (3) 6.0 g of glucose in 100 ml of distilled water to which has been added a small crystal of thymol. Each time the indicator solution is needed, mix equal parts of the three solutions in a test tube and boil in a cup of water until the color disappears. Place tube in anaerobic container immediately and begin process of securing anaerobic conditions. If the container is satisfactorily deoxygenated, the color in the solution should not reappear. If the blue color does return, it is a sign that the container leaks or has not been satisfactorily exhausted of oxygen. (In the vegetable tissue jar, to be described, the color may appear but will disappear with the development of anaerobiosis during the incubation period.)

Oxygen Removal by Combustion Using Laidlaw Principle

For laboratories which are engaged in problems where anaerobic plating is to be done frequently, it is advisable to plan for this and to purchase equipment accordingly. Although the systems discussed above may be adequate for this purpose, it is well to consider one of the jars which utilize, on the Laidlaw (1915) principle, combustion as a means of securing the anaerobic environment. These methods were designed especially for incubation of plates, but other culture vessels (flasks, tubes, bottles, etc.) may be used. Jars using this principle are those of Brewer (Brown and Brewer, 1938) and McIntosh and Fildes (Fildes and McIntosh, 1921).

Brewer Anaerobic Jar[1]

Materials for method of Evans, Carlquist, and Brewer (1948): (1) *Brewer jar* complete with electric cord, (2) source of illuminating gas or hydrogen, (3) plasticene (see footnote 2 on page 123), (4) vacuum pump for evacuation.

Method. Place plates in jar. Add tube of methylene blue solution. Include a tube of soda lime in the jar to absorb excess CO_2. Place roll of (warmed) plasticene around rim of jar. Put on lid and press down on plasticene to form seal. Add the lid clamp but tighten *only slightly. If used with illuminating gas,* attach the jar by the rubber tubing to the vacuum pump. Evacuate until the manometer or gauge reads approxi-

[1] Brewer jar. Baltimore Biological Laboratory, Baltimore, Maryland, and Fisher Scientific Company, Pittsburgh, Pennsylvania.

mately 20 cm or 8 in. After this degree of evacuation is reached, connect the rubber tube to the gas supply (a three-way stopcock facilitates this change without loss of vacuum). Attach the electric plug (110-volt alternating or direct current) and allow the gas and electric current to remain attached for 30 to 45 min. At the end of this time clamp the rubber tube tightly, remove the electric cord, and place the jar in the incubator. (Formation of water droplets on the inside walls of the jar indicates the proper functioning of the apparatus.) To open the jar, remove the clamp. Insert a knife blade between lid and rim of jar, using caution to avoid scoring the soft metal rim of the top and making subsequent leakage more likely. If used with hydrogen, attach the jar, without evacuation, to the hydrogen tank and admit the gas at a pressure of 1–2 psi. Attach the electric connection, and allow the current and gas both to remain on for 30 min. Then treat the jar as above. The jar may be used as a gas-replacement system by evacuating the jar on a water aspirator pump to 700 mm Hg negative pressure; fill with illuminating gas, repeating this process three times or more. Avoid dropping of agar from inverted plates due to excessive evacuation. This method may fail to produce anaerobiosis as adequate as that achieved by catalytic combustion as described above.

Advantages. Convenient system for incubation of a number of plates in experiments where speed of obtaining anaerobiosis is essential. Recommended for clinical laboratories. Inexpensive system after the initial outlay for apparatus. *Disadvantages.* Some possibility of explosion or cracking of jar. Initial expense of equipment is more than for other methods, but this may be a good investment if routine work is to be done over a period of time. Requires source of hydrogen or illuminating gas and electricity; while these are available in most laboratories, they are not available in others such as some mobile laboratory units, temporary laboratories in field surveys, etc.

Biological Methods for Oxygen Removal

Vegetable-tissue Jar

Materials for method of McClung, McCoy, and Fred (1935): (1) jar or other container which may be sealed airtight (recommended: 6- by 18-in. or 6- by 12-in. Pyrex cylinder[1]); (2) square (7 by 7 in.) of plate glass or a glazed plate; (3) plasticene,[2] ¼ lb; (4) glass tumbler; (5) supply of

[1] Pyrex cylinder. Pyrex Catalogue No. 850. Corning Glass Works, New York, or supply house.

[2] *Sealing materials.* Plasticene of a suitable grade is sold by Baltimore Biological Laboratory, Baltimore, Maryland, and by J. L. Hammet Co., Cambridge, Massachusetts. Care should be taken in choosing products for sealing, since some dry to a hard cake upon incubation. A silicone grease is preferred by many workers, since it gives a seal less likely to leak, and jar tops are more easily removed than with plasticine. A suitable product is marketed as Dow-Corning High Vacuum Grease by Dow Corning Corp., Midland, Michigan.

oats or other grain (other tissues, particularly chopped Irish potatoes, may be used but are less conveniently stored for occasional use, and in some cases produce objectionable odors which are evident when the jar is opened); (6) tap water.

Method. Place inverted tumbler (if plates are to be used) or other support in bottom of cylinder. Add oats to fill at least one-tenth of the capacity of the cylinder. Add sufficient tap water to moisten the oats. Stack plates or other cultures on support. Add tube of methylene blue solution (see above). Place layer of plasticene, previously softened by placing in incubator, on rim of cylinder. Push plate-glass square firmly against plasticene; using fingers, press the clay against both the square and the cylinder until a satisfactory seal is obtained. Place jar in incubator immediately. (A 40- to 48-hr incubation period is recommended.)

If plate cultures are employed, replace the ordinary petri dish cover with unglazed porcelain ("clay") tops[1] to absorb the moisture which collects within the cylinder. If porcelain tops are unavailable, add a petri dish lid containing $CaCl_2$ to absorb the moisture.

Advantages. The method is inexpensive and employs easily available materials. No special apparatus is required, an advantage in laboratories where anaerobic cultures are not usually prepared. It may be used at any incubation temperature without danger of explosion. It is particularly suitable in problems requiring large numbers of plate cultures. It is recommended especially for cultural and physiological studies of strains which have been purified by other methods. *Disadvantages.* Several hours may be needed for anaerobic conditions to become established, and therefore the method is not suitable when the results are required quickly. It is not recommended for routine clinical use where speed of isolation of pure culture is an important factor. With certain enrichments it is not suitable for purification of species contaminated with aerobic sporeforming bacteria because of the quick growth of these forms. In plate-culture experiments, as in the isolation of new strains, no one plate may be removed from the cylinder for observation until the end of the incubation period, for to do so would destroy the anaerobic conditions within the cylinder.

Use of Aerobe to Absorb Oxygen

Another biological method for oxygen removal utilizes the growth of an aerobic organism (usually *Staphylococcus aureus*, *Serratia marcescens*, or *Saccharomyces cerevisiae*). A wide variety of applications of this system have appeared in the literature. The technics suggested[2] involve the growth of the aerobic organism in pure culture on a medium separate from that on which the anaerobe is to be cultured.

[1] Unglazed porcelain ("clay") tops for petri dishes. The Coors porcelain dish, sold by Arthur H. Thomas Company, has been found to be more uniform in size and quality than others tested.

[2] These are similar to the Fortner method and are recommended in place of it. In the Fortner method the aerobe is streaked on one half of the plate and the anaerobe on the other half of the same dish.

Materials for method of Snieszko (1930): (1) two petri dishes of ordinary size; (2) paper tape, Scotch tape, adhesive plaster, or plasticene; (3) culture of *Serratia marcescens* or other fast-growing aerobic organism; (4) tube of nutrient agar.

Method. Select two petri dishes which have bottoms of exactly the same size, and sterilize these in position in their usual top sections. Pour nutrient agar into the bottom half of plate A, and after solidification streak the medium heavily (or flood across surface with 0.5 ml of broth culture) with the aerobic organism. (As an alternate method, seed the agar before pouring.) Pour into plate B, a medium suitable for the anaerobe (see Chap. III); when hard, streak with the sample or culture of the anaerobe (or seed with the latter prior to pouring).

Remove the two bottoms from their respective tops, and fit together at their rims. Use tape or other sealing device around the juncture to provide an airtight seal. Place plate in the incubator immediately. If thermophilic anaerobic cultures are to be made, replace the *S. marcescens* by a thermophilic aerobe, or before placing plates in thermophilic incubator, incubate for 18 hr at 32°C to allow *S. marcescens* to grow and to use the oxygen.

Advantages. No elaborate equipment is needed, since the method uses ordinary petri plates and other common materials. Thus it is available as an emergency method in almost any laboratory at any time. The technic is extremely simple and can be set up by inexperienced individuals. Since each set of plates is an individual unit, observation of the growth of each anaerobe may be carried out without destroying anaerobic conditions for other cultures. *Disadvantages.* The method is somewhat time-consuming when large numbers of cultures are to be made and is therefore not suitable in laboratories where routine platings of a number of cultures is not an unusual event. Anaerobiosis may not be attained promptly enough to prevent death of the inoculum of nonsporeforming species of vegetative cells of anaerobic sporeformers. Negative results on isolation attempts with this procedure are hence not reliable.

Chemical Methods for Oxygen Removal

Many of the methods proposed for removal of oxygen from the environment for anaerobic culture involve the initiation of a chemical reaction in which oxygen is consumed. Of the various systems which have been suggested, those which are recommended have been tested and used sufficiently to show their utility, and they do not require elaborate apparatus.

Phosphorus Jar

Materials. (1) Sticks of yellow (or white) phosphorus (which *must be kept under water in tightly stoppered, wide-mouth bottle;* the small sticks, 5/16-in.-diameter, are the most useful); (2) Pyrex cylinder, or any convenient jar or container which may be sealed airtight; (3) pair of long forceps or chemical tongs; (4) plasticene; (5) small amount of tap water.

Method. Place small amount of tap water in bottom of cylinder to remove the P_2O_5 which forms. Stack inoculated plates or tubes on support. Add tube of methylene blue solution (see page 122). Place small (50-ml) beaker on top of cultures. Remove two or three short (1½- to 2-in.) pieces of phosphorus from water *with forceps or tongs*, and place in beaker. Immediately put lid on jar and seal with plasticene. (Upon drying for a few minutes the phosphorus should ignite spontaneously and remain burning as long as there is oxygen present.) If experience shows that the phosphorus used does not ignite spontaneously but merely gives off a gray smoke, ignite it before the jar is sealed by a match *held with the forceps*. Since considerable heat is developed, place beaker, unless resistant glass is used, 3 in. from the top of the container and put a "blank" plate under the beaker rather than an inoculated plate. After the phosphorus ignites and the jar is tightly sealed, place it directly in the incubator. At the time the container is opened have available a crock or pan filled with water. As soon as the lid is taken from the jar, remove the beaker containing the phosphorus with the tongs and submerge under the water in the pan and save for later use. After this remove the cultures from the jar.

Advantages. Quick method of obtaining anaerobiosis. It is relatively inexpensive since the only materials are phosphorus and a container which may be sealed. *Disadvantages.* Care must be exercised to prevent accidental burns which are very painful. Inexperienced technicians should be cautioned concerning the dangers.

Alkaline Pyrogallol Methods

Another chemical method for removing oxygen in order to promote anaerobic growth is to utilize the oxygen absorptive capacity of the reaction between alkali and pyrogallic acid. Of the technics and devices reported which make use of this reaction two may be recommended as being especially useful. One of these concerns a technic applied to individual plate culture, and the other relates to a system for individual tube cultures.

SPRAY (OR BRAY) PLATE CULTURES

Materials. (1) Spray (1930) anaerobic dish;[1] (2) plasticene, or tape for sealing; (3) 20 per cent aqueous NaOH; (4) 40 per cent aqueous pyrogallic acid.

Note: The Spray dish consists of an ordinary glass petri-dish top and a special bottom which is deep and has a raised ridge across the center. The top of the bottom dish has a lip into which the top section of the dish fits. Although constructed of heat-resistant glass, in practice considerable breakage may be encountered during

[1] Spray anaerobic dish. Fisher Scientific Company, Pittsburgh, Pennsylvania, or E. H. Sargent Company, Chicago, Illinois.

sterilization and handling of the Spray dish. This is eliminated in the Bray[1] dish, which is Pyrex and essentially the same design as the Spray dish. In the Bray dish, however, the need for the lip is eliminated, since the top of the bottom section is slightly smaller in diameter than the remainder of the bottom section. This allows the top to fit down over the rim of the bottom section.

Method. Pour anaerobic medium in the top half of the dish, and after solidification, streak from sample or culture, or pour seeded plate. After inverting dish, place 10 ml of 20 per cent aqueous NaOH solution in one section of the bottom dish and 4 ml of 40 per cent aqueous pyrogallic acid in the other. Seal dish with plasticene or tape. Tilt dish to mix solutions and place in incubator.

Advantages. Anaerobiosis is attained quickly. It is a useful method for single plate culture. Since each plate is a single unit, observations may be made at any time and any particular plate of a series may be opened when visual inspection reveals growth to be at the desired stage. Recommended for clinical laboratory technicians seeking a quick method of purification of possible pathogenic types. *Disadvantages.* Preparation of individual dishes is time-consuming, and in work on any scale, anaerobic jars may be preferred. Loss of anaerobiosis due to leaks in seal is not uncommon. The excess of alkali present results in absorption of CO_2, which may be disadvantageous with some strains. In addition, a small amount of CO is given off during oxidation of the pyrogallate. Some strains are inhibited by this substance. Preliminary replacement of air by an inert gas before introduction of the pyrogallate will minimize this effect.

Anaerobic Jar with Alkaline Pyrogallol for Plate Culture

For each 100 ml of jar capacity 1 g of pyrogallic acid and 10 ml of 2.5N NaOH are used. Plates are stacked in the jar, together with anaerobic indicator tube, and pyrogallic acid is added to alkali in a large-diameter test tube. The jar is quickly sealed, using plasticine or silicone grease.

Advantages. General availability of materials and suitability for a variety of jars, which need be only sealable. *Disadvantages.* As noted above, absence of CO_2 and presence of CO may inhibit growth of some strains.

Tube Culture

Materials for method of Griffin (1932): (1) Two test tubes with approximately ⅝ in. diameter (one empty and the other containing a liquid or slant culture of the anaerobe), (2) two one-holed rubber stoppers to fit tubes, (3) short piece of small-diameter rubber tubing, (4) two short pieces of glass tubing of diameter to fit tightly in holes of rubber stoppers, (5) small glass vial, (6) dry pyrogallic acid, (7) strong aqueous NaOH.

[1] Bray anaerobic dish. Corning Glass Works, Corning, New York, Pyrex No. 3155, or dealer.

Method. Put a column of pyrogallic acid, approximately 1½ in. high, in the bottom of the empty tube. Stand empty vial in this acid. With pipet, fill vial two-thirds full of NaOH solution. Fashion a connecting unit from the rubber stoppers and rubber and glass tubing. Insert one of the stoppers in the tube with the chemicals. Push down cotton plug in culture tube to a level 1 in. above the medium. Insert second stopper in this tube. Tilt tube containing chemicals sufficiently to allow NaOH solution to spill over the acid.

Advantages. If a supply of chemicals is at hand, it is useful as an emergency system, when the special equipment required by other systems is not available. *Disadvantages.* Not suitable for large numbers of cultures, or at least, such use would be more time consuming than other methods.

Plating System Using Strongly Reducing Medium

The single plating device introduced by Brewer (1942) is an ingenious method which offers a means of plating cultures without added equipment. The dish is used with agar containing strong reducing agents and is designed so that at the periphery the top rests on the medium, forming a seal. The remainder of the top is slightly raised, so that a small amount of air is trapped over the surface. The oxygen entrapped is removed by the reducing agents in the medium.

Brewer Culture Dish[1]

Materials. (1) Brewer anaerobic culture dish; (2) regular petri dish with bottom either 15 or 10 mm deep; (3) infusion agar suitable for anaerobes which contains suitable reducing agents, such as the following: 0.2 per cent sodium thioglycollate, 0.1 per cent sodium formaldehyde sulfoxylate, and 0.0002 per cent methylene blue.

Method. Pour sterilized medium in bottom of regular petri dish (25 ml minimum in 10-mm dish and 40 ml minimum in 15-mm dish). Streak center area from sample or culture. Replace the lid of the regular dish with the Brewer anaerobic lid. (The lid at its periphery should touch the agar at all points in order that a perfect seal be obtained. In the successfully prepared dish, the agar in the center of the dish remains colorless while the blue color returns to the agar at the end of the dish because of oxygenation of the dye which serves as an oxidation-reduction-potential indicator.) Place plates in the incubator immediately after they are prepared, and examine as needed during the incubation period. When transfers are to be made from the plate, break the seal by a slight turn of the lid.

Advantages. A useful, quick method of single-plate culture. An extremely simple method which is easy to learn and use. The only trick in the technic is to have

[1] Brewer anaerobic dish. Baltimore Biological Laboratory, Baltimore, Maryland, and Kimble Glass Company, Vineland, New Jersey.

sufficient agar in the original dish so that a perfect seal is formed when the special lid is added. Recommended for routine use in hospital laboratories, and particularly for mobile laboratories, where anaerobic cultures for pathogens may be encountered. *Disadvantages.* Surface moisture may result in film formation in some instances; this may be reduced by using a porcelain top (see footnote 1 on page 124) on the regular dish prior to the Brewer anaerobic lid or by drying the plates in incubator before streaking. Some organisms apparently are inhibited by the reducing agents. This is not serious, since the reports indicate that all pathogenic types are easily cultured by this method. The Brewer anaerobic lids are, at the present time, relatively expensive.

TECHNICS FOR STUDY OF ANAEROBIC BACTERIA[1]

In the above section the various pieces of apparatus and methods for their use with anaerobic bacteria have been considered. Formulas for the particular media which are recommended may be found in Chap. III. The remainder of this chapter will be devoted to a discussion of the details of certain technics which should aid the worker who has not had previous experience with anaerobes.

It may not be amiss to insert here a precautionary note concerning the necessity of very careful inspection of the purity of cultures. There are instances on record, in the older literature, where two species grew symbiotically on plate culture with such constancy that recorded observations were made of the colony type of mixture, the investigator being unaware of the existence of more than one type. In all studies concerning obligate anaerobes, a check on the purity of the culture should be made with regard to aerobic contaminants. The following test is suggested: For most cultures, streak a glucose nutrient agar slant and incubate it at 37°C; but for anaerobic species having a lower or higher optimum temperature, incubate a second agar slant at the temperature which is optimum for the anaerobe. If the culture appears free of aerobic types, investigate the purity with respect to anaerobic contaminants. Make repeated platings, and scrutinize intensely the colonies which develop. For cultures which will grow on the egg-yolk agar (with 0.3 to 0.5 per cent glucose added for the butyric group) of McClung and Toabe (1947), contamination in cultures is more readily revealed on this medium than on media not containing egg yolk.

Preliminary Microscopic Examination

If the sample is suitable, one should make preliminary examination using the gram stain. The conventional method of staining a smear, heat

[1] In this chapter reference will be made to the "pathogenic group" and the "butyric-butyl group." The former term is used to designate such organisms as *Clostridium tetani*, *C. septicum*, *C. histolyticum*, *C. chauvoei*, *C. perfringens*, *C. parabotulinum*, *C. botulinum*, and *C. sporogenes*. In the butyric-butyl group are included *C. butyricum*, *C. beijerinckii*, *C. butylicum*, *C. pasteurianum*, *C. acetobutylicum*, *C. felsineum*, *C. roseum*, and *C. thermosaccharolyticum*.

fixed on a glass slide, should be used, except that the decolorizer should be either 95 per cent ethyl alcohol (*preferred*) or 25 parts of acetone and 75 parts of ethyl alcohol. The use of greater amounts of acetone must be avoided because of the ease with which anaerobes are decolorized. Decolorization of young cultures of clostridia is relatively common even when alcohol is used with caution. The presence of numerous gram-negative cells does not rule out clostridia. The usefulness of the gram method is limited in smears prepared from blood, fibrin, or albumin. In samples of pathologic material, large, gram-positive rods are likely to prove to be anaerobic bacilli, but a final diagnosis must not be based on microscopic observations unsupported by cultural tests. Of the strictly aerobic gram-positive species, *Bacillus anthracis* Koch is the only usual pathogen. The characteristic morphology of *Clostridium perfringens* (syn. *C. welchii*) and the regularity of its appearance in certain clinical conditions frequently combine to give presumptive evidence of value; similarly, the typical microscopic picture presented by a spore-bearing *C. tetani* culture should be remembered when such forms are encountered in pathologic material. All anaerobic species are non-acid-fast; therefore, this stain has no diagnostic importance.

Microscopic Examination of Pure Cultures

Gram stain. If the organism in question will grow within this period, apply the gram stain to a 16- to 18-hr culture and observe the same caution with reference to the decolorizer as noted above. Ordinarily the stain is satisfactory when prepared from any enrichment medium in which the organism will grow. In recording the gram reaction of a new species, state the medium from which the smear was made and the age of the culture.

Examination for motility. The majority of the sporeforming anaerobic bacilli are motile; the most important exception is *Clostridium perfringens* (*C. welchii*). The technic by which the motility examination is made is often of utmost importance in securing the correct results. *Unless the culture is known to be nonpathogenic, discard all cover slips and slides into a disinfectant solution or sterilize by steam before washing.* Use young cultures (12–18 hr) except as noted. Accept the results of hanging drop or wet-mount preparations under cover slips only if observation reveals positive motility. If motility is doubtful or appears to be negative, *initiate other procedures.* For example, use a flattened capillary tube sealed at each end. Heat glass tubing, of small diameter, and flatten a small area. Prepare a capillary tube from the flattened section. Draw a small amount of culture into this tube, and seal the tube in the flame on both sides of the drop of culture. Examine this preparation with the high-power objective. If the motility is still recorded as negative, make further observations on younger (4–6 hr) cultures. For these, examine the third or fourth tube of a serial passage series, using the medium which appears to give the best growth of the culture. Semisolid agar (0.5 per cent) may be inoculated by stab and observed for clouding due to motility. In some cases this procedure is complicated by breaking up of the agar as a result of gas production. Because of the relatively small number of species which are nonmotile, considerable caution should be exercised in reporting cultures which appear to be nonmotile. Naturally occurring nonmotile variants of motile species, however, have been encountered.

Flagella stain. For material for preparation of flagella stains use young cultures growing in the medium which is most favorable to the organism being studied. A temperature lower than that generally considered optimum is recommended (Leifson, 1951). If difficulty is encountered in securing positive slides from cultures known or thought to be motile, use the technic of Leifson, and consult the directions given by O'Toole (1942) for suggestions in technic which refer particularly to anaerobic bacteria.

Capsule stain. For the capsule stain one may use any of the conventional methods. The most important capsulated species is *Clostridium perfringens* (*C. welchii*). Material taken from artificially infected laboratory animals generally serves as the origin of smear preparations. If stains from *in vitro* cultures are desired, the medium of Svec and McCoy (Chap. III) is useful if other media prove unsuccessful.

Demonstration of spores. Cultures surviving 20 min heating at 80°C may be presumed to be sporeformers. It is, however, useful to demonstrate the spores microscopically. The exact method of making the spore stain is of little importance in comparison with other factors, as each of the common methods (Dorner, Moeller, and malachite green) appears satisfactory. One must, however, pay some attention to the medium in which one expects to induce sporulation. Media containing fermentable carbohydrates are not satisfactory, in general, for the pathogenic group. The media naturally containing carbohydrate (e.g., corn mash or potato infusion), on the other hand, appear ideal for most of the butyric-butyl group. For the pathogens one should use the deep brain, beef heart, or alkaline egg medium. In some instances spores may be demonstrated within 24–28 hr after inoculation, but if the culture is negative at this time, older cultures should be examined. Protection from evaporation must be given cultures which are to be incubated longer than one week. *Clostridium perfringens* (*C. welchii*) appears to be one of the most difficult species in which to demonstrate spores microscopically with regularity. If success is not attained using the above-mentioned media in cultures having the characteristics of this organism, one may use the medium recommended by Ellner (1956).

Since some taxonomic systems give considerable attention to the size and position of the spore, these characteristics should be recorded when the original laboratory examination is made. The characteristic appearance of *Clostridium tetani* spores has been noted above; these are round in shape and borne at the end of a slender vegetative rod. This is almost the only instance in which the picture of the spore and sporangium assumes importance in species diagnosis, and this observation must be supported by cultural or pathologic information as nontoxic organisms of similar microscopic characters occur.

Granulose reaction. The cells of certain species, particularly during the early stages of spore formation, store granulose. To test for this, add a drop of Lugol's iodine to a wet mount preparation. Cells containing granulose will stain blue or violet, while others will appear yellow.

Cultivation Technics[1]

Preliminary Enrichment Methods

Ordinarily the best method to be followed in initiating growth of an anaerobe from a sample is to inoculate one of the tubed media rather than to proceed directly to plate culture. Certainly this should be done if there is question concerning the possible success of the preliminary culture, and it is advised that parallel tube cultures be inoculated to serve as reserve cultures at the same time the plating is done, if the plating technic is favored. The medium to be used will be a matter of choice, as discussed in Chap. III, depending upon the nature of the sample. If aerobic contamination is suspected and the anaerobe is thought to be in the spore state, a duplicate primary culture should be heated briefly (boil for 1 or 2 min, or hold at 80°C for 20 min). This should be a duplicate culture, however, in case the anaerobic form is a nonsporeformer or is a sporeformer in the vegetative state. Almost all types of tubed media should have the dissolved oxygen driven off by boiling or heating in flowing steam.

The specimen of choice for bacteriological analysis of deep wounds consists of bits of tissue taken at the time of *débridement* during the initial surgery, or at the time of change of dressing. Alternate, though less desirable, samples include swab samples or tissue exudate obtained by aspiration. Such samples should be planted in heart-infusion broth with heart particles (often called chopped-meat medium) and in addition, streaked on blood or egg-yolk agar. (See Chap. III for details of preparation of these media.) Following incubation the heart-infusion cultures

[1] The use of petrolatum, mineral oil, or other materials as a seal at the surface of liquid media is not recommended.

should be plated for isolation of pure cultures. It should be remembered that infections from such wounds are commonly polymicrobic in origin.

Preliminary Purification Procedures

It is often difficult to isolate anaerobic bacteria from enrichments which also contain aerobic bacteria. It would be presumed that aerobic bacteria could ordinarily be eliminated merely by the anaerobic environment when this is introduced. Often in practice this is not the case, and other procedures must be instituted. It is of value frequently to attempt partial or complete elimination of the contaminants in tube culture using a liquid medium before plating is done. Materials derived from human or animal sources, other than feces, are usually contaminated with nonsporulating aerobic rods and cocci. Cultures derived from milk, soil, water, grains, feces, etc., contain, in addition, sporeforming aerobes. In fecal and perhaps other samples the contamination may include nonsporeforming anaerobes. If the nonsporeforming anaerobe is *wanted*, then anaerobic plating and picking of isolated colonies should be combined with optimum temperature and selective medium to secure the culture. In all cases the original enrichment tube should be preserved in the refrigerator, after growth is evident, until the purification routine is successfully completed. This will ensure a supply of starting material should something go wrong with the purification.

Generally one of the easiest practices to be followed to get rid of nonsporeforming types is as follows: Heat subcultures from the contaminated enrichment, retaining the original tube, of course, unheated. Heat the newly inoculated tubes 20 min at 80°C or a shorter time at higher temperatures. Take care to ensure the presence of the spores of the anaerobe. Use old cultures in a sugar-free medium as the best source of material to be heated, although other cultures may be satisfactory in special situations.

For enrichments contaminated with sporeforming aerobes the above procedure may not be satisfactory, owing to the heat resistance of the aerobic spores. In this case, one may employ dyes as bacteriostatic agents. Nearly all, if not all, aerobic sporeformers are inhibited by crystal violet, and most of the anaerobic types are relatively resistant. Two or three serial transfers may therefore be made in a medium containing this dye (approximately 1:100,000 final concentration) to eliminate the aerobe. The exact concentration of the dye to be used may vary with the medium and the conditions at hand. If the dye is used in some of the complex media, its effectiveness may be reduced during sterilization; therefore, the dye should be added to such media after sterilization. Either liquid or solid media may be used.

Sodium azide (0.02 per cent final concentration) together with chloral hydrate (0.01 per cent final concentration) may be used, singly or together, to inhibit gram-negative aerobic spreading organisms. Sorbic acid (York and Vaughn, 1944) in a final concentration of 0.12–0.15 per cent in thioglycollate broth will greatly aid in eliminating aerobic spore-forming bacilli, staphylococci, and many gram-negative contaminants. Initial enrichment broths are subcultured to sorbic acid thioglycollate broth, incubated 24 hr, then plated to an isolation plate. In the event that cultures are contaminated by *Pseudomonas* strains, polymyxin B may be added to the sorbic acid thioglycollate broth; a concentration of 10 μg per ml will usually inhibit these organisms and permit recovery of clostridia present (Lindberg, Mason, and Cutchins, 1954).

Another method for elimination of aerobic sporeformers utilizes the fact that while growth of the aerobe may take place in an anaerobic environment, the conditions for sporulation are unfavorable. Under such conditions the anaerobe will be expected to sporulate freely. Thus liquid cultures in tubes or plate cultures taken from an anaerobic jar are chosen for material for heating as in the case of the nonsporeforming contaminants.

Isolation Procedures

From a purely theoretical viewpoint, microscopic single-cell methods of isolation are ideal, but the low percentage of successes with these procedures excludes them from any uses except research. Several reports are in the literature indicating success with anaerobes using the Chambers micromanipulator or similar instruments, and wherever there is great need for strains of single-cell origin, the technic should be attempted. Because of the sensitivity of the vegetative cells toward oxygen, it is recommended that spores be picked rather than vegetative cells. One should use freshly exhausted media showing highly reducing activity for the subcultures, and naturally the medium should be suited to the organism being purified. If growth is not evident within the first 48 hr, the tubes may be protected from evaporation and incubated indefinitely. Reputable workers have reported dormancy of spores for six months' or longer duration.

In routine problems either plating or deep agar tube methods are available for purification of cultures from the original enrichment tubes. As stated above, the usual procedure in the isolation of anaerobes from samples in which contamination is excessive is best done by attempting partial purification in tube culture. This, however, need not be the case if the population of the sample is dominated by one species. In these the plating routine may be started without the preliminary enrichment procedure. Perhaps a few words should be included concerning details of technic. Since some of the anaerobes tend to spread rapidly over the surface of the agar, in many instances it will be found that "poured" agar plates are to be preferred to plates inoculated by streaking the surface. Two common methods are available for preparing these: (1) Melt tubes of the plating medium, cool, and inoculate before pouring; (2) place a small amount of sterile tap water in the culture dish, inoculate, and pour the agar into the dish immediately. If conditions warrant, use crystal violet in the agar. Place the plates in the anaerobic environment as soon as possible. (The size of inoculum to be used will vary so that some practice may be necessary to give a dilution sufficient so that well-isolated colonies will appear.) If difficulty is encountered in obtaining discrete colonies, decrease the agar concentration in the plating medium to 0.75–1.0 per cent.

Another method is available for colony isolation which may be preferred, particularly if the special apparatus needed for some of the plating methods is not at hand. This method involves the inoculation of a column of medium as mentioned in the opening pages of this chapter in the discussion of methods useful to determine whether or not a particular strain is an obligate anaerobe. For isolation purposes the fewer the number of colonies appearing in the medium the better. The percentage of fermentable sugar should be reduced to the lowest amount which gives good growth of the organism in order to prevent the production of gas, which may crack the medium. Assuming that we have available a deep tube of agar in which there appear several isolated colonies, two methods of isolation are available: (1) If soft glass tubes are used, cut the glass and break the tube at a short distance below the desired colony. Deposit the agar quickly in a sterile petri dish. Using a hot needle or small blade, cut across the plug of agar near the colony and transfer it to a suitable liquid medium. (This method is preferred if the tube shows aerobic contamination in the upper layers.) (2) If Pyrex tubes are used, eject the plug of agar into the sterile dish by applying a bunsen flame to the bottom end. Before this, heat the sides of the tube and sterilize the mouth of the tube in the flame. During the ejection step of the technic, hold the mouth of the tube so that it points directly into the sterile dish. After the column of agar is deposited in the dish, proceed as discussed above.

Inoculation Technics

The following points of culture transfer and other routine technics are sufficiently different from the procedures used with aerobes so that some note is needed:

Steam or boil most liquid media for a few minutes immediately prior to inoculation in order to drive off oxygen which may have been absorbed following sterilization. Attempt to deliver the inoculum to the *bottom* of the new tube of medium, for it is this portion of the medium which will stay reduced the longest. Although it is possible to initiate growth from a small number of cells, in routine studies use a more adequate inoculum. To facilitate the placing of the inoculum in the bottom of the tube with liquid and semisolid media, substitute a Wright or Pasteur pipet (used with small rubber bulbs) for the inoculation needle. By this means transfer a small drop (0.1 or 0.2 ml) of the culture to the new tube. Use the pipet also in the isolation of subsurface colonies particularly from media in which the concentration of agar is reduced. Prepare these pipets from 6- to 8-in. lengths of sterile 8- to 9-mm soft glass tubing (with cotton plug in each end) by applying heat to the center of the glass and pulling to form two capillary pipets.

In general, use a culture from 16–20 hr old. With the pathogenic types this time may be extended a few hours with no harm. With the butyl-butyric types, however, which sporulate readily in many media, there is a critical period in which the culture is not very satisfactory for transfer purposes. As the culture goes into the spore stage, it is less and less suitable until sufficient time elapses for the spores to mature. When spores are present in the inoculum, with these cultures and perhaps others as well, the new tube should be given a heat treatment (80°C for 20 min) *after* inoculation.

Generally, if an anaerobic sporeforming culture is desired in an experiment, inoculate a tube of a favorable medium from a stock culture which contains spores, heat-shock it, and use the resulting culture for the experiment rather than the inoculation of the latter tube or flask directly from the spore-containing culture. Maintain the stock culture in the spore state, follow the above transfer routine rather than carry the anaerobe in a serial passage, and use such cultures for sources of inoculum for experimental flasks or tubes. This is particularly true with the actively fermentative types, where serial passage may yield a culture of undesirable characters—even though it is descended in pure state from a culture that was satisfactory.

Other Methods of Value

Stock Culture Methods

The anaerobes are susceptible to freezing-drying technic as a means of preservation of cultures over a long period of time as shown by Roe (1940). This technic is unnecessary, however, as species of *Clostridium* are usually viable in spore state over a long period of time. For the pathogenic group, one should use beef-heart infusion, alkaline egg medium, and brain mash, with the last perhaps being the best. With the butyric-butyl group, use plain corn mash or potato infusion. Prepare the plain corn mash in a manner similar to the method given for corn-liver medium with the exception that the liver powder is omitted. Brain medium may be suitable also (see also Chap. III).

In any medium, after all gassing has subsided and spores have been demonstrated microscopically, the tube should be sealed in the flame or the stopper covered to protect the medium from evaporation and the tube placed in a cool room or refrigerator. Viable subcultures may be obtained from such tubes for months or even years in some instances. Another method which has been used with success is worthy of mention. This involves the storage of cultures on sterile soil: Dry fresh garden soil and sift through a fine-mesh screen; add 5 per cent of $CaCO_3$ to neutralize any acidity of the culture. Place soil in tubes in 2-in. columns, and autoclave overnight. Test each tube for sterility, using both aerobic and anaerobic media. If sterile, add 2 or 3 ml of a well-sporulated culture with a sterile pipet and dry the tube (preferably in a vacuum desiccator). To obtain an active culture from this stock (which may be stored at room temperature) transfer a small amount of the soil to an enrichment medium and heat-shock. By the soil stock method a relatively permanent source is available from which cultures may be revived as needed without destroying the stock culture.

Serological Reactions

The serological relationships of the sporeforming anaerobes have been reviewed (McCoy and McClung, 1938; Smith, 1955). The toxin-antitoxin reaction is of value as a taxonomic aid with certain species. In such an instance one takes advantage of the fact that relationships may be established by the success or failure of the reaction of antitoxin, prepared against the toxin of a known organism, with the toxin from the unidentified strain. In some instances the anaerobic species are monotypic with respect to toxin formation. In other species this is not true, and subgroups have been established within these species or species groups on the basis of non-cross-neutralization tests.

The problem of the complex relationships of the toxins produced by the clostridia is beyond the scope of this chapter, but those interested in the details beyond those presented by Smith (1955) should consult Oakley (1954) and Van Heyningen (1950, 1955).

REFERENCES

Barker, H. A. 1936. Studies upon the methane-producing bacteria. *Arch. Mikrob.*, **7,** 420–438.

Brewer, J. H. 1940. Clear liquid mediums for the "aerobic" cultivation of anaerobes. *J. Am. Med. Assoc.*, **115,** 598–600.

———. 1942. A new petri dish cover and technique for use in the cultivation of anaerobes and microaerophiles. *Science,* **95,** 587.

Brown, J. H., and J. H. Brewer. 1938. A method for utilizing illuminating gas in the Brown, Fildes, and McIntosh or other anaerobe jars of the Laidlaw principle. *J. Lab. Clin. Med.*, **23,** 870–874.

Dack, G. M. 1940. Non-spore-forming anaerobic bacteria of medical importance. *Bacteriol. Rev.*, **4,** 227–259.

Ellner, P. D. 1956. A medium promoting rapid quantitative sporulation in *Clostridium perfringens*. *J. Bacteriol.*, **71,** 495–496.

Evans, J. M., P. R. Carlquist, and J. H. Brewer. 1948. A modification of the Brewer anaerobic jar. *Am. J. Clin. Pathol.*, **18,** 745–747.

Fildes, P. 1931. Anaerobic cultivation. Chap. VI in "System of Bacteriology," vol. **9.** (Great Britain) Medical Research Council.

Fildes, P., and J. McIntosh. 1921. An improved form of McIntosh and Fildes anaerobic jar. *Brit. J. Exptl. Pathol.*, **2,** 153–154.

Griffin, A. M. 1932. A modification of the Buchner method of cultivating anaerobic bacteria. *Science*, **75,** 416–417.

Hall, I. C. 1922. Differentiation and identification of the sporulating anaerobes. *J. Infectious Diseases*, **30,** 445–504.

———. 1928. Anaerobiosis. Chap. XIII in "The Newer Knowledge of Bacteriology and Immunology," edited by E. O. Jordon and I. S. Falk. University of Chicago Press, Chicago.

———. 1929. A review of the development and application of physical and chemical principles in the cultivation of obligately anaerobic bacteria. *J. Bacteriol.*, **17,** 255–301.

Heller, H. H. 1921. Principles concerning the isolation of anaerobes. Studies in pathogenic anaerobes. II. *J. Bacteriol.*, **6,** 445–470.

Hewitt, L. F. 1950. "Oxidation-reduction Potentials in Bacteriology and Biochemistry," 6th ed. E. S. Livingstone, Ltd., Edinburgh.

Knight, B. C. J. G. 1931. Oxidation-reduction potential measurement in cultures and culture media. Chap. XIII in "System of Bacteriology," vol. **9.** (Great Britain) Medical Research Council.

Knorr, M. 1923. Ergebnisse neurer Arbeiten über krankheitserregende Anaerobien. I. Teil. Kranksheitserregende anaerobe Sporenbildner, ausschliesslich Tetanus und Botulinus. *Zentr. Gesam. Hyg.*, **4,** 81–100, 161–180.

———. 1924. Ergebnisse neuerer Arbeiten über krankheitserregende Anaerobien. II. Teil, 1: Botulismus. *Zentr. Gesam. Hyg.*, **7,** 161–171, 241–253.

Laidlaw, P. P. 1915. Some simple anaerobic methods. *Brit. Med. J.*, **1,** 497–498.

Lebert, F., and P. Tardieux. 1952. "Technique d'isolement et de détermination des bactéries anaérobies," 2d ed., 55 pp. Pacomhy, Paris.

Leifson, E. 1951. Staining, shape, and arrangement of bacterial flagella. *J. Bacteriol.*, **61,** 377–389.

Lindberg, R. B., R. P. Mason, and E. Cutchins. 1954. "Selective Inhibitors in the Rapid Isolation of Clostridia from Wounds." *Bacteriol. Proc.*, pp. 53–54.

McClung, L. S., E. McCoy, and E. B. Fred. 1935. Studies on anaerobic bacteria. II. Further extensive uses of the vegetable tissue anaerobic system. *Zentr. Bakteriol.*, II Abt., **91,** 225–227.

———, and E. McCoy. 1941. "The Anaerobic Bacteria and Their Activities in Nature and Disease: A Subject Bibliography," Suppl. 1: Literature for 1938 and 1939. *xxii* and 244 pp. University of California Press, Berkeley, Calif.

———, and R. Toabe. 1947. The egg yolk plate reaction for the presumptive diagnosis of *Clostridium sporogenes* and certain species of the gangrene and botulinum groups. *J. Bacteriol.*, **53,** 139–147.

———. 1956. The anaerobic bacteria with special reference to the genus *Clostridium*. *Ann. Rev. Microbiol.*, **10**, 173–192.

McCoy, E., E. B. Fred, W. H. Peterson, and E. G. Hastings. 1926. A cultural study of the acetone butyl alcohol organism. *J. Infectious Diseases*, **39**, 457–483.

———, ———, ———, and ———. 1930. A cultural study of certain anaerobic butyric acid-forming bacteria. *J. Infectious Diseases*, **46**, 118–137.

———, and L. S. McClung. 1938. Serological relations among the spore-forming anaerobic bacteria. *Bacteriol. Revs.*, **2**, 47–97.

———, and ———. 1939. "The Anaerobic Bacteria and Their Activities in Nature and Disease: A Subject Bibliography" (in two volumes). *xxiii* and 295 pp.; *xi* and 602 pp. University of California Press, Berkeley, Calif.

McIntosh, J. 1917. The classification and study of the anaerobic bacteria of war wounds. (*Gt. Brit.*) *Med. Research Council, Spec. Rpt. Ser.*, **12**, 1–58.

Meyer, K. F. 1928. Botulismus. In W. Kolle, R. Krause, und P. Uhlenhuth, "Handbuch der pathogenen Mikroorganismen," 3 Aufl., **4**, 1269–2364.

Molland, J. 1944. Oxidation-reduction potentials in cultures of anaerobic bacteria. *Acta Pathol. Microbiol. Scand.*, **21**, 673–712.

Oakley, C. L. 1954. Bacterial toxins. Demonstration of antigenic components in bacterial filtrates. *Ann. Rev. Microbiol.*, **8**, 411–428.

O'Toole, E. 1942. Flagella staining of anaerobic bacilli. *Stain Technol.*, **17**, 33–40.

Prévot, A.-R. 1950. "Manual de classification et de détermination des bacteries anaérobies," 2d ed., 290 pp. Masson et Cie, Paris.

———. 1955. "Biologie des maladies due aux anaérobies," 572 pp. Éditions Médicales Flammarion, Paris.

Reed, G. B., and J. H. Orr. 1941. Rapid identification of gas gangrene anaerobes. *War Med.*, **1**, 493–510.

———, and ———. 1943. Cultivation of anaerobes and oxidation-reduction potentials. *J. Bacteriol.*, **45**, 309–320.

Roe, A. F. 1940. Report on viability of 200 cultures of anaerobes desiccated for six years. *J. Bacteriol.*, **39**, 11–12.

Smith, L. DS. 1955. "Introduction to the Pathogenic Anaerobes," 253 pp. University of Chicago Press, Chicago.

Snieszko, S. 1930. The growth of anaerobic bacteria in petri dish cultures. *Centr. Bakteriol.*, II Abt., **82**, 109–110.

Spray, R. S. 1930. An improved anaerobic culture dish. *J. Lab. Clin. Med.*, **16**, 203–206.

———. 1936. Semisolid media for cultivation and identification of the sporulating anaerobes. *J. Bacteriol.*, **32**, 135–155.

Van Heyningen, W. E. 1950. "Bacterial Toxins," 133 pp. Charles C Thomas, Publisher, Springfield, Ill.

———. 1955. Recent developments in the field of bacterial toxins. *Schweiz. z. allgem. Pathol. u. Bakteriol.*, **18**, 1018–1035.

York, K., and R. H. Vaughn. 1944. Use of sorbic acid enrichment media for species of *Clostridium*. *J. Bacteriol.*, **68**, 739–744.

Zeissler, J. 1930. Anaërobenzuchtung. In W. Kolle, R. Krause, und P. Uhlenhuth, "Handbuch der pathogenen Mikroorganismen," 3 Aufl., **10**, 35–144.

Zeissler, J., and L. Rassfeld. 1928. Die anaerobe Sporenflora der europaischen Kriegsschauplätze 1917. *Veröffentl. Kriegs-Konstitutionspathol.*, **5**, Heft 2. 99 pp.

CHAPTER VII

Routine Tests for the Identification of Bacteria

H. J. Conn, M. W. Jennison, and O. B. Weeks

INTRODUCTORY

The Society of American Bacteriologists issues descriptive charts for use in characterizing bacterial species. The charts are blank forms to be recorded, at least one chart to be used for each culture studied. The "Manual of Methods for Pure Culture Study of Bacteria" was originally published to secure uniformity in the methods used for determining these characteristics. The present "Manual of Microbiological Methods" has become much broader than this, and practically all the methods covered in the first editions of the original manual are now included in this chapter.

The methods described in this chapter are intended primarily for aerobic saprophytes and cannot therefore be considered applicable in general to strict anaerobes, nutritionally "fastidious" organisms, and bacteria having other "special" cultural characteristics. Chapter VI must be consulted in studying the latter group, while Chap. IX gives methods specially applicable to animal pathogens. Special methods for plant pathogens are given in Chap. XII. In the case of other special groups, the investigator will therefore be forced to modify the methods or to use others more suited to the group in question.

THE DESCRIPTIVE CHARTS

There are two descriptive charts, each printed on 8½- by 11-in. sheets of heavy paper: the Standard Descriptive Chart and the Descriptive Chart for Instruction. The general plan of each is to have the body of it consist, under various headings, of a series of blanks to be completed and descriptive terms to be underlined as the various characteristics of the cultures are determined. In addition to this, there is a place on the mar-

Name of organism.......... Source.......... Studied by.......... Culture No..........
Date of isolation.......... Habitat.......... Optimum conditions: Media.......... Temp..........°C.
Is phase variation observed?.......... Phase on this Chart: S, R, M, G (smooth, rough, mucoid, gonidial).......... Phases recorded on other charts:..........

Underscore required terms.

VEGETATIVE CELLS: Medium used..........
Reaction (pH).......... Temp.......... Age..........d
Size of Majority..........
Ends, *rounded, truncate, concave, tapering*
MOTILITY: In broth.......... On agar..........
SPORANGIA and ENDOSPORES: *present, absent.*
Medium used.......... pH.......... Temp.......... Age..........d
Endospore Form: *spherical, ellipsoid, cylindrical*
IRREGULAR FORMS:
Present on.......... in.......... days at..........°C.

AGAR COLONIES: Temperature..........°C. Age..........d.
Form, *punctiform (i. e. under 1 m.m. diam.), circular (i. e. over 1 m.m. diam.), filamentous, irregular, rhizoid.*
Surface, *smooth, rough, concentrically ringed, radiately ridged.*
Edge, *entire, undulate, lobate, erose, filamentous, curled.*
Elevation of growth, *effuse, flat, raised, convex.*
Optical Characters, *opaque, translucent, opalescent, iridescent.*

GELATIN COLONIES: Temperature..........°C. Age..........d.
Form, *punctiform, circular, irregular, filamentous.*
Elevation, *flat, raised, convex, pulvinate, crateriform (liquefying).*
Edge, *entire, undulate, lobate, erose, filamentous, curled.*
Liquefaction, *cup, saucer, spreading.*
Surface, *smooth, contoured, rugose.*
Optical Characters, *opaque, translucent, opalescent, iridescent.*

AGAR STROKE: Temperature..........°C. Age..........d.
Growth, *scanty, moderate, abundant, none.*
Form of growth, *filiform, echinulate, beaded, spreading, arborescent, rhizoid.*
Lustre, *glistening, dull.*
Chromogenesis .... *photogenic, fluorescent.*
Odor, *absent, decided, resembling* ....
Consistency, *butyrous, viscid, membranous, brittle.*
Medium, *grayed, browned, reddened, blued, greened, unchanged.*

NUTRIENT BROTH: Temperature..........°C. Age..........d.
Surface growth, *ring, pellicle, flocculent, membranous, none.*
Clouding, *slight, moderate, strong, transient, persistent, none, fluid turbid, granular growth.*
Odor, *absent, decided, resembling*..........
Sediment, *compact, flocculent, granular, flaky, viscid.*
Amount of sediment, *abundant, scanty, none.*

GELATIN STAB: Temperature..........°C. Age..........d.
Growth, *uniform, best at top, best at bottom.*
Line of puncture, *filiform, beaded, papillate, villous, arborescent.*
Liquefaction, *none, crateriform, infundibuliform, napiform, saccate, stratiform* begins in..........d. complete in..........d.
Degree of liquefaction in..........days..........
Method used..........
Medium, *fluorescent, browned, unchanged.*

SKETCHES

Surface Colonies | Deep Colonies

Surface Colonies | Deep Colonies

Medium: .... Temperature ....°C. Age ....d.

Medium: .... Temperature ....°C. Age ....d.

Medium: .... Temperature ....°C. Age ....d.

FERMENTATION

Temperature..........°C.

Medium.......... containing.......... .... and:

	Monosaccharides							Disaccharides						Polysaccharides						Alcohols							Glucosides			
	Arabinose	Rhamnose	Xylose	Glucose	Fructose	Galactose	Mannose	Lactose	Sucrose	Maltose	Trehalose	Melibiose	Cellobiose	Raffinose	Melezitose	Starch	Inulin	Dextrin	Glycogen	Glycerol	Erythritol	Arabitol	Adonitol	Mannitol	Sorbitol	Dulcitol	Salicin	Aesculin	Coniferin	a-Methyl Gluc.
Gas in fermentation tube																														
Amt. CO₂ in Eldredge tube																														
Reaction (pH) after..........*d.*																														
Titrable acidity in ml. of N/.......... NaOH																														

BRIEF CHARACTERIZATION

As each of the following characteristics is determined, indicate in proper marginal square by means of figure, as designated below. In case any of these characteristics are doubtful or have not been determined, indicate with the letters U, V, and X according to the following code:
U, undetermined; V, variable; X, doubtful.

MORPHOLOGICAL

VEGETATIVE CELLS
- Form & arrangement: 1, streptococci; 2, diplococci; 3, micrococci; 4, sarcinæ; 5, rods; 6, commas; 7, spirals; 8, branched rods; 9, filamentous
- Diameter: 1, under 0.5μ; 2, between 0.5μ and 1μ; 3, over 1μ
- Gram stain: 0, negative; 1, positive
- Flagella: 0, absent; 1, peritrichic; 2, polar; 3, present but undetermined
- Capsules: 0, absent; 1, present
- Chains (4 or more cells): 0, absent; 1, present

SPORANGIA: 0, absent; 1, elliptical; 2, short rods; 3, spindled; 4, clavate; 5, drumsticks

ENDOSPORES: 0, absent; 1, central to excentric; 2, subterminal; 3, terminal

CULTURAL

AGAR STROKE
- Growth: 0, absent; 1, abundant; 2, moderate; 3, scanty
- Lustre: 1, glistening; 2, dull

AGAR COLONIES
- Form: 1, punctiform; 2, circular (over 1 mm. diameter); 3, rhizoid; 4, filamentous; 5, curled; 6, irregular
- Surface: 1, smooth; 2, contoured; 3. rugose

GELATIN COLONIES
- Form: 1, punctiform; 2, circular (over 1 mm.); 3, irregular; 4, filamentous
- Surface: 1, smooth; 2, contoured; 3, rugose

PHYSIOLOGICAL

- Biologic relationships: 1, pathogenic for man; 2, for animals but not for man; 3, for plants; 4, parasitic but not pathogenic; 5, saprophytic; 6, autotrophic
- Relation to free oxygen: 1, strict aerobe; 2, facultative anaerobe; 3, strict anaerobe, 4, microaerophile
- In nitrate media: 0, neither nitrite nor gas; 1. both nitrite and gas; 2, nitrite but no gas; 3, gas but no nitrite
- Chromogenesis: 0, none 1, pink; 2, violet; 3, blue; 4, green; 5, yellow; 6, orange; 7, red; 8, brown; 9, black
- Other photic characters 0. none; 1, photogenic, 2, fluorescent; 3, iridescent
- Indole: 0, negative; 1, positive
- Hydrogen sulfide: 0, negative; 1, positive
- Hemolysis: 0, negative; 1, positive
- Methemoglobin: 0, negative; 1, positive

PROTEIN LIQUEFACTION OR DIGESTION
- Gelatin: 0, negative; 1, positive
- Casein: 0, negative; 1, positive
- Egg albumin: 0, negative; 1, positive
- Blood serum: 0, negative; 1, positive

INDICATOR REDUCTION
- Litmus: 0, negative; 1, positive
- Methylene blue: 0, negative; 1, positive
- Janus green: 0, negative; 1, positive

Rennet production: 0, negative; 1, positive

FIG. 2. Standard Descriptive Chart. Back of chart is shown on page 142. Size of actual chart is $8\frac{1}{2}$ by 11 in. Distributed by the Society of American Bacteriologists.

SUPPLEMENTARY DATA

TEMPERATURE RELATIONS

Medium ...... pH ......
Optimum temperature for growth ...... °C.
Maximum temperature for growth ...... °C.
Minimum temperature for growth ...... °C.
THERMAL DEATH POINT: Time 10 minutes: ...... °C.
Medium ...... pH ......
THERMAL DEATH TIME:
Medium ...... pH ......

Temp.	Time	Temp.	Time
......°C.	......min.	......°C.	......min.
......°C.	......min.	......°C.	......min.
......°C.	......min.	......°C.	......min.
......°C.	......min.	......°C.	......min.
......°C.	......min.	......°C.	......min.

CHROMOGENESIS

Gelatin ......
Agar ......
Potato ......
......

OTHER PHOTIC CHARACTERS

Photogenesis on ......
Iridescence on ......
Fluorescence in ......
......

RELATION TO REACTION (pH) OF MEDIUM

Medium ......
Optimum for growth: *about pH* ......
Limits for growth: *from pH* ...... *to* ......

RELATION TO FREE OXYGEN

Method ......
Medium ...... Temp ...... °C.
Aerobic growth: *absent, present, better than anaerobic growth, micro-aerophilic*
Anaerobic growth: *absent, occurs in presence of glucose, of sucrose, of lactose, of nitrate; better than aerobic growth*
Additional data: ......
......
......

MILK

Temperature ...... °C.

Reaction:	......d. ......;	......d. ......;	......d. ......
Acid curd:	......d. ......;	......d. ......;	......d. ......
Rennet curd:	......d. ......;	......d. ......;	......d. ......
Peptonization:	......d. ......;	......d. ......;	......d. ......

LITMUS MILK

Temperature ...... °C.

Reaction:	......d. ......;	......d. ......;	......d. ......
Acid curd:	......d. ......;	......d. ......;	......d. ......
Rennet curd:	......d. ......;	......d. ......;	......d. ......
Peptonization:	......d. ......;	......d. ......;	......d. ......

Reduction of litmus *begins in* ...... *days, ends in* ...... *days*

ACTION ON ERYTHROCYTES

Cells: ......
Method: *plate, broth, filtrate*
Hemolysis: *negative, positive*
Methemoglobin: *negative, positive*

PRODUCTION OF INDOLE

Medium ......
Test used ......
Indole *absent, present in* ...... *days*

PRODUCTION OF HYDROGEN SULFIDE

Medium ......
Test used ......
H_2S *absent, present in* ...... *days*

ACTION ON NITRATES

Medium ...... Temp ...... °C.
Nitrite: ......d. ......; ......d. ......; ......d. ......; ......d. ......
Gas (N_2): ......d. ......; ......d. ......; ......d. ......; ......d. ......
Medium ...... Temp ...... °C.
Nitrite: ......d. ......; ......d. ......; ......d. ......; ......d. ......
Gas (N_2): ......d. ......; ......d. ......; ......d. ......; ......d. ......
Ammonia production (in amino-N-free nitrate medium): *negative, positive*
Complete disappearance of nitrate in ...... medium: *negative, positive*
Disappearance of 2 p.p.m. nitrite in ...... medium: *negative, positive*

REDUCTION OF INDICATORS

Medium ...... pH ...... Temp ...... °C.

Indicator	Conc.	Reduction:	
......	......%	......hr. ......;	......hr. ......
......	......%	......hr. ......;	......hr. ......
......	......%	......hr. ......;	......hr. ......
......	......%	......hr. ......;	......hr. ......

STAINING REACTIONS

Gram: ......d. ......; ......d. ......; ......d. ......; ......d. ......
Method ......
Spores: Method ......
Capsules: Method ......
Medium ......
Flagella: Method ......
Special Stains: ......
......
......

ADDITIONAL TESTS

Methyl red: *negative, positive*
Voges-Proskauer: *negative, positive*
Growth in sodium citrate: *absent, present*
Growth in uric acid: *absent, present*
Hydrolysis of starch: *complete (iodine colorless); partial (iodine reddish-brown); none (iodine blue)*
Nitrogen obtained from the following compounds: ......
......
......
......

PATHOLOGY

ANIMAL INOCULATION

Medium used ...... Age of culture ...... Amount ...... Incubation period ......

		Whole culture			Cells			Filtrate		
Animal										
Type of Injection	Subcutaneous	*								
	Intraperitoneal									
	Intravenous									
	Per os									

*In each instance where pathogenicity is observed, indicate location of lesion, and type, e. g. edema, histolysis, gas, hemorrhage, ulcer, diphtheritic, etc.

ANTIGENIC ACTION

Animal ...... Medium used ...... Age of culture ......
Type injection ...... Number of injections ......
Culture causes production of *cytolysins, agglutinins, precipitins, antitoxin.*
Specificity: Antibodies produced effective against other antigens as follows ......
......
Immune sera from ......
...... effective against this organism as antigen

SPECIAL TESTS

This DESCRIPTIVE CHART presented at the annual meeting of the SOCIETY OF AMERICAN BACTERIOLOGISTS, Dec. 28, 1934, by the Committee on Bacteriological Technic. Prepared by a sub-committee consisting of M. W. Jennison and H. J. Conn.

gin for recording the most important characteristics by a system of numerical notation.

The special feature of the Standard Descriptive Chart is that all the most important characteristics of an organism may be recorded on the front of the sheet, partly in the margin, partly in the larger section at the right, while the fermentative reactions are to be entered at the bottom. By the use of right-hand margin and bottom edge, a long series of charts may be compared, one on top of the other, by glancing only at these two edges. The back of the Standard Chart is now reserved largely for supplementary data, nearly all of which is summarized on the front. (See Fig. 2.)

The increasingly large number of tests called for in the study of bacteria has resulted in making a somewhat complicated chart. Although all these tests may be needed in some research work, they plainly are not needed in the use of the chart for instruction purposes. To meet the demand for a simpler chart for use in teaching, a new form known as the Descriptive Chart for Instruction was published in 1939. This chart is designed to fit a standard notebook for 11- by 8½-in. sheets. (See Fig. 3.) In numerous research laboratories, also, this chart is proving more useful than the Standard Chart because of its flexibility and the amount of space available for special tests.

DETERMINING OPTIMUM CONDITIONS FOR GROWTH

Before beginning the study of any pure culture, it is important to know something about the growth requirements of the organism. If the organism in question does not grow in ordinary media, because it requires the complete absence either of oxygen or of organic matter (or has other "special" requirements), it obviously cannot be studied by the methods called for on the Descriptive Chart. For such organisms the investigator must use his own methods of study and may record the results in the blank space at the bottom of the back of the chart. For those organisms that grow on ordinary media, methods must be varied according to whether the organisms grow better in liquid or in solid media and at high temperature or low temperature. It is important, therefore, that before an unknown culture which is able to grow in laboratory media is studied, these two points in regard to growth requirements be determined. (As pointed out in Chap. III, many such media are now available in dehydrated form.)

Bacteria may not grow well upon ordinary laboratory media when they are first isolated. In some instances cultures may be adapted to the media used for routine testing of cultural properties through a series of transfers. Such adaptations may be necessary to permit a study of an unknown culture in terms of previously described cultures.

After these growth peculiarities are determined, it is possible to proceed with the study of an organism under optimum conditions. Space is left on the chart under all the procedures listed where the medium used and the temperature of incubation can be recorded. As far as possible the same uniform set of conditions should be used throughout the entire study of one organism. If, for example, one set of tests is made on solid media at 25°C, the other tests should be made likewise. Leaving out those organisms referred to above which require special conditions of study and other organisms of "special" growth requirements, such as the thermophilic bacteria, there are four different sets of conditions that will suit practically all bacteria, namely, liquid media at 37°, solid media at 37°, liquid media at 21–25°, and solid media at 21–25°.

Space is provided on the Standard Chart for recording optimum medium and temperature. This does not ordinarily mean that one must determine the one best medium for the growth of the culture or the exact degree of temperature at which it grows most rapidly. In the first blank one may record such terms as "organic, solid"; "organic, liquid"; "inorganic, solid"; etc., unless it be known that there is one particular medium specially adapted to the organism in question. Under the second blank one may record temperature in general terms, as: "20–25°," "35–40°," "45–50°," or "over 55°."

It is also important to remember that certain organisms (frequently facultative anaerobes) which do not grow in either solid or true liquid media will grow in a "semisolid" medium (that is, a nutrient solution in which 0.05–0.1 per cent of agar has been dissolved). It is, of course, important that such organisms be studied under optimum conditions, and for their study the procedures given in this manual should ordinarily be modified by using media containing 0.05–0.1 per cent of agar instead of the usual liquid or solid media.

Thermal death point, as called for under "Temperature Relations" on the back of the chart, is undoubtedly best determined with the use of capillary tubes. Short pieces of thin-walled tubing having an internal diameter of 1–1.5 mm are filled with the culture (consisting mostly of spores if it is a sporeformer) and are heated for varying periods of time at the temperatures under investigation. After heating, each tube is broken into a tube of a medium in which the organisms grow well. A tabulation of results gives a good idea of the thermal death point. This procedure requires careful attention to detail, and one should consult the description of it by Magoon (1926). Results are most valuable if the length of time before death is recorded, in which case, this becomes a test for *thermal death time.*

Name of organism............ Studied by............ Culture No............

Source............ Habitat............ Date............

Descriptions (*Underscore required terms.*)	Sketches
CELL MORPHOLOGY Medium: Vegetative cells: Age: Form and arrangement: *streptococci, diplococci, micrococci, sarcinae, rods, commas, spirals, branched rods, filaments.* Motility in broth: Flagella: Size: Irregular forms: Sporangia: *none, rods, spindles, elliptical, clavate, drumstick.* Age: Endospores: Shape: *spherical, ellipsoid, cylindrical.* Position: *central to excentric, terminal, subterminal.*	Temp. °C.
STAINING CHARACTERISTICS Gram: Age: Method: Special stains:	
AGAR STROKE Age: Amount of growth: *scanty, moderate, abundant.* Form: *filiform, echinulate, beaded, spreading, rhizoid.* Consistency: *butyrous, viscid, membranous, brittle.* Chromogenesis: ; *fluorescent, iridescent, photogenic.*	Temp. °C.
AGAR COLONIES Age: Form: *punctiform, circular, filamentous, rhizoid, irregular.* Elevation: *effuse, flat, raised, convex.* Surface: *smooth, contoured, radiate, concentric, rugose.* Margin: *entire, undulate, erose, filamentous, curled.* Density: *opaque, translucent.*	Temp. °C.
NUTRIENT BROTH Age: Surface growth: *none, ring, pellicle, flocculent, membranous.* Subsurface growth: *none, turbid, granular.* Amount of growth: *scanty, moderate, abundant.* Sediment: *none, granular, flocculent, viscid, flaky.*	Temp. °C.
GELATIN STAB Age: Liquefaction: *none, crateriform, infundibuliform, napiform, saccate, stratiform.* Rate: *slow, moderate, rapid.*	Temp. °C.
OTHER MEDIA Age:	Temp. °C.

FERMENTATION Temp. °C.					
Medium: Carbohydrate: % Indicator:	Glucose	Lactose	Sucrose		
Acid in days					
Acid in days					
Gas in days					
Gas in days					

ACTION ON MILK Temp. °C.						
Indicator:	Days					
Reaction						
Acid curd						
Rennet curd						
Peptonization						
Reduction (before coagulation)						

FIG. 3. Descriptive Chart for instruction. Back of chart is shown on page 146. Size of actual chart is 8½ by 11 in. Distributed by the Society of American Bacteriologists.

ACTION ON NITRATES

Medium: Temp. °C.

Nitrite: ....d. ; ....d. ; ....d.

Gas (N): ....d. ; ....d. ; ....d.

HYDROGEN SULFIDE PRODUCTION

Medium: Age:

H_2S: *present, absent.* Temp. °C.

TEMPERATURE RELATIONS

Growth in refrigerator (°C.): *present, absent.*
Growth at room temperature (°C.): *present, absent.*
Growth at 37° C.: *present, absent.*
Growth at 50° C.: *present, absent.*

INDOLE PRODUCTION

Medium: Age:

Method: Temp. °C.

Indole: *present, absent.*

RELATION TO FREE OXYGEN

Medium: Age:

Method: Temp. °C.

Aerobic growth: *absent, present, better than anaerobic growth, poorer than anaerobic growth.*

Anaerobic growth: *present, absent.*

ADDITIONAL TESTS

This simplified Descriptive Chart recommended for instruction purposes by the Committee on Bacteriological Technic of the Society of American Bacteriologists and put on sale provisionally during 1939.

INCUBATION

Cultures should be incubated at or near the optimum temperature of the organism or organisms under investigation. As a rule it is not necessary, however, to know the exact optimum temperature of each organism. If the laboratory is equipped with a series of incubators running at 20, 25, 30, and 37°C, the temperature requirements of practically all bacteria except the thermophilic forms can be very satisfactorily met. Room temperature is sometimes used in place of 25° but is not to be recommended because of its uncontrollable variations.

Length of incubation varies and is specified on the chart under many of the tests. In cases where it is not specified, one should observe the following general rule: On the day when good growth first appears, the proper descriptive terms on the chart should be underlined. Any changes occurring and noted in subsequent study should also be recorded on the chart. The meaning of the terms given in this section of the chart will in general be made clear by consulting the glossary (Chap. XIII).

VARIATION

In using these methods it must be remembered that among bacteria, the individual members of any species may differ from one another in respect to both physiology and morphology, thus making it difficult to define the limits of the species; also that any individual culture in repeated examinations may produce variable results in connection with some test even when studied under apparently constant conditions. For these reasons it is important that single determinations shall never be used for characterizing any culture that has been studied or much less for characterizing any species or type that is being described. Determinations must be repeated at different times and under different conditions in order to learn definitely the physiological characteristics of a culture. Whenever possible, an effort should be made to correlate the variations in physiology and serology with colony type and to list separately the physiological characteristics of the "smooth," "rough," "mucoid," "opaque," "translucent" strains, etc. When an organism shows any tendency to "dissociate" into "phase variants," its description is incomplete if it applies to only one phase or to a culture containing a mixture of two phases or more. In such case the phase variants should be separated by plating methods or otherwise and a separate chart should be used for each individual strain studied. The individual charts may be filed for the investigator's information, but it must be insisted that results of such work should not be published for the use of other bacteriologists until repeated determinations have been made and, if possible, have been shown to bear some relation to the phase indicated by colony type.

MICROMETHODS

Attention is called to the type of methods for determining biochemical characteristics known as "quick" methods or micromethods. The methods in question depend on making mass inoculations into media preheated to 37°C and making readings after very brief periods of incubation. The principle involved is that the enzymes produced act so quickly that if mass inoculations of vigorous cultures are made, action of the enzymes can be studied almost like a chemical reaction without waiting for further bacterial growth to occur. A method depending on this principle, for determining rennet production (see page 166), was given in several editions of the old manual, but it is not considered to be so well standardized as those more recently worked out by Weaver in the United States and Cowan in England, together with their respective associates. The quick tests prove quite effective for determining nitrate reduction, indole, H_2S, and acetyl-methyl-carbinol production and in some cases for sugar fermentations as well as for the methyl red test. Studies by Cowan (1953*a* and *b*) have shown, however, that some micromethods are not readily suited to routine work. Determinations such as the methyl red test and fermentations of carbohydrates give variable results unless factors such as buffer concentrations, substrate concentration, and quantity of cells are carefully controlled. In addition there is always the possibility that a cultural test, micro or macro, may be made so sensitive that the property being tested loses its diagnostic significance. An illustration of this is the Batty-Smith procedure for detecting acetylmethyl carbinol (Shaw *et al.*, 1951) and several methods for the determination of H_2S (Clarke, 1953). This point is discussed by Clarke and Cowan (1952).

CULTURAL CHARACTERISTICS

Space is provided on both charts for recording appearance of colonies, growth on agar stroke, in broth and gelatin stab. In addition to the space provided for sketches, various terms are listed in order that those which apply may be underlined. The meaning of all the terms is given in the glossary (Chap. XIII).

As some of the terms, especially in regard to shape and structure of colonies, are more easily described graphically than verbally, the diagram in Fig. 4 is included here to assist the student in understanding the appropriate terms.

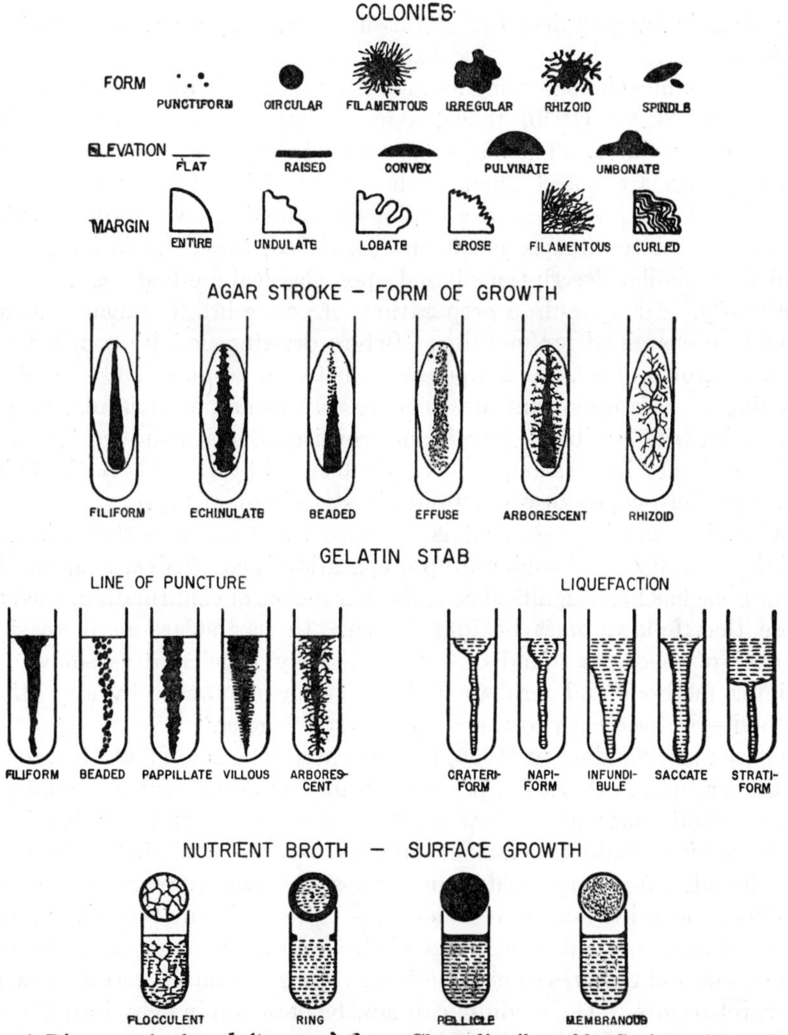

FIG. 4. Diagram of cultural characteristics. Chart distributed by Society of American Bacteriologists.

STUDY OF CELL MORPHOLOGY

The routine study of morphology should include examinations of stained, dried preparations and of unstained organisms in hanging drop. Stained preparations to show the vegetative cells should be made, preferably from agar slant cultures, from a few hours to two days old, according to the rapidity of growth. The medium and temperature used and the age of the culture should be recorded. The examination of unstained organisms in hanging drop is a useful supplementary procedure too often neglected.

Studies of morphology using methods such as those described by Knaysi (1951) and Bisset (1950) reveal a more natural morphology than do classical procedures. The majority of morphological descriptions of bacterial species are based upon studies of air-dried, heat-fixed, stained smears. If morphological descriptions obtained from improved cytological procedures (see page 30) are published, these should be accompanied by similar descriptions based upon classical methodology.

Motility. Hanging-drop preparations of young broth or agar cultures should be examined for motility. Before drawing definite conclusions, cultures grown at several temperatures between 20 and 37°C should be examined. It is important not to confuse Brownian or molecular movement with true motility. The former consists of a "to-and-fro" motion without change in position, except as influenced by currents in the fluid. A phase microscope can prove useful in studying motility.

When interpreting results it is important to remember that whereas definite motility in a hanging-drop preparation is conclusive, weak motility or none has little significance and other means of confirmation, such as those that follow, must be undertaken. In particular, an increasing number of cases are found of organisms fully flagellated as shown by staining methods and serology but absolutely nonmotile by any other method—bacteria with so-called "paralyzed flagella."

Tittsler and Sandholzer (1936) have proposed the use of stabs in a semisolid agar (meat extract 0.3 per cent, peptone 0.5 per cent, agar 0.5 per cent). Motile organisms show a diffuse zone of growth spreading from the line of inoculation; nonmotile cultures do not. For this test, incubation should be for 6 days at 30°C unless positive results are secured sooner. For gram-negative nonsporeformers, 12- to 18-hr incubation gives more clear-cut results. This test is a good check on the hanging-drop method but is slow and requires some experience before one can be certain how to interpret results. The medium can now be obtained in dehydrated form under the name Motility Test Medium.

Conn and Wolfe (1938), moreover, have recommended a flagella stain even on cultures that do not appear motile upon examination in hanging drop. The modification of the Bailey flagella stain given in Chap. II is simple and quick enough to be employed for routine examinations; positive results cannot be misinterpreted and show the arrangement of flagella as well as the mere presence or absence of motility. A few further refinements of the method, making it more adaptable to routine use on bacteria of various types, published by Fisher and Conn (1942), are also given in Chap. II. The Leifson (1951) procedure is also well suited for routine use.

Presence of endospores. Routine examinations should be made on agar slant cultures a week old, employing methylene blue or dilute crystal violet to stain the vegetative rods and leave spores unstained. If spore-like bodies are present whose exact nature is uncertain, one of the spore stains recommended in Chap. II should be employed.

In most cases there is little trouble in finding spores if the organism produces them. All rather large rods, however (0.8 μ or more in diameter), should be regarded as possible spore producers even if microscopic examination does not show spores. Such bacteria should be mixed with sterile broth or physiological saline solution and heated to 85°C for 10 min; if still alive, endospores may be regarded as probably present. One should also make repeated transfers of the culture onto agar and examine at various ages. A culture of a large rod should not be recorded as a nonsporeformer unless all these tests are negative.

Acid-fast staining. Various methods have been proposed for determining if an organism is "acid fast." They are all essentially modifications of the same general procedure and are similar to the spore stains of Moeller (1891) and Foth (1892). The committee is as yet unprepared to recommend any one of them in particular. Several are listed in Chap. II.

Capsules. An organism should not be recorded as having capsules unless they have been actually stained by one of the methods of capsule staining described in bacteriological textbooks. Four of the common methods of capsule staining, namely those of Anthony, of Hiss, of Huntoon, and of Churchman, are given in Chap. II. The committee has obtained good results with Anthony's and Hiss' methods. Capsules do not appear in all media; the medium of choice should be milk serum slants or exudates from infected animals.

Irregular forms. Forms that differ from the typical shape for the organism, such as branching forms, clubs, spindles, or filaments, should be noted and sketched. Simple observation is enough to show that these irregular forms occur quite uniformly in certain cultures; hence their existence must not be ignored. The *interpretation* of these forms is at present under dispute, and the decision as to their significance must be awaited. The committee recommends that the microscopic study of any culture include an examination of the growth on various media and at various ages upon each medium, with sketches of all the shapes that occur.

Gram stain. The gram stain was until recently an entirely empirical procedure for distinguishing between two groups of organisms, the actual significance of which was not understood. Since 1940, however, the work of Henry and Stacey (1943), Bartholomew and Umbreit (1944), and others has shown that a positive reaction is dependent upon the presence of ribonucleic acid in the outer layers of the cells, which can be removed by treatment with ribonuclease and replated on them by treatment with magnesium ribonucleate. Thus gram-positive organisms can be artificially converted to gram-negative ones and then restored to their gram-positive state.

In addition to this fact, it is also true that many bacteria are neither definitely positive nor negative; some organisms are gram-variable and may appear either negative or positive according to conditions. Other organisms contain granules which resist decolorization and may cause misinterpretation. The importance of taking such variations into account has been repeatedly emphasized. Such organisms should be recorded as gram-variable rather than made to appear either positive or negative by some modification of technic. To determine if an organism belongs to this variable group, it is necessary that it be stained at two or three different ages by more than one procedure. If an organism changes from positive to negative or vice versa during its life history, this change should be recorded, with a statement as to the age of the culture when the change was first observed. It is often practical to record such an organism as prevailingly positive or prevailingly negative; obviously, however, this cannot be done without a very considerable series of determinations. Tests must therefore be made after 1 and 2 days' incubation, sometimes also in even older cultures. It must, moreover, be recognized that *gram-variable* organisms are not necessarily ones that show *uneven* gram staining; the latter should be recorded as staining unevenly, not as gram-variable.

The two methods at present recommended are the ammonium oxalate method (Hucker) and Kopeloff and Beerman's modification of the Burke technic. In the former the manipulation is more simple, but the latter gives better results if the organism is growing in a medium that may be of acid reaction (e.g., exudates) and is claimed to distinguish better between true and false positive reactions. These two procedures are given in Chap. II.

RELATION TO FREE OXYGEN

In relation to free oxygen, organisms are generally classified as strict aerobes, facultative anaerobes, or strict anaerobes. A fourth group of microaerophiles may also be recognized. None of these distinctions is clear cut, but the following method gives a rough grouping of bacteria in regard to their oxygen requirements.

Agar shake culture affords a good routine method of determining the oxygen requirements of an organism. A tube of deep agar medium containing glucose or some other available carbon source is inoculated while in fluid condition at 45°C with an inoculum not too heavy to permit discrete colonies, rotated to mix the inoculum with the medium, and cooled.

Some bacteriologists prefer to pour or to pipet the inoculated medium into another sterile tube to ensure thorough mixing.

Upon incubation, strict aerobes will be found to grow upon the surface and in the upper layers only, microaerophiles will grow best just a few millimeters below the surface, facultative anaerobes will grow throughout the medium, and strict anaerobes will grow only in the depths, if at all.

ACTION ON NITRATES

Nitrate reduction should be indicated by complete or partial disappearance of nitrate accompanied by appearance of nitrite, ammonia, or free nitrogen. As quantitative nitrate tests are too time-consuming for routine pure culture work, one must ordinarily be satisfied with tests for the end products only.

The following routine procedure is recommended: Inoculate into nitrate both and onto slants of nitrate agar (containing 0.1 per cent KNO_3 plus beef extract and peptone as usual). Test the cultures on various days as indicated on the chart. On these days examine first for gas as shown by foam on the broth or by cracks in the agar. Then test for nitrite with the following reagents.

1. Dissolve 8 g of sulfanilic acid in 1 liter of 5*N* acetic acid (1 part of glacial acetic acid to 2.5 parts of water), or in 1 liter of dilute sulfuric acid (1 part of concentrated acid to 20 parts of water).
2. Dissolve 5 g of α-naphthylamine in 1 liter of 5*N* acetic acid or of very dilute sulfuric acid (1 part of concentrated acid to 125 parts of water). Or dissolve 6 ml of dimethyl-α-naphthylamine in 1 liter of 5*N* acetic acid. This latter reagent has recently been recommended by Wallace and Neave (1927), and by Tittsler (1930), as it gives a permanent red color in the presence of high concentrations of nitrite.

Put a few drops of each of these reagents in each broth culture to be tested and on the surface of each agar slant. A distinct pink or red in the broth or agar indicates the presence of nitrite. It is well to test a sterile control which has been kept under the same condition to guard against errors due to absorption of nitrous acid from the air.

A rapid method, calling for less than an hour's incubation, has been devised by Bachmann and Weaver (1947), and a rapid microtechnic by Brough (1950) and by Clarke and Cowan (1952).

Among the micromethods, Brough's is detailed in a concise, easily followed form. The culture is grown 18–24 hr at optimum temperature on a nutrient agar giving good growth. The growth is washed off and sufficient to give high turbidity is added to 1 ml (in a small test tube) of 0.1 per cent KNO_3 in nutrient broth; Clarke and Cowan show that the solution to which the KNO_3 is added may be pH 6.8 buffer instead of broth. The medium should be preheated in a water bath to 37°C before inoculation; after inoculation it needs to be incubated only 15 min at that temperature before adding the reagents of the nitrite test.

Presence of nitrite shows the nitrate to have been reduced, and the presence of gas is a strong indication that reduction has taken place. A negative result does not prove that the organism is unable to reduce nitrates; in such a case further study is necessary as follows:

In case the fault seems to lie in poor growth, search should be made for a nitrate medium in which the organism in question does make good growth by means of the following modifications: increasing or decreasing the amount of peptone, changing the amount of nitrate, altering the reaction, adding some readily available carbohydrate, adding 0.1–0.5 per cent of agar to a liquid medium to furnish a semisolid substrate. The appearance of nitrite in any nitrate medium whatever (while it is absent in a sterile control) should be recorded as nitrate reduction.

Absence of nitrite in the presence of good growth may indicate complete consumption of nitrate or its decomposition beyond the nitrite stage as well as *no* reduction at all. Test, therefore, for nitrate by adding a pinch of zinc dust to the tube to which the nitrite reagents have been introduced and allowing it to stand a few minutes. If nitrate is present, it will be reduced to nitrite and show the characteristic pink color. Confirmation of the test may be obtained by placing a crystal of diphenylamine in a drop of concentrated sulfuric acid in a depression in a porcelain spot plate and touching with a drop of the culture (or of the liquid at the base of the slant if agar cultures are used). The test will be more delicate if the culture is first mixed with concentrated sulfuric acid and allowed to cool. A blue color indicates presence of nitrate, provided nitrite is absent, but as nitrite gives the same color with diphenylamine, this test must not be used when nitrite is present in the same or greater order of magnitude.

If none of these tests indicate utilization of the nitrate, the organism probably does not reduce nitrate, but to be certain of the fact further investigation is necessary. It must be understood, however, that for routine diagnostic work a determination of nitrite on standard nitrate broth or agar is ordinarily sufficient; this is because most descriptions in the literature containing the words "Nitrates not reduced" mean merely that no nitrite is produced on this medium. *But in recording such results the student should be careful to state only the observed fact, i.e., that nitrite is or is not found in the nitrate medium employed.*

CHROMOGENESIS

Color production should be recorded if observed in broth or on beef-extract agar, gelatin or potato or if noticed to a striking extent on any other medium (e.g., starch media). In the margin the space devoted to chromogenesis refers to the color produced on beef-extract agar. This is merely because chromogenesis on such a medium is referred to in the literature more frequently than on the other media above mentioned. As a matter of fact, agar containing peptone is a very poor medium for the purpose. With some organisms, synthetic media are necessary to bring out color.

Minor variations in pigmentations when a culture is grown under different conditions should not be given too much emphasis, as they often merely reflect ability to grow on the medium in question. Some species, however (cf. Harrison, 1929), do show qualitative differences when the environment is changed, and such should be noted. Similarly, differences in the presence or absence of air should be recorded. Frequently it is well to note the final H-ion concentration of the culture, as some pigments vary in their hue with reaction of the substrate. Cultures should be allowed to develop completely before recording chromogenesis, not making a final reading till after several days at room temperature. Frequently pigmentation deepens when a culture is held in diffuse light.

INDOLE PRODUCTION

During the last 40 years, results of investigations on the indole test have been published by Zipfel (1912), Frieber (1921), Fellers and Clough (1925), Goré (1921), Holman and Gonzales (1923), Koser and Galt (1926), and Kovács (1928). The two important points brought out in these papers are that the medium be of correct composition and that the test used be specific for indole.

The important consideration in regard to the medium is that a peptone be employed containing tryptophane, which is not always present in bacteriologic peptones. Peptones are ordinarily digests of lean meat, but for the indole test a casein digest which contains tryptophane is apparently more satisfactory.

The medium used should, therefore, contain 1.0 per cent of casein digest. No other ingredients need be added if the organism under investigation will grow in a solution of it alone. If the organism is not able to grow in such a medium, add such ingredients as are needed to assure its growth. If necessary, add agar and perform the test on agar slants.

If the organism produces good growth, 1–2 days' incubation is ordinarily sufficient. In fact, with rapid-growing organisms, the reaction may be positive in 24 hr but negative the following day. Therefore both 24- and 48-hr tests are recommended. Arnold and Weaver (1948), however, with confirmation by Clarke and Cowan, (1952) show that 6–120 min incubation is enough by the following "quick" method:

The culture is grown on tryptone agar at optimum temperature till good growth is obtained; then heavy inoculation (to give high turbidity) is made into preheated (to 37°C) tryptophane broth, 1 ml to the tube. It is incubated at 37° for not over 2 hr, the test made with the Kovács reagent, as given below.

The test for indole may be performed by the technic of Ehrlich-Böhme, by either the Goré or the Kovács modification of the same or by the Gnezda technic. The Kovács method is especially simple and convenient. These procedures are as follows:

Böhme (1905) called for the following solutions:

Solution 1

Para-dimethyl-amino-benzaldehyde	1 g
Ethyl alcohol (95%)	95 ml
Hydrochloric acid, concentrated	20 ml

Solution 2

Saturated aqueous solution of potassium persulfate ($K_2S_2O_8$)

To about 10 ml of the culture fluid add 5 ml of solution 1, then 5 ml of solution 2, and mix well by rotating between the hands; a red color appearing in 5 min indicates a positive reaction. This test may also be performed (and sometimes more satisfactorily) by first mixing up the culture with ether and adding solution 1 (Ehrlich's reagent) dropping down the side of the tube so that it spreads out as a layer between the ether and the culture fluid. After this method of applying, solution 2 seems to be unnecessary.

The Goré (1921) test uses these same solutions, but the method of application is as follows: Remove the plug of the culture tube (which must be of white *absorbent* cotton), moisten it first with 4–6 drops of solution 2, then with the same amount of solution 1. Replace the plug and push down until 1 or 1½ in. above the surface of the culture. Place the tube upright in a boiling-water bath, and heat for 15 min without letting the culture solution come in contact with the plug. The appearance of a red color on the plug indicates the presence of indole.

The Kovács (1928) test is a simplification of that of Böhme, using only one solution; it is now the method of choice in many laboratories:

Para-dimethyl-amino-benzaldehyde	5 g
Amyl or butyl alcohol	75 ml
Hydrochloric acid, concentrated	25 ml

This reagent may be used as in the Böhme test, but no solution 2 is required. The red color appears in the alcohol layer.

The Gnezda (1899) oxalic acid test is made as follows: dip a strip of filter paper in a warm saturated solution of oxalic acid; on cooling, this is covered with crystals of the acid. Dry the strip of paper thoroughly (sterilization by heat seems unnecessary), and insert into the culture tube under aseptic conditions, bent at such an angle that it presses against the side of the tube and remains near the mouth. Reinsert the plug, and incubate the culture. If indole is formed, the oxalic acid crystals take on a pink color.

It is recommended that the Goré or the Kovács test be used in a routine way. In interpreting the results obtained it must be remembered that when the reagents are added directly to the medium, they react with alpha-methyl-indole as well as with indole itself, but as the former compound is nonvolatile, it cannot react to the Goré or Gnezda tests. Hence the Ehrlich test unmodified is less specific for indole than the Goré modification or the Gnezda test.

Some samples of para-dimethyl-amino-benzaldehyde and of amyl and butyl alcohol have been found unsatisfactory for the indole test. It is well, therefore, to check new supplies of these chemicals against samples known to be satisfactory.

THE PRODUCTION OF HYDROGEN SULFIDE

Hydrogen sulfide is generally detected in bacterial cultures by observing the blackening which it produces in the presence of salts of certain metals, such as lead, iron, or bismuth, owing to the dark color of the sulfide of these metals. Two methods have been utilized for employing these tests: one by incorporating the metallic salt in the medium and the other by using a test strip of filter paper impregnated with the metallic salt in question.

In early editions of the manual four media containing either lead or iron salts were given. The lead salt media, however, were discredited some time ago because of the toxic properties of these salts, and Hunter and Crecelius (1938) show the superiority of bismuth media over iron media. Zobell and Feltham (1934), moreover, have shown distinct advantages from the use of lead acetate test strips, without any of these metallic salts in the media. The advantage of the test strip technic is that it is more sensitive and does not introduce the possibility of inhibiting the bacterial growth if the concentration of metallic salt in the medium is too great. It is important, as emphasized by Hunter and Crecelius, that the indicator and method employed be stated when results are given. Untermohlen and Georgi (1940) suggest use of nickel or cobalt salts but specially emphasize the variations in results with different media and indicators.

When using the test-strip technic the bacteria may be grown in ordinary broth, peptone solution alone, or a peptone agar suitable to the organism in question. One must be certain that the peptone contains available sulfur compounds. This can be determined by running a check tube inoculated with a slow hydrogen sulfide producer. For this procedure the test strip should be prepared by cutting white filter paper into strips approximately 5 by 50 mm, soaking them in a saturated solution of lead acetate, sterilizing them in plugged test tubes, and drying in an oven at 120°C. One of these strips should be replaced in the mouth of the culture tube before incubation in such a position that one-quarter to one-half of the strip projects below the cotton plug. These tubes should be incubated at about the optimum temperature of the organism under investigation and examined daily to notice whether or not blackening of the test strip has occurred.

Because of the inconvenience of the test-strip technic, media in which iron salts are incorporated are now generally preferred. A dehydrated medium of such composition is available and has been found quite satisfactory.

For a "quick" method, Morse and Weaver (1947) recommend using small tubes containing 0.8-ml portions of 2 per cent thiopentone solution preheated to 37°C in a water bath. These are inoculated from a 6-hr agar slant culture using sufficient quantity to give a turbidity indicating 2,100 millions per liter. A strip of paper saturated with lead acetate solution is inserted in the tubes, and they are reincubated for 30–45 min. Clarke (1953) proposes a slightly different test; she calls attention to the fact that very few of the cultures which she tested, using a micromethod, failed to produce H_2S from cysteine. Thiosulfate appeared to be a more useful substrate than cysteine, since not all the cultures studied were able to reduce the compound to H_2S. When micromethods are used, the importance of recording the conditions of the test cannot be overemphasized.

LIQUEFACTION OF GELATIN

The conventional method of determining liquefaction, which has been given with but slight modification in all the reports on methods, is as follows:

Make a gelatin stab (plain 12 per cent gelatin), and incubate 6 weeks at 20°C, provided the organism under investigation will grow at that temperature. Care must be taken to observe whether the organisms produce rapid and progressive liquefaction or merely slow liquefaction not extending far from the point of inoculation. In the latter case the liquefaction may be due merely to endo-enzymes that are released from the cell after death and may not be what is generally called "true liquefaction" (that is, the process resulting from the action of enzymes diffusing out of actively growing cells). Some slow liquefiers are true liquefiers, however, and the distinction between slow and rapid liquefaction must be regarded as very artificial.

In early editions of the manual the Frazier (1926) method was given, but it was omitted from later editions as not proving practicable. A recent modification of it by Smith (1946), however, proves useful and has two advantages over the gelatin stab method: (1) It does not require low temperature incubation; (2) it is more sensitive in the case of weak liquefiers. The procedure is as follows: Streak culture on a plate of nutrient agar containing 0.4 per cent of gelatin. Incubate at 28°C for 2–14 days according to rate of growth. Cover plate with 8–10 ml of a solution of 15 g of $HgCl_2$ in 100 ml of distilled water and 20 ml of concentrated HCl. This reagent forms a white opaque precipitate with the unchanged gelatin, but a liquefier is surrounded by a clear zone.

There is another method recommended for organisms that do not grow at 20°C. By this technic an inoculated tube of gelatin is incubated at 37°C, or whatever temperature may be the optimum, and then after incubation the tubes are placed in a cold-water bath or in a refrigerator to determine whether or not the gelatin is still capable of solidifying. Suitable uninoculated controls must always be run in parallel, especially if the optimum growth conditions for the organism necessitate prolonged exposure of the gelatin to hydrolysis by mild acid, alkali, or heat. In addition, precautions should always be taken to prevent evaporation of moisture which might conceivably tend to obscure a slow liquefaction. This method has the advantage of rarely giving positive results except in case of "true liquefaction." On the other hand, it may well fail to detect cases of real liquefaction that have proceeded so slowly that the gelatin can still set even after several week's incubation. The significance of this test can be increased by using weaker than normal gelatin, 4 per cent gelatin, for example, or even less.

CLEAVAGE OF SUGARS, ALCOHOLS, AND GLUCOSIDES

Fermentable substance to employ. Quite a wide range of pure alcohols and carbohydrates is available for use in fermentation tests. In routine work the choice is often limited to the more common and less expensive substances, but in special research work economy is of less importance. The three sugars glucose, sucrose, and lactose and the alcohols glycerol and mannitol are most widely employed because they are readily available. Whether these compounds give valuable information depends upon the group of organisms being studied. If the group, like the colon group, is capable of fermenting nearly all these substances, these readily fermented sugars and alcohols may have very little value in separating the species one from another; one must then employ one or more of the rarer compounds if carbohydrate fermentation is considered to have any diagnostic significance in the group under study. In other words the selection is based upon the group of bacteria under investigation.

Attention must always be paid to the sterilization of heat labile fermentable substrates (see Chap. III).

Basal medium. The compound to be tested must be added to some basal medium suited to the group of organisms under investigation. For routine work it is best to employ two such basal media, namely, beef-extract peptone broth and beef-extract peptone agar, selecting one or the other according to whether the organisms under investigation grow better in liquid or solid media. These media should be prepared as directed in Chap. III. It should be noted that some commercial peptones contain fermentable sugars (Vera, 1949); hence care must be exercised in regard to the peptone selected, and controls must be run.

One should notice particularly whether or not good growth is obtained in any or all of these media after adding the fermentable substance under investigation. If poor growth or none is obtained in the broth and on the agar, one should vary the basal medium employed.

If a culture is to be studied in liquid, the media should be sterilized in fermentation tubes; if on solid media, agar slants should be used—see

Conn and Hucker (1920). Agar slants may be inoculated either on the surface alone or partly on the surface and partly in a stab at the base. It has been found in practice that if much gas is produced, it may occur at the very base of the column of agar even when all the growth seems to occur on the surface, but if there is reason to suspect that gas production is being overlooked, shake cultures may be used in addition to the agar slant.

Demonstration of cleavage. Utilization of the sugar (or other fermentable substance) may be indicated by a chemical determination showing its partial or complete disappearance or by the demonstration of the end products of fermentation. These end products are generally organic acids, sometimes accompanied with the evolution of gases, e.g., free hydrogen, carbon dioxide, or occasionally methane. Determinations of the amount of sugar remaining or of the nature of the organic acids produced are very valuable in discriminating investigations but require time-consuming chemical work that is difficult to employ in the routine examination of large numbers of cultures. These chemical methods are referred to in more detail elsewhere (Chap. VIII). In many instances, however, a sufficient amount of information is obtained merely by demonstrating an increase in acid or the presence of gas.

For routine work in the case of organisms concerning which little advance information is at hand, the use of indicators is especially valuable in determining whether or not production of acid has occurred. It must be remembered, however, that in many instances more useful and significant information can be obtained by means of titration. For rapid tests, useful with some organisms, see Hannan and Weaver (1948). Cowan (1953*a* and *b*) has developed a rapid method to replace an earlier micromethod which proved unreliable.

When the indicator method is employed, the indicators may be incorporated with the media in the first place or may be added subsequently when the final reaction is being determined. If they are added when determining final reaction, the color obtained should be compared with color standards (see Chap. IV) in order to secure accuracy. The use of indicator media is less accurate but is a much more rapid procedure; when the cultures are growing on agar, moreover, it is the only satisfactory procedure.

When using indicator media, make them up according to the directions given on page 53 of Chap. III. The indicator most commonly added is bromcresol purple, but with organisms producing considerable acid, bromcresol green or even bromphenol blue may be employed. When studying a series of unknown organisms it is often advisable to inoculate all onto the prescribed sugar medium with bromcresol purple; later those that show acid may be reinoculated onto the same medium with bromcresol green; and subsequently those positive to this indicator upon the same medium with bromphenol blue. If it is decided to observe the production of alkalinity as well as acidity, one may employ bromthymol blue or, better, a mixture of bromcresol purple with cresol red; in a solid medium this practice is often of value as it may show the production of acid in one part of the tube and of alkalinity in another.

TABLE 15. THE SENSITIVE RANGES OF THE THREE INDICATORS RECOMMENDED FOR USE IN INDICATOR MEDIA

pH:	7.0 6.0 5.5 5.0 4.0 3.0
Br. cres. purple:	purple|←sensitive range→|yellow..............................
Br. cres. green:	..................blue|←sensitive range→|yellow..............
Br. phenol blue:	..........................blue|←——sensitive range——→|yellow

With indicator media it is difficult to learn the exact reaction by reference to color standards, but a good estimate as to hydrogen-ion concentration can be obtained by inspection, particularly when three tubes are used, one with each of the three indicators recommended above. For this purpose Table 15, showing the relation of the ranges of these three indicators to one another, will be found useful.

After some experience a bacteriologist can usually devise some method for recording on the chart, by a system of numerals or + signs, the strength of reaction observed with each indicator employed; such a system often proves practical for comparative purposes but gives no very definite information as to final H-ion concentration.

Gas production in liquid media can be measured in percentage of gas in the closed arm of the Smith or the Durham fermentation tube. The Durham tube, which is most commonly used, consists of a small test tube (e.g., 75 by 10 mm) inverted in a large tube (e.g., 150 by 18 mm). In the case of solid media it is recorded as present or absent according to whether or not bubbles or cracks are present in the agar. This test is especially valuable if the organism is tested in a shake culture, but the presence of gas can usually be detected in an ordinary agar slant. These tests for gas production are chiefly useful if the organism produces primarily hydrogen; if the gas is all carbon dioxide, little or none will accumulate in the fermentation tube because of the great solubility and rapid diffusion into the air.

Interpretation of results. In case an organism produces gas or considerable increase in acidity in either broth or beef-extract peptone agar in the presence of some fermentable substance and this does not occur in the basal medium without the addition of the fermentable substance, it may safely be concluded that cleavage of this substance has occurred. Very often for routine diagnostic purposes such information is enough. To understand the true action of the organism on any carbon compound, however, much more investigation must be made as explained elsewhere (see Chap. VIII). This is particularly necessary in the case of organisms that produce a small amount of acid in some tubes but not in others containing the same carbon source and in cases where the addition of some carbon source results in a distinctly improved growth without the appearance of demonstrable acid or gas. In routine work, accordingly, one should record as positive only those organisms that produce considerable acid or gas from a given compound and as negative only those that consistently fail to show any acid or gas or any increase of growth when supplied with the carbon compound under investigation. All others should be regarded as border-line cultures, calling for further investigation.

It is especially important to recognize that the question whether or not cleavage of a carbohydrate occurs depends greatly on the cultural characteristics. Clarke and Cowan (1952) remark that tests of fermentative ability are often tests of the ease with which the buffer capacity of a medium is overcome. Accordingly some bacteria fail to produce an acid reaction in a beef-extract-peptone medium but will do so from the same carbohydrate in a synthetic or semisynthetic medium. It must accordingly be recognized that although tests for fermentative ability in some medium may have diagnostic usefulness, they do not necessarily indicate actual ability to metabolize the carbohydrate therein.

HYDROLYSIS OF STARCH

The breaking down of starch is rather more complicated than that of sugars because of the extensive hydrolysis that is necessary before it can be utilized by the bacteria. The first stage of this process is generally known as diastatic action because of the similarity to that brought about by the enzyme diastase. The final end result is usually an increase in acid, so one may obtain good evidence as to the utilization of starch by substituting it for sugar in the above methods (pages 159ff.) and determining acid produced or increase in H-ion concentration. It is often desirable, however, to secure evidence as to the intermediate products and as to whether the starch has been entirely consumed or not, and various methods have been proposed for this purpose.

This test may be made on raw starch, dissolved by boiling, or on the so-called "soluble starch." The latter is a partly hydrolyzed product, but it is often used as "starch" in this test because its iodine reaction is like that of true starch and different from that given by typical dextrins. If soluble starch is used, its true nature must be taken into account, but at the same time it must be remembered that true starch is partly hydrolyzed when sterilized in culture media, and even cultures growing in such a substratum are not furnished with raw starch as the sole carbohydrate. When such media are filtered, possibly "soluble starch" is all that remains.

A satisfactory method has been proposed by Eckford (1927) for learning the type of action on starch brought about by organisms capable of making good growth in broth. The same method may be adapted to organisms which prefer some other liquid medium by substituting it for broth in Eckford's method. The procedure, however, is not well adapted to those bacteria that fail to grow well in liquid medium. The technic is as follows:

Add 0.2 per cent soluble starch to broth and incubate cultures a week to 10 days. Examine on second, fourth, seventh and tenth days for hydrolysis of starch, production of acid, and reduction of Fehling's solution. For this test a drop is placed in a depression on a porcelain plate and a larger quantity in a serological test tube. The latter is tested for acid production with an indicator of the proper pH range. To the drop on the plate add a drop of dilute iodine solution and read reaction as follows: if blue, no hydrolysis; if reddish brown, partial hydrolysis with production of erythrodextrin; if clear, hydrolysis complete, with production of dextrin or perhaps glucose. The tubes showing complete hydrolysis may be tested for reducing sugar with Fehling's solution.

For bacteria that do not grow well in liquid media, no better method has yet been proposed than the plate technic given in all previous editions of the manual with little modification. This method has its disadvantages but is often useful; it is as follows:

Use beef-extract agar containing 0.2 per cent of soluble starch. Pour it into a petri dish, and after hardening make a streak inoculation on its surface. Incubate at optimum temperature for the organism under investigation. Observations are to be made on the second day for rapidly growing organisms but not until the seventh day for the more slowly growing ones. To make the test, flood the surface of the petri dishes with Lugol's iodine or with a saturated solution of iodine in 50 per cent alcohol. The breadth of the clear zone outside the area of growth indicates the extent of starch destruction. By means of a simultaneous inoculation on another plate containing the same medium with bromcresol purple as an indicator one may at the same time learn whether or not acid is produced as an end product.

THE METHYL RED AND VOGES-PROSKAUER TESTS

Special tests as to cleavage of glucose are commonly made in the differentiation of the organisms of the colon-aerogenes group. The medium ordinarily employed is as follows: 5 g of proteose peptone (Difco, Witte's, or some brand recognized as equivalent), 5 g of cp glucose, 5 g of K_2HPO_4 in 1,000 ml of distilled water. The dry potassium phosphate should be tested before using in dilute solution to see that it gives a distinct pink color with phenolphthalein. According to Smith (1940), the K_2HPO_4 in this medium should be replaced with the same amount of NaCl, if the tests are to be carried out on aerobic sporeformers.

Tubes should be filled with 5 ml each, and each culture should be inoculated into duplicate (or triplicate) tubes for each of the two tests. Incubation should be at optimum temperature of the organism under investigation, and tubes should be incubated 2–7 days, according to the rate of growth of the organism in question. Although the same medium is used for both the methyl red and the Voges-Proskauer tests, they must be performed in separate tubes. The latter test depends upon the production of acetyl-methyl-carbinol from the glucose. Fabrizio and Weaver (1947) show the possibility of a rapid test for the production of this compound; Cowan (1953) agrees as to its practicability. A similar microtest for the methyl red reaction proves more difficult.

A positive methyl red reaction is regarded as being present when the culture is sufficiently acid to turn the methyl red (0.1 g dissolved in 300 ml of 95 per cent ethyl alcohol and diluted to 500 ml with distilled water) a distinct red; a yellow color with the methyl red indicator is regarded as a negative reaction, while intermediate shades should be considered doubtful.

For the Voges-Proskauer reaction, according to the "Standard Methods" of the APHA (1946), to 1 ml of culture add 0.6 ml of 5 per cent α-naphthol in absolute alcohol and 0.2 ml of 40 per cent KOH. It is important to shake for about 5 sec after addition of each reagent. The development of a crimson to ruby color in the mixture from 2 to 4 hr after adding the reagents constitutes a positive test for acetyl-methyl-carbinol. Results should be read not later than 4 hr after addition of the reagents.

Various other tests have been suggested for this reaction, both to obtain results more quickly and because some organisms apparently give different results with different tests. In any case, weakly positive reactions may be obscured by the color of the reagent. A procedure which has given excellent results with many thousand cultures run by a member of the committee (CAS) is the creatine test of O'Meara, as modified by Levine, Epstein, and Vaughn (1934). In this procedure the test reagent added to the culture is 0.3 per cent creatine in 40 per cent KOH. This reagent deteriorates rapidly at temperatures over 50°C but may be kept 2 weeks at room temperature (22–25°C) or for 4 to 6 weeks in a refrigerator.

A recent modification of Coblentz (1943) is similar to the APHA method but uses a massive inoculum in broth from an infusion-agar slant culture, followed by incubation of the broth for 6 hr. Also, the 40 per cent KOH has 0.3 per cent of creatine added to it to intensify the reaction. After addition of the reagents (α-naphthol and KOH-creatine) the culture is shaken vigorously for 1 min; a positive reaction is characterized by an intense rose-pink color developing in a few seconds to 10 min.

The microtest of Fabrizio and Weaver (1947) calls for inoculation with a loopful of growth from a 6- to 12-hr infusion agar slant into 0.5 ml of infusion medium with 1 per cent trypticase and 0.5 per cent NaCl, placed in small tubes and preheated in a 30° water bath. It is then incubated at 30°C in a water bath for 90 min. It is then tested for acetyl methyl carbinol by the same method as given above, except that smaller quantities of the reagent (0.15 and 0.05 ml, respectively) are added.

ACID PRODUCTION IN MILK

Acid production in milk may be determined very simply, but the opacity of the milk must be taken into account if accurate determinations are desired. The milk must be considerably diluted before adding indicator for comparison with a buffer standard.

Indicator milk is often useful. Litmus has been used most frequently, as it indicates reduction as well as pH changes (although roughly). Neutral litmus milk (about pH 6.8) has a lavender color, which becomes red with acid production or blue with production of alkalinity. Reduction is indicated by a partial or complete fading of the color. The use of litmus milk has been seriously criticized because of the inaccurate nature of litmus as a pH indicator; nevertheless the differences it brings out have enough practical value so that it has not yet been superseded by any other indicator in milk.

The use of bromcresol purple, as was recommended by Clark and Lubs (1917), does not show changes in O/R potential.

TABLE 16. DEGREES OF ACIDITY EASILY RECOGNIZED IN MILK

Acidity	Indicator, reaction, etc.	Approximate pH value
"Neutral"	Same color with bromcresol purple as sterile milk, i.e., blue to gray-green	6.2–6.8
"Weak"	Color with bromcresol purple lighter than in sterile milk, i.e., gray-green to greenish yellow	5.2–6.0
"Moderate"	Yellow with bromcresol purple. Not curdled	4.7–6.0
"Strong"	Curdled. Blue or green to bromphenol blue	3.4–4.6
"Very strong"	Yellow to bromphenol blue	Under 3.4

It is possible to recognize the five degrees of acidity listed in Table 16 by the use of bromcresol purple (either in the milk before inoculation or added after incubation), the subsequent addition of bromphenol blue, and observation as to the presence of curdling. This is only a rough method of measurement, but in the routine study of milk cultures it will often be found valuable.

H. C. Brown (1922) proposed condensed milk diluted with 4 parts of water containing phenol red. The reaction is adjusted by addition of alkali until first appearance of a brick red. Subsequent changes of reaction in either direction can be observed.

RENNET PRODUCTION

The production of the enzyme rennet (lab) can sometimes be recognized in litmus milk by noticing the occurrence of coagulation without the appearance of acid. It is often obscured by simultaneous digestion, however, and two other methods have been proposed which often show rennet production with cultures that fail to show it when inoculated directly into milk.

Conn (1922) grows bacteria in milk sterilized in the usual manner; after the appearance of whey or peptonized milk, 0.5 ml is transferred to 10 ml of unsterilized milk and placed in a 37° incubator. Examinations are made every 5 min for the first half hour and at less frequent periods thereafter for a few hours longer. First appearance of coagulation is noted. Undoubtedly this test could be standardized, if desired, so as to be performed like the other microtests given here.

Gorini (1932) obtains vigorous growth on an agar slant, then covers the growth with milk, fractionally sterilized at temperatures not over 100° so as not to alter the color of the milk. The growth is mixed with the milk by use of a platinum needle, and the tube is incubated at 37° until coagulation occurs.

Although the committee is not prepared to recommend either method, it is felt that by a combination of the two a good indication of rennet production can be obtained.

REFERENCES

American Public Health Association. 1946. "Standard Methods for the Examination of Water and Sewage," 9th ed., published by the Association, New York.

Arnold, W. M., Jr., and R. H. Weaver. 1948. Quick microtechniques for the identification of cultures. I. Indole production. *J. Lab. Clin. Med.*, **33**, 1334–1337.

Bachmann, Barbara, and R. H. Weaver. 1947. A quick microtechnique for the detection of the reduction of nitrates to nitrites by bacteria. (Abstract) *J. Bacteriol.*, **54**, 28.

Bartholomew, J. W., and W. W. Umbreit. 1944. Ribonucleic acid and the Gram stain. *J. Bacteriol.*, **48**, 567–578.

Bisset, K. A. 1950. "The Cytology and Life-history of Bacteria." The Williams & Wilkins Company, Baltimore.

Böhme, A. 1905. Die Anwendung der Ehrlichschen Indolreaktion für bacteriologische Zwecke. *Centr. Bakteriol.*, I Abt. Orig., **40**, 129–133.

Brough, F. K. 1950. A rapid microtechnique for the determination of nitrate reduction by microörganisms. *J. Bacteriol.*, **60**, 365–366.

Brown, H. C. 1922. Use of phenol red as an indicator for milk and sugar media. *Lancet*, **202**, 842.

Clark, W. M., and H. A. Lubs. 1917. A substitute for litmus for use in milk cultures. *J. Agr. Research*, **10**, 105–111.

———, and S. T. Cowan. 1952. Biochemical methods for bacteriology. *J. Gen. Microbiol.*, **6**, 187–197.

Clarke, P. H., and Cowan, S. T. 1953. Hydrogen sulphide production by bacteria *J. Gen. Microbiol.*, **8**, 397–407.

Coblentz, J. M. 1943. A rapid test for acetyl methyl carbinol production. *Am. J. Public Health*, **33**, 815.

Conn, H. J., and G. J. Hucker. 1920. The use of agar slants in detecting fermentation. *J. Bacteriol.*, **5**, 433–435.

———. 1922. A method of detecting rennet production by bacteria. *J. Bacteriol.*, **7**, 447–448.

———, and Gladys E. Wolfe. 1938. Flagella staining as a routine test for bacteria. *J. Bacteriol.*, **36**, 517–520.

Cowan, S. T. 1953*a*. Fermentations: Biochemical micro-methods for bacteriology. *J. Gen. Microbiol.*, **8**, 391–396.

———. 1953*b*. A micromethod for the methyl red test. *J. Gen. Microbiol.*, **9**, 101–109.

Eckford, Martha O. 1927. Thermophilic bacteria in milk. *Am. J. Hyg.*, **7**, 201–221. (See p. 208.)

Fabrizio, Angelina, and R. H. Weaver. 1947. A quick microtechnique for the detection of acetylmethylcarbinol by bacteria. (Abstract) *J. Bacteriol.*, **54**, 69.

Fellers, C. R., and R. W. Clough. 1925. Indol and skatol determination in bacterial cultures. *J. Bacteriol.*, **10**, 105–133.

Fisher, P. J., and Jean E. Conn. 1942. A flagella staining technic for soil bacteria. *Stain Technol.*, **17**, 117–121.

Foth, 1892. Zur Frage der Sporenfärbung. *Centr. Bakteriol.*, **11**, 272–278.

Frazier, W. C. 1926. A method for the detection of changes in gelatin due to bacteria. *J. Infectious Diseases*, **39**, 302–309.

Frieber, W. 1921. Beitrage zur Frage der Indolbildung und der Indolreacktionen sowie zur Kenntnis des Verhaltens indolnegative Bacterien. *Centr. Bakteriol.*, I Abt. Orig., **87**, 254–277.

Gnezda, J. 1899. Sur les reactions nouvelles des bases indoliques et des corps albuminoides. *Compt. rend. acad. sci.*, **128**, 1584.

Goré, S. N. 1921. The cotton-wool plug test for indole. *Indian J. Med. Research*, **8**, 505–507.

Gorini, C. 1932. The coagulation of milk by *B. typhosus* and other bacteria considered inactive on milk. *J. Pathol. Bacteriol.*, **35**, 637.

Hannan, John, and R. H. Weaver. 1948. Quick microtechniques for the identification of cultures. *J. Lab. Clin. Med.*, **33**, 1338–1341.

Harrison, F. C. 1929. The discoloration of halibut. *Can. J. Research*, **1**, 214–239.

Henry, H., and M. Stacey. 1943. Histochemistry of the Gram-staining reaction for microorganisms. *Nature*, **151**, 671.

Holman, W. L., and F. L. Gonzales. 1923. A test for indol based on the oxalic acid reaction of Gnezda. *J. Bacteriol.*, **8**, 577–583.

Hunter, C. A., and H. G. Crecelius. 1938. Hydrogen sulphide studies. I. Detection of hydrogen sulphide in cultures. *J. Bacteriol.*, **35**, 185–196.

Knaysi, G. 1951. "Elements of Bacterial Cytology," 2 ed. Comstock Publishing Associates, Inc., Ithaca, N.Y.

Koser, S. A., and R. H. Galt. 1926. The oxalic acid test for indol. *J. Bacteriol.*, **11**, 293–303.

Kovács, N. 1928. Eine vereinfachte Method zum Nachweis der Indolbildung durch Bakterien. *Z. Immunitätsforsch.*, **55**, 311–315.

Leifson, Einar. 1951. Staining, shape, and arrangement of bacterial flagella. *J. Bacteriol.*, **62**, 377–389.

Levine, Max, S. S. Epstein, and R. H. Vaughn. 1934. Differential reactions in the colon group of bacteria. *Am. J. Public Health*, **24**, 505–510.

Magoon, C. A. 1926. Studies upon bacterial spores. *J. Bacteriol.*, **11**, 253–283. (See pp. 261–264.)

Moeller, H. 1891. Über eine neue Methode der Sporenfärbung. *Centr. Bakteriol.*, **10**, 273–277.

Morse, M. L., and R. H. Weaver. 1947. A quick microtechnique for the detection of hydrogen sulfide production. (Abstract) *J. Bacteriol.*, **54**, 28–29.

Shaw, C., J. M. Stitt, and S. T. Cowan. **1951.** Staphylococci and their classification. *J. Gen. Microbiol.*, **5**, 1010–1023.

Smith, N. R. 1940. Factors influencing the production of acetyl-methyl-carbinol by the aerobic spore-formers. *J. Bacteriol.*, **39**, 757.

———, Gordan, R. E., and Clark, F. E. 1946. Aerobic mesophilic sporeforming bacteria. *U.S. Dept. Agr. Misc. Publ.* 559.

Tittsler, R. P. 1930. The reduction of nitrates to nitrites by *Salmonella pullorum* and *Salmonella gallinarum*, *J. Bacteriol.*, **19**, 261–267.

———, and L. A. Sandholzer. 1936. The use of semi-solid agar for the detection of bacterial motility. *J. Bacteriol.*, **31**, 575–580.

Untermohlen, W. P., and C. E. Georgi. 1940. A comparison of cobalt and nickel salts with other agents for the detection of hydrogen sulfide in bacterial cultures. *J. Bacteriol.*, **40**, 449–459.

Vera, H. D. 1949. Accuracy and sensitivity of fermentation tests. *Soc. Am. Bacteriologists, Abs., 49th*, p. 6.

Wallace, G. I., and S. L. Neave. 1927. The nitrite test as applied to bacterial cultures. *J. Bacteriol.*, **14**, 377–384.

Zipfel, H. 1912. Zur Kenntnis der Indolreaktion. *Centr. Bakteriol.*, I Abt. Orig., **64**, 65–80.

Zobell, Claude E., and Catherine B. Feltham. 1934. A comparison of lead, bismuth and iron as detectors of hydrogen sulphide produced by bacteria. *J. Bacteriol.*, **28**, 169–176.

CHAPTER VIII

Physiological and Biochemical Technics

R. D. DeMoss and R. C. Bard

INTRODUCTORY

Even the bacteriologist not experiencing repeated contacts with microbial biochemistry is aware of the vast accumulation of knowledge in this field during recent years. Indeed, the methods and interpretations of microbial biochemistry have attained leading roles in modern microbiology, and as a result of this position, the importance of these phases of study in the training of new workers and in current research can hardly be overemphasized.

The main guiding principle during the preparation of this chapter has been the concept of environmental adaptation. The extreme adaptability and response of bacteria allow the choice of specific growth conditions in order to understand more easily and more completely a given biochemical situation. Thus, the premise has been adopted which demands a thorough study of physiological conditions relating to a specific biochemical process prior to a detailed investigation of that biochemical process. Accordingly, the following technics apply to physiological studies, often requiring a measurement of growth, and to both general and specific biochemical reactions. The material presented will deal primarily with bacteria, although investigations with algae, molds, protozoa, and even viruses along similar lines are being pursued vigorously and a unity of concept is being formulated.

The methods described in this chapter are not in general applicable to routine work in the sense of taxonomic investigations (see Chap. VII). In some cases these technics represent an extension of the routine tests, while the majority of the methods, because of the requirement for more precise and detailed measurements, are based upon different principles and usually depend upon instrumentation in addition to visual observation.

GROWTH MEASUREMENTS

In routine tests for taxonomic purposes, semiquantitative measurements of growth are usually deemed satisfactory. However, for research problems in the physiological and biochemical aspects of bacteriology, quantitative methods are generally necessary. Either the rate or extent of growth is used for quantitative purposes, and the extent of growth generally represents the most accurate expression, particularly when related to small experimental differences. Microbiological assay methods for the determination of amino acids, vitamins, and other growth factors, as well as antibiotics, are usually dependent upon growth measurements either directly or indirectly. Investigations into the nutritional requirements of bacteria, although often satisfied by qualitative methods, are more definitively answered by quantitative growth determinations. Within the area of nutritional studies is included the elucidation of biochemical pathways by means of biochemical mutants. Various mutants must be defined on the basis of nutritional requirements before becoming useful in such a study.

Several methods for the quantitative estimation of growth are described in detail below. Since most investigators publish results in terms of the milligrams of N or milligrams of cells (dry weight) used, the method employed for measuring is generally related by means of a standard curve to one of these unitages. For example, using a given suspension of bacterial cells, one may determine turbidity, dry weight, and total Kjeldahl nitrogen and then construct curves of turbidity as a function of dry weight and of nitrogen. With subsequent suspensions, only the turbidity needs to be determined, and the dry weight may then be read from the previously constructed curve. However, it should be pointed out that the margin of error will vary with cells harvested at different stages of growth. In addition, the turbidity–dry-weight relationships determined for one species do not necessarily apply to another species or under different conditions of growth.

Quantitative Methods

Cell counts. *Indirect method.* Dispense sterile saline solution (0.85 per cent; w/v) in 9.0-ml amounts in sterile, cotton-plugged test tubes. Serially dilute 1.0 ml of a culture or suspension of the organism in the saline to give 1:10, 1:100, 1:1,000, etc., dilutions, respectively, of the culture.

Dispense 10- to 15-ml quantities of melted nutrient agar in sterile plugged tubes, and hold at 40–45°C. To each agar tube, add 1.0 ml of a suitable dilution of bacterial suspension, mix well, pour into a sterile petri dish, allow to harden, and incubate. The incubated agar plate should contain 30–300 colonies for accuracy in counting, and all dilutions should be plated at least in triplicate. After the colonies have been counted, the number of cells in the original suspension is calculated. Alternately, the inocula may be spread over the surface of agar plates in the case of obligate aerobes.

In the latter procedure, plates are first poured and allowed to solidify. To remove excess water from the surface of the agar, the plates are dried overnight at 37°C or for 2 hr at 50–55°C. The sample to be counted is deposited (0.1 ml in 3–4 separate drops) on the agar surface and spread uniformly by means of a sterile bent glass rod. Since excess surface water has been removed, difficulties due to colony spreading are obviated.

A culture in the maximum stationary phase of the growth curve will often contain about 10^9 cells per ml; thus, the suitable dilution for plating will be found in the seventh serial dilution tube for a fully grown culture. The sixth and eighth tubes should also be plated in order to allow for errors in estimation of the original culture population.

It should be obvious that not all bacterial species will grow on nutrient agar and will therefore require a different agar medium in order to develop colonies. Anaerobic species must, of course, be incubated under anaerobic conditions. Furthermore, some species will require different diluent fluids, since saline may allow lysis of some cells, thereby resulting in low counts.

The method is based upon the assumptions that each cell develops into one colony and that each colony is derived from only one cell. These assumptions are not necessarily valid for those species which grow in chains or clumps. Since not all cells are capable of reproduction, each cell will not, as assumed, yield a colony. For these reasons, the estimate of number of cells will nearly always be lower than the true value. In addition, the proportion of nonreproducing cells may vary at different stages of growth; thus the error will not be constant. Inconstancy of error also will obtain when the size of the clump or the length of the chain varies with age of the culture.

For a discussion of the application of statistical method to bacterial enumeration procedures, see the review by Stearman (1955) and the previous reports cited therein.

Direct microscopic method. A counting chamber of the hemocytometer type is employed in this method. The chamber consists of a ruled slide and a cover slip constructed in a manner such that a definite known volume is delimited by the cover slip, slide, and ruled lines. Detailed description and instructions for use are furnished with the commercially available apparatus. The Petroff-Hauser or Helber bacterial counting chamber gives results superior to those obtained with the ordinary hemocytometer. The Petroff-Hauser slide is ruled in 0.05-mm squares, with the distance between the cover slip and slide equal to 0.02 mm.

According to Wilson and Knight (1949) the bacterial suspension should contain about 10^7–10^8 cells per ml for accurate counting. This cell concentration results in 3–12 cells per square of the counting chamber. A total count of about 400 cells per slide results in a final count correct to within 10 per cent. For best results, the direct count is made with dark-field illumination.

Results obtained with the counting chamber will always be higher than those from the colony-count method, since those cells which are not capable of reproduction are also counted. The direct-count method is, however, more rapid.

Cell volume. This technic involves the hematocrit principle of blood-cell determination. Unlike blood cells, however, bacterial cells occupy a smaller proportion of the volume of the suspending medium, and larger samples must be employed. A tube similar to the Hopkins vaccine tube, which holds 5- to 10-ml samples containing 0–0.05 ml of cell volume, is satisfactory. Using the Hopkins tube, the cell volume may be estimated directly, and if desired, the number of cells may be estimated by calculation from the known cell size. Schmidt and Fischer (1930) centrifuged cell suspensions in capillary tubes and then measured the height of the columns of sedimented bacteria.

A number of disadvantages accompany this method. Unless standard conditions are rigidly followed, the final cell volume observed will be variable, since the cells can be packed loosely or tightly, depending upon the rate and time of centrifugation. The average volume per cell also changes during the course of growth. Thus, the problem of relating total cell volume to other growth measurements becomes complex. Any variation in suspending medium will probably affect the water content of the cells, thus causing variable results in the total volume observed. In this method the differences observed are often small, thus allowing much greater opportunity for error.

Dry weight. Using distilled water, wash and resuspend the bacterial cells. Add aliquots of varying volume to small tared watch glasses or weighing bottles, and dry in an oven overnight at 85°C. After accurate weighing on an analytical balance, the dry weight of the bacterial mass in the original suspension is calculated.

If salt or other solutions are used in place of distilled water, it is necessary to determine the dry weight of the solution constituents. The determination, as calculated from such results, will contain an error introduced by the wet volume of the bacterial cells. On the other hand, the use of water for washing and resuspending the cells tends to extract salts or other soluble compounds from the cells, resulting in a final weight which is lower than the true weight.

Since the turbidity of a cell suspension is more easily and rapidly determined, a standard curve is often constructed, relating dry weight to optical density (see below).

Turbidity. Any of the available photometers are suitable for turbidity determination, although the type (Evelyn) which utilizes the sample tube as a lens for concentrating the transmitted light on a single phototube is probably the more accurate. Turbidity is measured most accurately at 420 mμ, providing the suspending fluid or medium is colorless. For yellow or brownish media, a wavelength of 660 mμ is employed. Turbidity changes during growth are often expressed simply as changes in optical density. However, at the higher values of optical density (0.4–2.0) turbidity is not a linear function of other expressions such as dry weight, total nitrogen, or cell numbers. Therefore, the culture or suspension should be suitably diluted before measurement, in order to fall within the range 0.0–0.4 optical density unit.

If turbidity is selected as the method of determination of growth, number of cells, dry weight, or cell N, a standard curve relating optical density to the desired unitage is constructed. Determinations of both optical density (optical density $= 2 - \log_{10}$ per cent transmission) and, for example, dry weight are performed on aliquots of a given cell suspension. The standard curve then consists of a plot of optical density vs. dry weight. Since the optical density of a given cell suspension varies with the wavelength employed, separate standard curves must be constructed for each wavelength which is to be used.

The measurement of turbidity during growth may be accomplished by growing the culture directly in colorimeter tubes or in specially fabricated flasks fitted with side arms which may be inserted in a colorimeter (Wiame and Storck, 1953).

Production of acid or alkali. The growth of several acid-forming bacteria may be followed by simple titration of the culture, using standard alkali. For example, nutritional studies or microbiological assays with certain lactic acid bacteria are commonly recorded in terms of milliliters of 0.01*N* NaOH required to titrate the culture to the bromthymol blue end point. Conversely, during growth of some *Mycobacterium* species, acid disappears from the medium and growth may be recorded as milliliters of 0.01*N* HCl required to reach the indicator end point.

One limitation of this method is obvious. After maximum growth is attained, metabolism (e.g., acid formation) continues; thus, the proper incubation time must be selected in order to determine true growth rather than true acid production. In addition, some species of bacteria accumulate acid in the early stages of growth but subsequently metabolize the acid.

Total nitrogen. Add a 2.0-ml aliquot (10–100 μg of total N) of the washed bacterial suspension to a 30-ml micro-Kjeldahl flask. Add 2.0 ml of a digestion mixture (500 ml of H_2SO_4, sp gr 1.84; 75 g of Na_2SO_4; 2.0 g of $CuSeO_3$; and 500 ml of H_2O), and boil gently in a hood or on a Kjeldahl rack for 1 hr longer than is necessary to clarify the solution. Allow the mixture to cool; add a small amount of antifoam agent (caprylic or oleic acid, or silicone) and 5.0 ml of 10*N* NaOH. Immediately connect the flask to a small distillation apparatus fitted with a graduated receiver tube containing 2.0 ml of 0.05*N* H_2SO_4. The delivery tube should extend well below the surface of the acid in the receiver. Gently boil the sample for 5 min; during the last minute, raise the delivery tube slightly above the surface of the acid. This procedure allows all the distilled ammonia to drip free of the delivery tube and prevents drawing the receiving solution back into the distilling flask when the flame is removed from the latter. Dilute the receiving solution to contain 10–15 μg per ml of ammonia-nitrogen. In a colorimeter tube, mix 2.0 ml of diluted distillate, 2.0 ml of Nessler's reagent (contains per liter: 4.0 g of HgI, 4.0 g of KI, and 1.75 g of gum ghatti), and 3.0 ml of 2*N* NaOH. Allow to stand for 15 min at room temperature, and read in the colorimeter at 540 mμ. The amount of ammonia-nitrogen is determined from a curve previously constructed using known samples of ammonia-nitrogen.

It should be remembered that this method cannot be used with aliquots of a culture taken directly from the medium, since the sample will include medium constituents and will yield falsely high results. The nesslerization procedure can also be replaced by direct titration of the distillate when the receiver acid solution is measured quantitatively.

Discussion and Recommendations

It is obvious that application of any of the methods described above depends upon the experimental approach and the type of apparatus and materials available. The technic selected also will depend to a certain extent upon the investigator's definition of growth. If growth is viewed as a simple increase in size, then a cell count will not reveal the extent of growth. On the other hand, if growth is recognized as increase in both the amount of protein and the number of cells, then expressions of total N, turbidity, and cell count will each represent growth, although not precisely, and will also be in some measure of disagreement, since each is based upon different physical and chemical attributes of the cells. Thus, where possible, the method selected should be used throughout a complete study, since comparisons among experiments will then be valid.

PREPARATION OF CELLS AND EXTRACTS

Growth and Harvest of Metabolically Active Cells

It is impossible to suggest specific growth media and conditions of incubation suitable for the production of microbial cells demonstrating all types of high metabolic activity. In practice it is known that the metabolism of a given organism reflects many contributing factors: the nature and age of the organism, its nutritional requirements, and reactions to physical and chemical conditions, etc. (see Werkman and Wilson, 1951). A medium yielding a large crop of cells is not necessarily one which will produce cells of high metabolic activity. The reader is referred to Chap. III for some details of media composition, to Snell (1951) for a discussion of bacterial nutrition, and to Gale (1943) and Stanier (1951) for consideration of the many factors affecting the enzymatic activities of cells. Prior to a biochemical study of a given organism, it is desirable and often necessary to investigate closely the relationship between the conditions of growth and of metabolic activity. Cultivation of the organism under conditions which enhance the quantitative presence of the biochemical reaction of interest (e.g., adaptive enzyme and apoenzyme formation) is a powerful tool available to the microbial biochemist, and once these conditions are recognized, their control is relatively simple.

Harvesting of cells from a liquid medium is achieved by centrifugation, the type of machine employed depending primarily on the speed desired and the volume of medium involved. For large volumes the Sharples type of ultracentrifuge is employed, while for smaller volumes many types of angle head centrifuges are available. Although refrigeration during centrifugation may be desirable and even obligatory in some cases, in most instances room-temperature centrifugation is adequate. Harvesting of cells from a solid medium is done simply by washing the surface of the medium with water or saline, using a spatula or glass rod and collection of the resultant suspension.

Preparation of Resting Cell Suspensions

After centrifugation the medium is discarded and the packed cells resuspended in a suitable menstruum and washed once or twice to remove nutrients absorbed to the cell surface. The menstruum employed varies, although distilled water, saline, or buffer is most commonly used. The washed cell paste is resuspended, and since it is often necessary in biochemical work to relate a given metabolic activity to cell number or cell mass, the cell suspension density is usually controlled and measured. Determination of cell number is performed in the usual manner: dilution and plate count or total microscopic count. However, metabolic activity is not necessarily related to viability or total cell number, a measure of cell mass or protoplasm being more directly related to biochemical activity. If dry weight of cells is related to the turbidity of a cell suspension, a quick and accurate relationship of cell mass is obtained (see the previous section).

Such a suspension is referred to as a resting cell suspension because the organisms are washed free of nutrients and are suspended in a nutrient-free liquid, thus rendering them nonproliferative. In actual use, the endogenous activity of the cell suspension must be determined as well as the activity upon the addition of a specific substrate. When the endogenous activity is low, the experimental values can be readily corrected by subtraction of the endogenous rate. However, if the endogenous rate is high, certain complications may arise which may best be avoided by attempts to lower this rate. Bubbling air at room temperature into the suspension for an hour or more may result in lowered endogenous values by the oxidation of nutrients still present within the cells.

Some of the advantages of the resting cell technic include (1) elimination of growth in nonproliferating cells, (2) ease of preparation with good reproducibility of data, (3) uniformity of suspensions for quantitative pipetting, (4) excellent spectrum of enzymes stable under these conditions, and (5) maintenance of activity at refrigerator storage temperature, the time depending on the lability of the particular enzyme system. The major difficulties are (1) selective permeability of the living cell which prevents entrance of many types of compounds into the cell and (2) the number of reactions which may be competing for the substrate or the products or both.

Dried Cells

The use of dried preparations of microbial cells has certain advantages over the resting cell technic. The permeability barrier is markedly reduced and in some cases eliminated; the number of reactions occurring is reduced; the preparations are often stable for months or even years, thus allowing excellent reproducibility of results. Although some enzyme systems appear quite labile to drying, persistent efforts with some modification in technic have resulted in stable preparations of many enzymes.

There are two methods of drying in general use: vacuum and acetone drying. Vacuum drying consists of drying the cell suspension *in vacuo* over a suitable desiccant, such as Drierite or phosphorus pentoxide, in a desiccator. A minimum volume of water is used to prepare the cell paste, since thin layers of cell paste present the maximum surface to the desiccant. A good high-vacuum pump and an amount of desiccant capable of absorbing the volume of water present are required. Acetone drying consists of the dropwise addition of a heavy cell paste to about 10 vol of ice-cold acetone with constant stirring. The cells precipitate and are removed immediately by vacuum filtration, the acetone being removed by suction. The dried cell preparations and, in most cases, the cell-free enzyme extracts below should be stored at 0 to $-20°C$ to prevent destruction of enzyme activity.

Cell-free Enzyme Extracts

For many types of studies, for example those involving reaction mechanisms, cell-free or soluble enzyme extracts are desirable. There are numerous technics available, none of which works in all cases. In principle, all methods attempt to disrupt the cell structure, thus allowing the liberation of intracellular material. The bacterial debris and any whole cells left are removed by high-speed centrifugation, usually in the cold, leaving the soluble elements in the supernatant liquid. If the desired enzymatic activity is not located in the supernatant liquid, it is possible that such activity may be found in the sedimented debris, and the latter should be examined before the particular procedure employed to obtain the cell-free extract is discarded as unworkable.

Brief descriptions of these methods follow. Autolysis: cell suspension made to pH 7.5–8.5 with phosphate buffer and kept at 37°C under toluene for 12–24 hr. The degree of autolysis depends on the type of organisms. Although autolysis increases with time, so does inactivation of enzymes. Lysis: most commonly employed with *Micrococcus lysodeikticus* using the enzyme lysozyme to disrupt the cells (see, for example, McManus, 1951). Freezing and thawing: enzymes are often released by rupture of the cell membrane with alternate freezing and thawing (see, for example, Koepsell and Johnson, 1942). Wet grinding: (1) hand grinding of frozen cell paste in a mortar with powdered glass, alumina, carborundum, or other suitable abrasives (see, for example, McIlwain, 1948); (2) semimechanical method of Kalnitsky *et al.* (1945) which involves the passage of a paste of bacteria and powdered glass between concentric cones of heavy glass, the inner cone revolving by connection to a motor and the outer cone held firmly in place; (3) a wet-crushing mill described by Booth and Green (1938) but not generally available; (4) shearing action of minute glass beads in a high-speed blender (Lamanna and Mallette, 1954). Dry grinding: consists of slow rotation *in vacuo* for several hours of dried powders of bacteria mixed with glass beads. The shearing action of the beads appears to be the important factor. Pressure: A large instantaneous pressure forces the frozen cell material between two machined surfaces, with resultant cell disruption. The apparatus has been referred to as the "Hughes press" (see Hughes, 1951, and a modification by Gest and Nordstrom, 1956). Sonic vibration: this technic is becoming more popular because of the ease of manipulations. The bacterial cells are disrupted by the sonic energy (see, for example, Shropshire, 1947, and Stumpf *et al.*, 1946). The cells of various bacterial species differ in their susceptibility to sonic disruption. Gram-positive cells are more resistant than gram-negative cells; cocci are more resistant than rod forms; large cells are more resistant than small cells. Thus, the time of exposure in the sonic oscillator required for cell disruption will vary for different species. In addition, some enzymes are more labile than others under the conditions of sonic disruption. Although maximum protein concentration in the extract may be obtained after 50 min exposure, for example, maximum enzyme yield may be achieved after only 20 min and the total activity of the particular enzyme may decrease after that period (see Table 17). The nature of the suspending fluid markedly influences the

TABLE 17. EFFECT OF DISINTEGRATION TIME ON ENZYME YIELD

Time, min	10	20	30	40	50
Enzyme, units per ml	2.0	3.2	2.0	1.8	1.0
Protein, mg per ml	61.7	96.3	111.6	145.8	154.8
Specific activity, units per mg protein	0.03	0.03	0.02	0.01	0.003

composition of the extract obtained by sonic disruption; alkaline buffers (pH 8–9) often give better results. Thus, the conditions used for the preparation of cell-free extracts must be carefully investigated in order to obtain adequate results, reflecting the particular biochemical reaction of interest to the investigator.

Combination of methods may be employed: extraction of dried cells at low temperatures, oscillation of suspensions prepared from dried cells or frozen cell pastes, etc.

Protein Fractionation

The cell-free extracts of bacteria, prepared according to the foregoing methods, contain many of the bacterial enzymes. It is often the case that a given biochemical reaction cannot be easily measured in such an extract owing to the presence of interfering reactions catalyzed by other enzyme systems which are also present. Therefore, it often becomes necessary to separate the desired enzyme system from some or all of the other enzyme systems present. This result may be achieved either by isolation of enzyme protein or by isolation of enzyme activity, using enzymatic activity as a measure of the process. Since bacterial enzymes are protein in nature, most of the problems encountered in enzyme protein isolation are similar to those of animal protein fractionation. The isolation of enzyme activity is often accomplished by utilizing fortuitous differences in physical properties (e.g., sensitivity to heat or pH) of the desired active protein. A useful example is the purification of the enzyme myokinase from rabbit-muscle extract (Colowick and Kalckar, 1943). In contrast to other enzymes, the enzymatic activity of myokinase is relatively stable to heat and acids. The protein is thus easily separated from interfering reactions by heating the extract at 90°C for

2 min in the presence of 0.5*N* HCL. The majority of the other proteins are denatured and precipitated by this treatment and may be removed by centrifugation. The myokinase remains in the supernatant fluid in a nearly pure condition. It should be noted that not all denatured proteins are precipitated by this and other methods. Thus, it would be as useful in isolating an enzyme activity to denature all other activities yet allow the spurious proteins to remain in solution with the desired enzyme.

As pointed out above, the preparation of bacterial protein extracts often requires somewhat more extensive methods than those needed for animal protein solutions. In addition, certain special problems arise which are peculiar to the fractionation of bacterial protein mixtures. The cell-free extracts obtained by most of the methods cited above contain relatively large amounts of nucleic acid. Because nucleic acid extends the range of precipitation of a given protein from a mixture, it is highly desirable to remove as much nucleic acid as possible from the mixture before proceeding with the common fractionation procedures. Either one of two methods is commonly used for precipitating nucleic acid. After the crude cell-free extract is prepared, the solution is dialyzed for at least 4 hr at 0–5°C against a solution (distilled water or a suitable buffer at about 0.01 *M* and pH 8 is often used) in order to remove as many small molecules as possible. After dialysis, the protein solution is adjusted to pH 6.0 by the dropwise addition of *M* CH_3COOH. Nucleic acid is precipitated from the adjusted mixture by the dropwise addition of either (1) 0.05 vol of *M* $MnCl_2$ or (2) protamine sulfate. One milligram of protamine sulfate (17 mg per ml, pH 5.0) is added for each 100 mg of protein in the original cell-free extract. Protamine sulfate has a negative temperature solubility coefficient, and the above concentration represents an approximately saturated solution at room temperature. The additions are made at 0°C with gentle stirring. After 20 min, the mixture is centrifuged and the supernatant solution, with a greatly decreased concentration of nucleic acid, may be fractionated with conventional procedures. For these and many other specific protein fractionation procedures, the treatise of Colowick and Kaplan (1955) is recommended.

Protein Estimation

The optical method of Warburg and Christian (1941) for the estimation of protein is simple and rapid but is unfortunately less precise when protein solutions contain high concentrations of nucleic acid. Thus, this technic cannot be used reliably with crude cell-free bacterial extracts. The method depends upon the relative optical densities of the protein solution at 260 mμ (due to the purine and pyrimidine components of nucleic acid) and at 280 mμ (due to the aromatic amino acids of the protein). Table 18 gives the factors necessary for calculation. For clarity, a sample calculation is given below.

TABLE 18. PROTEIN ESTIMATION BY U.V. ABSORPTION

Ratio 280/260	% nucleic acid	Factor for 1.0-cm cell at 280 (factor × 280 reading = mg protein/ml)*
1.75	0	1.105
1.60	0.25	1.075
1.50	0.50	1.05
1.40	0.75	1.025
1.30	1.00	0.995
1.25	1.25	0.975
1.20	1.50	0.95
1.15	2.00	0.91
1.10	2.50	0.87
1.05	3.00	0.835
1.00	3.50	0.8
0.96	3.75	0.785
0.92	4.25	0.755
0.88	5.00	0.715
0.86	5.25	0.705
0.84	5.50	0.69
0.82	6.00	0.67
0.80	6.50	0.645
0.78	7.25	0.615
0.76	8.00	0.59
0.74	8.75	0.565
0.72	9.50	0.54
0.70	10.75	0.51
0.68	12.0	0.48
0.66	13.5	0.45
0.65	14.5	0.43
0.64	15.25	0.415
0.62	17.5	0.38
0.60	20.0	0.35
0.49	100	

* For 0.5-cm cells, multiply values in last column by 2.

Example. 0.2 ml of a protein solution (after protamine sulfate treatment) and 2.8 ml of distilled water are pipetted into a cuvette with a 1-cm light path. The reference cuvette contains distilled water.

Readings obtained: 280 mμ O.D. = 0.273
260 mμ O.D. = 0.248

$$\frac{280}{260} \text{ ratio} = \frac{0.273}{0.248} = 1.10$$

In the first column find 1.10.
From the third column, find the factor 0.87.

Calculation of protein concentration:
0.87 × 0.273 = 0.237 mg of protein per milliliter in the cuvette
0.237 × 15 (dilution factor) = 3.585 mg of protein per milliliter of original protein solution.

The sample may be recombined with the stock solution after estimation, since no protein is denatured during the procedure. This feature of the optical method is particularly advantageous at high degrees of enzyme purification when the enzyme solution often contains a very small amount of protein.

The protein concentration of extracts containing large quantities of nucleic acid (e.g., before protamine treatment) may be estimated by any of several colorimetric procedures. The method of Lowry *et al.* (1951) utilizing the Folin-Ciocalteau phenol reagent is recommended because of its high sensitivity and is described here.

Reagents:

1. 40 g of Na_2CO_3 in 500 ml of distilled water
2. 0.3 g of $CuSO_4$ and 0.6 g of sodium tartrate in 500 ml of distilled water
3. 1 part of Folin-Ciocalteau phenol reagent (available commercially) and 2 parts of distilled water

Procedure. To a protein sample containing 7–70 μg of protein, add sufficient water to make a total volume of 0.5 ml. Add 5.0 ml of carbonate-copper-reagent (equal volumes of reagents 1 and 2 mixed just before use), mix, and incubate at 37°C for 30 min. After cooling to room temperature, add 0.5 ml of the diluted phenol reagent, mix, let stand 20 min at room temperature, and read at 660 mμ. For preparation of the standard curve, a satisfactory standard protein solution may be prepared using crystalline bovine albumin (Armour and Co., Chicago). The dilute albumin solution is subject to surface denaturation and should be freshly prepared for each determination.

BIOCHEMICAL TECHNICS

Manometric Technics

Manometric methods are employed widely because of accuracy and speed in the quantitative analysis and rate measurements of reactions involving either the evolution or uptake of certain gases. In addition, acid production may be measured by following CO_2 evolution in a bicarbonate buffer in equilibrium with a CO_2 gas mixture. The reader is referred to two excellent books for detailed discussions of the methods and types of apparatus available (Dixon, 1943; Umbreit *et al.*, 1949). There are two major types of manometers in use, i.e., the constant-pressure type commonly referred to as the Barcroft respirometer and the constant-volume type or Warburg instrument. Inasmuch as the Warburg type is by far the most widely employed in this country, further comments will be concerned with this apparatus.

The fundamental principle involved is that quantitative changes in the amount of a gas can be measured by determining pressure changes as long as the temperature and volume of the gas are kept constant. The apparatus consists of a flask or vessel having one or more side arms equipped with stoppers and commonly containing a small center well sealed to the bottom. The flask is attached to a manometer having one closed and one open arm and usually containing Krebs' solution (Krebs, 1951) or Brodie's solution (density 1.03) rather than mercury (density 13.6), thus increasing the sensitivity of the pressure changes. The vessel is placed in a water bath at constant temperature, and the system shaken in order to increase the rapidity of gas exchange between the liquid and gaseous phases. The level of the liquid in the closed arm of the manometer is brought to a previously calibrated point (usually 150 or 250 mm), and the level in the open arm recorded. In this manner the system is always read at constant volume, and the recorded values represent pressure changes. Inasmuch as the manometers are influenced by barometric changes in pressure by virtue of having one end open to the atmosphere, a thermobarometer (flask containing water attached to a manometer) is included in all experiments. The pressure changes (positive or negative) recorded for the thermobarometer are used to correct all readings of the manometers attached to an experimental vessel.

The prime purpose of the side arm of the flask is to allow separation of the components of the system under investigation so that one may control the initial reaction time. For example, a bacterial suspension in a suitable buffer may be placed in the main compartment of the flask and the substrate in the side arm. After temperature equilibration, the substrate is added by removing the manometer from the water bath, closing the open end of the manometer with one's finger, and tipping the system at an angle sufficient to allow the material in the side arm to pour into the main compartment. The manometer is then returned to the water bath, and readings taken at suitable intervals, depending primarily on the speed of the particular reaction under study. The stopper of the side arm may be a venting plug, thus allowing introduction of gases of known composition other than air. The center well is used primarily to hold alkali to absorb CO_2 as, for example, in the measurement of O_2 uptake or H_2 output.

In actual operation certain information is required in order to make possible the quantitative calculation of the gas evolved or taken up.

One must know at a given temperature the gas volume of the flask and manometer (to the liquid reference level in the closed arm), the volume of fluid in the flask, the gas being exchanged, the solubility of that gas, and the density of the manometric fluid.

The Warburg technic is commonly used to measure rates of O_2 or H_2 uptake, CO_2 or H_2 production, and respiratory quotients with a variety of substrates. When properly used this method provides one of the most versatile tools for the microbial physiologist. It is pertinent here to point out that although the manometric apparatus itself allows only the measurement of gas exchange, it is often desirable to subject the contents of the vessel to chemical analysis at the conclusion of the manometric experiment. After deproteinization and centrifugation, the supernatant liquid may be analyzed for a variety of common intermediates and end products of metabolism, for many of which excellent micromethods are available. It is quite possible with some bacteria and with certain substrates to obtain an accurate dissimilation balance with the usual volume of contents (about 3 ml) employed in the Warburg flask.

Analytical Procedures

The material in this section is usually presented under the heading of "Fermentation Analysis." The less restricted title used above, however, is considered more appropriate because fermentation represents only one of the dissimilatory mechanisms found in microorganisms. The term "fermentation," or intramolecular oxidation, is properly reserved for those anaerobic processes in which the hydrogen acceptor originates from the substrate. Other mechanisms include anaerobic oxidation or intermolecular oxidation wherein carbonate, nitrate, sulfate, or an organic compound acts as the hydrogen acceptor and respiration or aerobic oxidation wherein oxygen serves as the hydrogen acceptor. Coupled reactions between two substrates have also been described as, for example, the Stickland reaction between pairs of amino acids (Stephenson, 1949). These distinctions are important if a valid mechanistic interpretation of the results is to be obtained and used to describe the probable intermediary pathways of dissimilation. The majority of the studies published to date deal with the products of carbohydrate, polyalcohol, and organic acid dissimilation. The principles in all cases are the same.

Measurement of dissimilatory products must include both their qualitative identification and quantitative determination. The quantity of substrate dissimilated, as well as any substances reacting with the substrate during its breakdown (e.g., oxygen), must be accounted for quantitatively by the products formed. In some instances, particularly during aerobic processes, part of the substrate disappearing is assimilated by the organisms while at the same time the rest of the substrate is broken down to various products (Clifton, 1946, 1951). Use of a growth medium during substrate dissimilation permits maximum assimilation and the elaboration of adaptive metabolic mechanisms, each with resultant effects upon the nature and quantity of end products. Thus, use of washed cell suspensions is desirable, employing a medium containing only the substrate to be dissimilated plus those other substances required for the process: phosphate, buffer, metallic ions, etc. Care must be exercised in sterilizing the reaction medium; heat-labile compounds are sterilized by filtration, usually of a concentrated solution, and added aseptically to the rest of the reaction medium. Sterilization of the reaction mixture is unnecessary if resting cell suspensions are employed, since short-term dissimilation (0.5–3 hr) is generally adequate.

In order to obtain a quantitative accounting of the dissimilation process or a "balance," it is necessary to determine the gases formed, used, or both, as well as the dissolved products. An atmosphere of defined composition must therefore be employed in contact with the reacting solution. Both these objectives may be realized by shaking the solution in a closed system (e.g., Warburg apparatus) or by bubbling nitrogen continuously through the solution (provided nitrogen fixation does not occur), the emitted gases being passed through a suitable absorption train. At the end of the chosen period of time, the reaction is stopped by the addition of enough nonvolatile mineral acid to bring the pH to 2–3; this also releases bound CO_2 which is included in the gas measurements. The solution is then freed of cells and proteinaceous material by treatment with a suitable agent (barium or zinc hydroxide or trichloracetic acid) and centrifugation, and the clarified solution analyzed. A small sample of the reaction mixture may be removed just before the reaction is stopped to check for bacteriological purity; this is necessary for reactions conducted in culture media and which proceed for long periods of incubation.

Owing to space limitations, it is not possible to describe here in detail the methods employed in dissimilatory products analysis. Reference is made to the following books: Johnson *et al.* (1949), McNair (1947), Neish (1952), and Umbreit *et al.* (1949), which contain detailed descriptions of individual methods and apparatus. Sensitive micromethods have been described and continue to appear which are particularly useful for analysis of Warburg flask contents, eliminating the need in many cases for more cumbersome equipment (Black, 1949; Bueding and Yale, 1951; Kennedy and Barker, 1951; Neish, 1952). The methods described by Neish (1952) are most generally applicable to the problems discussed herein, although slight modifications may be advisable for the particular problems which are under investigation. In any case, the appropriate analytical method should be thoroughly tested and proved under the particular conditions of application.

In general terms, portions of the clarified solution referred to above are subjected to several treatments. Disappearance of the substrate is measured, as are the types of products listed in Table 19, according to the procedure described below. The pH of the cleared solution containing the dissolved dissimilatory products is adjusted to pH 7–8 and distilled

TABLE 19

Neutral volatile products, pH 7–8	Volatile acids, pH 2–3	Nonvolatile products
Acetone	Formic	2,3-Butanediol*
Diacetyl	Acetic	Glycerol*
Acetaldehyde	Propionic	
Ethanol	Butyric	Lactic acid†
Isopropanol	*n*-Valeric	Pyruvic acid†
n-Butanol	Isovaleric	Fumaric acid†
Acetylmethylcarbinol	*n*-Caproic	Succinic acid†

* Ether-extractable at pH 7–8.
† Ether-extractable at pH 2–3.

to obtain a fraction containing the neutral volatile products. The residue is adjusted to pH 2–3 and steam-distilled to obtain the volatile acid fraction. The nonvolatile products remain in the residue.

Each of the products thus obtained is measured, taking into consideration the possible effects of other products upon the particular analysis. Partition chromatography may be used to separate and measure many of the components of the mixtures; the volatile acids may also be measured by Duclaux distillation (Neish, 1952).

After identification and quantitative determination of the dissimilatory products, two types of balances are drawn up to check the accuracy of the analysis. These are the carbon and oxidation-reduction (O/R) balances. An example of such balances, taken from Johnson *et al.* (1949), is given in Table 20.

As a result of dissimilation, substrate carbon appears in the products formed, and complete carbon recovery is to be expected if all products have been identified and if the quantitative determinations are not in error. Usually the analytical data are calculated on the basis of millimoles of product per 100 millimoles of substrate dissimilated; micromoles are used in microanalysis. To calculate the carbon balance, millimoles of any product are multiplied by the number of the carbon atoms in the molecule and indicated as millimoles of C. The total C millimoles of the products must equal the millimoles of the substrate similarly calculated or must be within the range of the experimental error of the analytical methods employed. Thus, the carbon-recovery figure equals 100 per cent theoretically; the value of 97.4 per cent in the balance presented below represents, however, very good carbon recovery.

To obtain the figures for the calculated amounts of CO_2 expected on a theoretical basis, a knowledge of the dissimilatory mechanism in operation must be available or certain assumptions concerning its operation

TABLE 20. CARBON AND OXIDATION-REDUCTION BALANCES

	Compound	Carbon	Calc CO_2	Oxid value	Oxid prod	Red prod
	mM	mM	mM		mM	mM
Glucose utilized	100.0	600.0	....	0		
Lactic acid	96.2	288.6	0.0	0		
Glycerol	6.8	20.4	0.0	−1	.....	6.8
Ethyl alcohol	85.9	171.8	85.9	−2	.....	171.8
Acetic acid	7.3	14.6	7.3	0		
Carbon dioxide	89.3	89.3	....	+2	178.6	
Totals		584.7	93.2		178.6	178.6

$$\text{Carbon recovery} = \frac{584.7}{600.0} = 97.4\% \quad \frac{\text{obs } CO_2}{\text{calc } CO_2} = \frac{89.3}{93.2} = 0.958$$

$$\text{O/R} = \frac{\text{oxidation value}}{\text{reduction value}} = \frac{178.6}{178.6} = 1.00$$

are made. In the balance presented above, the data suggest the operation of a mechanism of glucose dissimilation similar to the Embden-Meyerhof-Parnas scheme of glycolysis (Baldwin, 1947). The hexose glucose is split into two trioses which, by oxidation and reduction, are converted to the end products lactic acid and glycerol. Thus, on this basis, no CO_2 is expected to be liberated during the formation of these end products. However, a number of the triose molecules originating from glucose are decarboxylated, at the pyruvic acid level, to a C_2 compound and CO_2. By reduction, some of the C_2 compound is converted to ethanol while the rest of the C_2 compound is oxidized to acetic acid. Thus, for each mole of ethanol and of acetic acid formed, a mole of CO_2 is also liberated. On this basis, the theoretical yield of CO_2 is calculated. When this figure is compared with the actual experimental yield of carbon dioxide, a ratio of 1 is obtained if the reasoning and analyses are correct. In the balance presented above, the ratio actually obtained is 0.958, which is another indication of a very good balance.

A further check of analytical accuracy is the O/R balance. The O/R balance, or redox index, is the ratio of the number of equivalents of oxidation and reduction occurring during dissimilation. Since in any chemical reaction these must be equal, the O/R ratio should be 1. Oxidation-reduction balances are calculated by multiplying the millimoles of product by a value expressing its degree of oxidation or reduction compared with the general carbohydrate formula CH_2O. The values are positive if the compound is more highly oxidized than CH_2O and negative if more reduced. Compared with one oxygen atom with a value of $+1$, two hydrogen atoms have a value of -1. Thus, CO_2 has an oxidation value of $+2$; pyruvic ($C_3H_4O_3$) and formic (CH_2O_2) acids, $+1$; glucose ($C_6H_{12}O_6$) and lactic acid ($C_3H_6O_3$), 0; ethanol (C_2H_6O) and acetylmethylcarbinol ($C_4H_8O_2$), -2 (also known as reduction value if negative); etc. It is obvious that the addition or removal of water during the fermentation will not affect the O/R balance, since the O/R value of water is 0. However, the introduction of oxygen into the system will alter the O/R balance, because of the positive O/R value involved. The balance presented above indicates an O/R index of 1.00, which is in excellent agreement with the remaining analytical data.

The methods of calculation discussed above also may be applied to oxidations involving oxygen or to intermolecular fermentations (e.g., amino acid fermentation such as the Stickland reaction), always providing that the quantities of the participating substrates and products are known.

The theoretical calculations presented above are based on the assumption that the dissimilatory mechanism is known. The existing knowledge concerning the pathways of microbial dissimilation is, indeed, limited, although data are rapidly accumulating (Elsden, 1952; Gunsalus *et al.*, 1955). However, since the results of analyses are often employed to postulate the probable dissimilatory pathway, caution must be exercised in such interpretation.

More mechanistic studies of substrate dissimilation may be made. Individual stepwise enzymatic reactions may be observed, using dried cells, cell-free enzyme extracts, or purified enzyme preparations. Such reactions involve the breakdown of a given substrate, usually a phosphorylated compound, to specific end products and may be examined by a variety of technics (Lardy, 1949; Sumner and Somers, 1947; Sumner and Myrback, 1950–1952). Such studies represent some of the final steps in the description of the mechanism of substrate dissimilation; recent advances are discussed contemporaneously in the annual review series for biochemistry and for microbiology (published by Annual Reviews, Inc., Stanford, California).

Technics for Isolated Reactions

Reactions Involving Oxidation Reduction

Thunberg technic. The Thunberg technic for estimation of dehydrogenase activity is especially applicable to cell suspensions and to preparations which are too dense for spectrophotometric measurement (see page 190). In some instances, the Thunberg technic, observed instrumentally in a colorimeter, provides an economical substitute for a spectrophotometer.

In general, a tube fitted with an outlet for evacuation and a side arm is used. Buffer, substrate, and methylene blue are placed in the tube, and the bacterial suspension introduced into the side arm. The type and concentration of buffer and substrate depend on the nature of the system studied, while the methylene blue is commonly employed in a final concentration of 1/20,000 to 1/60,000 (1.3×10^{-4} *M* to 4.4×10^{-5} *M*). Methylene blue in the oxidized state is blue-colored; in the fully reduced state, the dye is white (leuco). The system is made anaerobic to prevent the autooxidation of the methylene blue by molecular oxygen; this is generally achieved by evacuation using a water aspirator or vacuum pump and then filling with nitrogen. The tube is placed in a constant-temperature bath, the bacterial suspension tipped in after temperature equilibration, and the rate, time, or both, required for dye reduction measured by visual inspection or by the aid of a photoelectric colorimeter. When methylene blue reduction is measured with the unaided eye, a standard (90 per cent reduction) is usually employed. This is prepared as the experimental tubes with two exceptions: adding one-tenth the amount of methylene blue and using a heat-inactivated cell suspension. As a control, a tube without substrate is included so that the rate of endogenous dye reduction may be ascertained; this endogenous rate is often significant, since the organism may contain a considerable quantity of hydrogen donors. A dehydrogenase is considered present when the methylene blue reduction time in the presence of added substrate is less than the endogenous time. A more complete discussion of this method is presented in Umbreit *et al.* (1944).

Many other dyes have been substituted for methylene blue in the Thunberg method, including indophenol derivatives and, more recently, triphenyltetrazolium chloride (Kun and Abood, 1949; Gunz, 1949; Smith, 1951). In each case, however, the general principle is the same; i.e., electrons are transferred from the substrate to the acceptor.

Acid formation. The availability of pH meters presents a quick, simple method for following acid production during the course of a reaction. When weakly buffered or unbuffered media are used, the time course of pH change is a relative measure of the progress of a reaction in which hydrogen ions are produced. The method has limited application, since the activity of a given system is usually dependent on pH and the linear relationship of time and pH is of short duration. In addition, pH change is not a simple function of progress of the reaction. This method has been applied by Sable and Guarino (1952) to the purification of glucono-kinase from yeast.

The formation of acid during a reaction or during a sequence of reactions may be followed manometrically, using essentially the same technics mentioned above in the section on manometric technics. This method takes advantage of the buffering capacity of the $HCO_3^- + H^+ \rightleftharpoons CO_2 + H_2O$ system. The formation of each acid is accompanied by release of a hydrogen ion into solution. As a result, CO_2 is evolved from the reaction mixture and may be quantitatively measured. A complete account of the buffer theory involved in this method, as well as a chart describing the relationships of pH, temperature, and bicarbonate concentration, may be found in Umbreit *et al.* (1949).

The method is applicable to the fermentation or oxidation of substrates which yield lactic or other acids as end products. If CO_2 is also an end product, a control flask is used containing a buffer other than bicarbonate (e.g., phosphate) at the same pH, allowing the determination of the true CO_2 value. By difference, the CO_2 evolved in the bicarbonate flask as a result of acid formation is calculated.

In an isolated reaction involving the reduction of di- or tri-phosphopyridine nucleotide, a hydrogen ion is produced; thus, with opaque systems which cannot be measured in a spectrophotometer (see below), the reaction may be followed manometrically as acid production. In this case, a stoichiometric quantity of the pyridine nucleotide must be added or a second system must be included which is capable of oxidizing the reduced pyridine nucleotide. For the second system, several reagents are available which accomplish the oxidation of reduced pyridine nucleotide, the best probably being $K_3Fe(CN)_6$. When coupled with the ethanol-acetaldehyde system, for example, the reactions are written as follows:

$$CH_3CH_2OH + DPN^+ \rightleftharpoons CH_3CHO + DPNH + H^+$$
$$2Fe(CN)_6^{---} + DPNH \rightleftharpoons 2Fe(CN)_6^{----} + H^+ + DPN^+$$
$$2H^+ + 2HCO_3^- \rightleftharpoons 2H_2O + 2CO_2$$
$$\text{Sum:}\quad CH_3CH_2OH + 2Fe(CN)_6^{---} + 2HCO_3 \rightleftharpoons CH_3CHO + 2Fe(CN)_6^{----} + 2H_2O + 2CO_2$$

Thus, under these conditions two molecules of CO_2 are evolved for each molecule of ethanol oxidized and two molecules of ferricyanide are required to oxidize one molecule of diphosphopyridine nucleotide. Since the oxidation of reduced diphosphopyridine nucleotide by ferricyanide is a relatively slow process in low concentrations of the latter, ferricyanide is added to the reaction mixture in an amount at least twice that required for oxidation of the substrate present.

A slight difference in stoichiometry exists for the case of simultaneous acid production and diphosphopyridine nucleotide reduction.

$$CH_3CHO + DPN^+ + H_2O \rightleftharpoons CH_3COO^- + DPNH + 2H^+$$

Here, three molecules of CO_2 are produced per molecule of substrate oxidized when measured by the ferricyanide manometric method.

Spectrophotometric measurements. The course of a given reaction is often followed spectrophotometrically, advantage being taken of the fact that the change in concentration of one of the products or reactants may be observed as a change in light absorption at a given wavelength. The classic examples of this technic are reactions which involve di- or triphosphopyridine nucleotide. These nucleotides exhibit an absorption peak at 340 mμ when in the reduced state, while the oxidized forms do not absorb an appreciable amount of light at that wavelength. Therefore, during the course of a reaction which results in reduction of pyridine nucleotide, light absorption at 340 mμ increases. The measurement may consist of readings taken at definite intervals after initiation of the reaction, in which case the time course of the reaction is plotted, or readings may be taken at the beginning and end of the reaction, in which case only the extent of the reaction is determined.

Other applications also involve changes in optical density (absorption) at a particular wavelength of light, such as the decrease in absorption at 265 mμ during the deamination of adenosine-5-phosphate or the increase in absorption at 220 mμ during formation of certain keto compounds.

The method obviously requires previous knowledge of the spectrum of the compound to be measured. In addition, the extinction coefficient for the compound at the particular wavelength employed must be known if the measurement is to be completely quantitative.

Although several types are available, the most commonly used spectrophotometer is the Beckman model DU. A complete instruction manual is supplied with this instrument.

Nitrogen Metabolism

Proteins. Proteins are estimated quantitatively by the methods previously described (see pages 179–180). The results of total nitrogen determination (pages 173–174) are converted to protein values when multiplied by 6.25. This conversion factor is merely an average value based on proteins containing 16 per cent of nitrogen and thus leads to some error.

Although considerable attention has been given to proteolysis in animals, there are comparatively few studies involving microorganisms. It is believed that native proteins probably do not enter the bacterial cell and that protein utilization reflects the ability of the cell to produce extracellular enzymes capable of hydrolyzing the protein to amino acids, the latter entering the cell during metabolism. Comparatively few bacterial species produce proteolytic enzymes. The simplest test for proteolytic activity is to ascertain the ability of a given bacterial species to liquefy certain proteins such as solidified gelatin, coagulated casein, or coagulated serum. This may be done either by incorporating the protein into the growth medium or by testing the cell-free filtrate of a culture on the protein itself. One may also determine the degree of proteolysis by measuring the amino nitrogen changes either by the Sorensen formol titration (Brown, 1925), which depends on the increase in acidity when neutralized formaldehyde is added to a solution containing ammonia, primary amines, amino acids, or polypeptides, or by the Van Slyke manometric technic (Peters and Van Slyke, 1932), which depends upon the production of gaseous nitrogen when nitrous acid acts on an aliphatic amine.

Amino acids. The three methods most commonly employed for the quantitative estimation of amino acids are chromatography and ion exchange, microbiological assay, and enzymatic technic. All have their merits and disadvantages. Chromatography and microbiological assay allow estimation of essentially all the amino acids, while the enzymatic methods are limited to a few.

The technics of chromatography depend upon the differential distribution of amino acids between two phases using some form of supporting column. The paper-partition method is perhaps the simplest and most commonly employed, for it accomplishes both separation and estimation of the various amino acids. A two-dimensional system is usually employed with phenol water and *n*-butanol-acetic water as the developing solvents. Most amino acid spots are made visible with the ninhydrin spray or dipping technic. The amino acid spots are estimated quantitatively by (1) measuring the spot area (Berry *et al.*, 1951) or (2) colorimetric measurement after elution of the spot (Housewright and Thorne, 1950).

Complete discussion of paper chromatographic theory and methods, including solvents, spray reagents, and Rf values, particularly as applied to amino acids, has been published (Block *et al.*, 1955; Berry *et al.*, 1951; Williams and Synge, 1950).

Ion-exchange columns are useful for batch and analytical scale separations of amino acid (Hirs *et al.*, 1952; Moore and Stein, 1954) but, although quite precise, require more time and attention than the paper method.

Microbiological assay is employed widely and depends upon strains of microorganisms (commonly lactic acid bacteria) of exacting but known nutrition, capable of growing in chemically defined media provided the amino acid to be assayed is supplied, and responding to graded amounts of this amino acid in a regular fashion. The limits and specificity of the growth response must be checked before unknown samples are assayed. Total growth is determined usually by acidity or turbidity measurements. These methods do not allow separation of amino acids but do permit the quantitative measurement of one amino acid in a mixture. Technics and organisms are available for the microbiological assay of all the amino acids (Snell, 1946; Dunn, 1949; Barton-Wright, 1955).

Although several enzymatic methods are available, the specific decarboxylase method is the one most widely used because of simplicity, specificity, accuracy, stability of dried bacterial preparations, and ease of manipulations. The method depends upon the manometric measurement of CO_2 production from the amino acid by enzymes from selected strains of microorganisms. Six amino acids may be determined quantitatively in this manner: arginine, lysine, tyrosine, histidine, glutamic acid, and ornithine (Gale, 1948; Umbreit and Gunsalus, 1945; Archibald, 1946).

The methods employed in the study of three of the major reactions of amino acids will be considered briefly.

The enzymatic removal of the carboxyl group of an amino acid with the formation of the corresponding amine and CO_2 is known as decarboxylation. Decarboxylase activity may be determined by measuring either the disappearance of the amino acid acted upon by one of the methods listed above or the resulting products. In most instances, CO_2 evolution is measured manometrically. In the case of aspartic acid decarboxylase, CO_2 measurements cannot be used because of the low activity. Billen and Lichstein (1949) have therefore employed a microbiological assay method for the β-alanine produced.

The enzymatic removal of the amino group of an amino acid with the formation of ammonia and most commonly but not exclusively the corresponding keto acid is known as deamination. Although the activity may be measured by determining the disappearance of the amino acid acted upon, the most common method is to measure the production of ammonia by nesslerization.

The enzymatic transfer of the amino group of an α-amino acid to an α-keto acid resulting in the formation of another α-keto acid, the former corresponding in structure to the original α-keto acid and the latter to the original α-amino acid, is known as transamination. The technics employed to study the transaminases are varied, but all measure the appearance or disappearance of the keto or amino acid. Green *et al.* (1945) discussed keto acid measurements, Cohen (1940) aspartate measurement by the chloramine-T reaction, Lichstein *et al.* (1945) amino acid measurement by decarboxylase method, and Feldman and Gunsalus (1950) amino acid measurement by paper-partition chromatography. The symposium on amino acid metabolism edited by McElroy and Glass (1954) is recommended as a source of reference to specific analytical procedures.

Nucleic acids, purines, and pyrimidines. Studies of nucleic acid metabolism necessitate the measurement of these complex substances themselves as well as the constituent purines, pyrimidines, carbohydrates, and phosphorus. A preliminary step involves the removal of nucleic acids from the rest of the cell material, followed by separation of the two types of nucleic acids, desoxypentose and pentose, and their quantitative analysis. The nucleic acids are then hydrolyzed, and the quantities of the constituent parts measured.

The treatise of Chargaff and Davidson (1955) contains a complete consideration of methods and results of nucleic acid research. Several useful general reference sources also are cited: Cold Spring Harbor Symposia (1947); Symposia of the Society for Experimental Biology (1947); Symposia on Biochemistry of Nucleic Acids (1951); as well as ion-exchange methods (Cohn, 1951) and paper chromatography methods (Hotchkiss, 1948; Carter, 1950) for purines and pyrimidines.

Phosphorus Metabolism

Although many phosphorylated compounds of biochemical importance have been described, isolated, and characterized, most of the work has dealt with animal tissues, yeast being the only microorganism that has received any appreciable attention in the past. However, it is probable that similar studies with microorganisms will yield analogous results, and indeed, work with the latter is accumulating. Thus, it appears valid to include here a brief description of phosphorylated compounds, regardless of their origin.

The manual of Umbreit *et al.* (1949) presents methods for the analysis of phosphorylated intermediates as well as technics for their isolation and synthesis. The isolation and purification of phosphorylated compounds derived from cellular material are usually accomplished by fractional precipitation of their metallic salts, but more recently, technics have been described employing the principle of solvent distribution (Plaut *et al.*,

1950) and two-dimensional chromatography (Bandurski and Axelrod, 1951). Most of the procedures described to date, however, leave room for improvement, and no single, universally satisfactory technic is known. Studies of phosphorus metabolism are now being pursued in which only a few biochemical steps are involved, employing dried cells or enzyme preparations, so that complexity of analysis is greatly reduced.

The uptake or appearance of inorganic phosphorus is readily measured by the method of Fiske and Subbarow (1925). The usual investigation leads from this type of observation to the finding that a large number of phosphorylated compounds, largely carbohydrate derivatives, are formed or broken down during cellular metabolism. Thus, the amount of inorganic phosphorus released by hydrolysis in *N* HCl at 100°C in 7 min (Δ7 P) from such compounds as adenosine di- and tri-phosphate, glucose-1-phosphate, etc., is measured and reflects the presence and quantity of such phosphorylated compounds. Similarly, hydrolysis under the same conditions for 180 min helps to characterize other phosphorylated compounds: phosphopyruvate, triose phosphate, etc., but care must be exercised in drawing conclusions from these results, since partial phosphorus liberation occurs even during shorter periods of hydrolysis and other compounds may be involved. Therefore, analysis is also made for the non-phosphorus moiety of the compound: pentose, glucose, fructose, etc. Many of the Δ7 P compounds are referred to as "high-energy phosphate" compounds, since, on hydrolysis, 10,000–15,000 cal are liberated compared with the 2,000–4,000 cal released upon hydrolysis of the more stable phosphate esters.

Lipmann and Tuttle (1945) have described a specific method for acyl phosphate to form the corresponding hydroxamic acid; addition of ferric ions results in the formation of a purple complex which is a measure of the acyl phosphate present. Another type of phosphorylated intermediate which is readily measurable includes the "alkali-labile" triose phosphates, glyceraldehyde-3-phosphate and dihydroxyacetone phosphate. Inorganic phosphorus is released from these compounds when placed in *N* NaOH or KOH for 20 min at room temperature (Meyerhof and Lohmann, 1934) and is measured as increase in inorganic phosphorus. Fresh alkali must be used, since silicates, originating from the walls of the glass container, analyze as inorganic phosphorus.

Mention should be made of the commercial availability of many phosphorylated compounds of biochemical importance. Such preparations, albeit of stated limited purity in many cases, are useful in metabolic studies as substrates as well as standards for analytical procedures. A useful recent compilation of the role of phosphorus in the metabolism of organisms has appeared (McElroy and Glass, 1951, 1952), including references to many analytical procedures.

Other Methods

Microbiological assay has become a powerful analytical tool and is widely used. Methods for amino acid assay have been mentioned above (Nitrogen Metabolism); for methods used in vitamin analysis, the compilation of György (1951) is useful.

Isotopes are being used at an ever-increasing pace in the study of microbial biochemistry. Several commercial organizations offer ready assistance with reference to the type of equipment best suited for specific purposes. Numerous centers exist for the training of new workers in this field, such as the laboratories of the Atomic Energy Commission. The literature in this field is rapidly accumulating, and several compilations of methods exist (Kamen, 1951; Wilson, 1948), with descriptions of new and improved procedures often appearing.

Bacterial Genetics

The use of biochemical mutants and of organisms adapted to metabolize specific substrates has played an important role in advancing biochemistry in the past decade. Such tools are being employed with great ingenuity, and methods for their use have appeared (Catcheside, 1951; Lederberg, 1951; Luria, 1950; Spiegelman and Landman, 1954).

BIBLIOGRAPHY

Archibald, R. M. 1946. In Amino acid analysis of proteins. *Ann. N. Y. Acad. Sci.*, **47**, 181–186.

Bandurski, R. S., and B. Axelrod. 1951. The chromatographic identification of some biologically important phosphate esters. *J. Biol. Chem.*, **193**, 405–410.

Barton-Wright, E. C. 1952. "The Microbiological Assay of the Vitamin B-complex and Amino Acids," 179 pp. Pitman Publishing Corporation, New York.

Berry, H. K., H. E. Sutton, L. Cain, and J. S. Berry. 1951. Development of paper chromatography for use in the study of metabolic patterns. *Univ. Texas Publ.* 5109, pp. 22–55.

Billen, D., and H. C. Lichstein. 1949. Studies on the aspartic acid decarboxylase of *Rhizobium trifolii*. *J. Bacteriol.*, **58**, 215–221.

Black, S. 1949. A microanalytical method for the volatile fatty acids. *Arch. Biochem.*, **23**, 347–359.

Block, R. J., E. L. Durrum, and G. Zweig. 1955. "A Manual of Paper Chromatography and Paper Electrophoresis," 484 pp. Academic Press, Inc., New York.

Booth, V. H., and D. E. Green. 1938. A wet-crushing mill for microorganisms. *Biochem. J.*, **32**, 855–861.

Brown, J. H. 1923. The formol titration of bacteriological media. *J. Bacteriol.*, **8**, 245–267.

Bueding, E., and H. W. Yale. 1951. Production of α-methylbutyric acid by bacteria-free *Ascaris lumbricoides*. *J. Biol. Chem.*, **193**, 411–423.

Carter, C. E. 1950. Paper chromatography of purine and pyrimidine derivatives of yeast ribonucleic acid. *J. Am. Chem. Soc.*, **72**, 1466–1471.

Catchside, D. G. 1951. "The Genetics of Microorganisms," 223 pp. Pitman Publishing Corporation, New York.

Chargaff, E., and J. N. Davidson. 1955. The nucleic acids in "Chemistry and Biology," vols I and II, 692 and 576 pp. Academic Press, Inc., New York.

Cohen, P. P. 1940. Transamination with purified enzyme preparations (transaminase). *J. Biol. Chem.*, **136,** 565–584.

Cohn, W. E. 1951. Some results of the application of ion-exchange chromatography to nucleic acid chemistry. *J. Cell. Comp. Physiol.*, **38** (suppl. 1), 21–40.

Cold Spring Harbor Symposia on Quantitative Biology. 1947. Vol. 12, Nucleic acids and Nucleoproteins, 279 pp. George W. Banta Publishing Company, Menasha, Wis.

Colowick, S. P., and H. M. Kalckar. 1943. The role of myokinase in transphosphorylation. I. The enzymatic phosphorylation of hexoses by adenyl pyrophosphate. *J. Biol. Chem.*, **148,** 117–126.

———, and N. O. Kaplan. 1955. "Methods in Enzymology," vol. 1, 835 pp. Academic Press, Inc., New York. (Vol. 4 is in process of publication.)

Dixon, M. 1943. "Manometric Methods," 2d ed., 155 pp. The Macmillan Company, New York.

Dunn, M. S. 1949. Determination of amino acids by microbiological assay. *Physiol. Revs.*, **29,** 219–259.

Elsden, S. R. 1952. Bacterial fermentation. In "The Enzymes," vol. 2, pt. 2, pp. 791–843. Edited by L. B. Sumner and K. Myrbäck. Academic Press, Inc., New York.

Feldman, L. I., and I. C. Gunsalus. 1950. The occurrence of a wide variety of transaminases in bacteria. *J. Biol. Chem.*, **187,** 821–830.

Fiske, C. H., and Y. Subbarow. 1925. The colorimetric determination of phosphorus. *J. Biol. Chem.*, **66,** 375–400.

Gale, E. F. 1948. In "Eiweiss-Forschung, Abhandlungen aus dem Gebiet der Proteins," pp. 145–155. H. A. Keune Verlag, Hamburg, Germany.

Gest, H., and A. Nordstrom. 1956. "A Plunger-hammer for the Hughes Microbial Press." In press.

Green, D. E., L. F. Leloir, and V. Nocito. 1945. Transaminases. *J. Biol. Chem.*, **161,** 559–582.

Gunsalus, I. C., B. L. Horecker, and W. A. Wood. 1955. Pathways of carbohydrate metabolism in microorganisms. *Bacteriol. Revs.*, **19,** 79–128.

Gunz, F. W. 1949. Reduction of tetrazolium salts by some biological agents. *Nature*, **163,** 98.

György, P. 1951. "Vitamin Methods," 2 vols, 571 pp., 740 pp. Academic Press, Inc., New York.

Hirs, C. H. W., S. Moore, and W. H. Stein. 1952. Isolation of amino acids by chromatography on ion exchange columns; use of volatile buffers. *J. Biol. Chem.*, **195,** 669–683.

Hotchkiss, R. D. 1948. The quantitative separation of purines, pyrimidines and nucleosides by paper chromatography. *J. Biol. Chem.*, **175,** 315–332.

Housewright, R. D., and C. B. Thorne. 1950. Synthesis of glutamic acid and glutamyl polypeptide by *Bacillus anthracis*. I. Formation of glutamic acid by transamination. *J. Bacteriol.*, **60,** 89–100.

Hughes, D. E. 1951. A press for disrupting bacteria and other microorganisms. *Brit. J. Exptl. Pathol.*, **32,** 97–109.

Johnson, M. J., and associates. 1949. "Laboratory Experiments in Fermentation Biochemistry," 42 pp. Department of Biochemistry, University of Wisconsin. Madison, Wis.

Kalnitsky, G., M. F. Utter, and C. H. Werkman. 1945. Active enzyme preparations from bacteria. *J. Bacteriol.*, **49**, 595–602.

Kamen, M. 1951. "Radioactive Tracers in Biology, an Introduction to Tracer Methodology," 2d ed., 429 pp. Academic Press, Inc., New York.

Kennedy, E. P., and H. A. Barker. 1951. Paper chromatography of volatile acids. *Anal. Chem.*, **23**, 1033–1034.

Koepsell, H. J., and M. J. Johnson. 1942. Dissimilation of pyruvic acid by cell-free preparations of *Clostridium butylicum*. *J. Biol. Chem.*, **145**, 379–386.

Krebs, H. A. 1951. Improved manometric fluid. *Biochem. J.*, **48**, 240–241.

Kun, E., and L. G. Abood. 1949. Colorimetric estimation of succinic dehydrogenase by triphenyltetrazolium chloride. *Science*, **109**, 144–146.

Lamanna, C., and M. F. Mallette. 1954. Use of glass beads for the mechanical rupture of microorganisms in concentrated suspensions. *J. Bacteriol.*, **67**, 503–504.

Lardy, H. A. 1949. "Respiratory Enzymes," 290 pp. Burgess Publishing Co., Minneapolis, Minn.

Lederberg, J. 1951. "Papers in Microbial Genetics," 303 pp. University of Wisconsin Press, Madison, Wis.

Lichstein, H. C., I. C. Gunsalus, and W. W. Umbreit. 1945. Function of the vitamin B_6 group: pyridoxal phosphate (codecarboxylase) in transamination. *J. Biol. Chem.*, **161**, 311–320.

Lipmann, F., and L. C. Tuttle. 1945. A specific micromethod for the determination of acyl phosphates. *J. Biol. Chem.*, **159**, 21–28.

Lowry, O. H., N. J. Rosenbrough, A. L. Farr, and R. J. Randall. 1951. Protein measurement with the Folin phenol reagent. *J. Biol. Chem.*, **193**, 265–275.

Luria, S. E. 1950. Genetics of micro-organisms. In "Methods in Medical Research," vol. 3, pp. 1–75. Year Book Publishers, Inc., Chicago.

McElroy, W. D., and B. Glass. 1951. "Phosphorus Metabolism," vol. 1, 762 pp. Johns Hopkins Press, Baltimore.

———, and ———. 1952. "Phosphorus Metabolism," vol. 2, 930 pp. Johns Hopkins Press, Baltimore.

———, and ———. 1955. "A Symposium on Amino Acid Metabolism," 1,048 pp. Johns Hopkins Press, Baltimore.

McIlwain, H. 1948. Preparation of cell-free bacterial extracts with powdered alumina. *J. Gen. Microbiol.*, **2**, 288–291.

McManus, I. R. 1951. A study of CO_2 fixation by *M. lysodeikticus*. *J. Biol. Chem.*, **188**, 729–740.

McNair, J. B. 1947. "The Analysis of Fermentation Acids," 290 pp. Westernlore Press, Los Angeles.

Meyerhof, O., and K. Lohmann. 1934. Über die enzymatische Gleichgewichtsreaktion zwischen Hexosediphosphosäure und Dioxyacetonphosphorsäure. *Biochem. Z.*, **271**, 89–110.

Moore, S., and W. H. Stein. 1954. Procedures for the chromatographic determination of amino acids on four per cent cross-linked sulfonated polystyrene resins. *J. Biol. Chem.*, **211**, 893–906.

Neish, A. C. 1952. Analytical methods for bacterial fermentations, 69 pp. *Natl. Research Council Can., Rept.* 46-8-3, 2nd revision, NRC 2952.

Peters, J. P., and D. D. Van Slyke. 1932. "Quantitative Clinical Chemistry," vol. 2, Methods, 957 pp. The Williams & Wilkins Company, Baltimore.

Plaut, G. W. E., S. A. Kuby, and H. A. Lardy. 1950. Systems for the separation of phosphoric esters by solvent distribution. *J. Biol. Chem.*, **184**, 243–249.

Sable, H. Z., and A. J. Guarino. 1952. Phosphorylation of gluconate in yeast extracts. *J. Biol. Chem.*, **196**, 395–402.

Schmidt, H., and E. Fischer. 1930. Die Bestimmung der Keimzahl von Bakteriensuspensionen mittels des Capillarzentrifugierverfahrens. *Z. Hyg. Infektionskrankh.*, **111**, 542–553.

Shropshire, R. F. 1947. Turbidimetric evaluation of bacterial disruption by sonic energy. *J. Bacteriol.*, **53**, 685–693.

Smith, F. E. 1951. Tetrazolium salt. *Science*, **113**, 751–754.

Snell, E. E. 1946. In Amino acid analysis of proteins. *Ann. N.Y. Acad. Sci.*, **47**, 161–179.

———. 1951. Bacterial nutrition-chemical factors. In "Bacterial Physiology," edited by C. H. Werkman and P. W. Wilson, pp. 214–255. Academic Press, Inc., New York.

Spiegelman, S., and O. E. Landman. 1954. Genetics of microorganisms. *Ann. Rev. Microbiol.*, **8**, 181–236.

Stanier, R. Y. 1951. Enzymatic adaptation in bacteria. *Ann. Rev. Microbiol.*, **5**, 35–56.

Stearman, R. L. 1955. Statistical concepts in microbiology. *Bacteriol. Revs.*, **19**, 160–215.

Stumpf, P. K., D. E. Green, and F. W. Smith, Jr. 1946. Ultrasonic disintegration as a method of extracting bacterial enzymes. *J. Bacteriol.*, **51**, 487–493.

Sumner, J. B., and K. Myrbäck. 1950–1952. "The Enzymes, Chemistry and Mechanism of Action," vol. 1, 1,361 pp.; vol. 2, 1,440 pp. (each in two parts). Academic Press. Inc., New York.

———, and G. F. Somers. 1947. "Chemistry and Methods of Enzymes," 2d ed., 445 pp. Academic Press, Inc., New York.

Symposia of the Society for Experimental Biology. 1947. "Nucleic Acid," 290 pp. Cambridge University Press, New York.

Symposium on Biochemistry of Nucleic Acids. 1951. *J. Cellular Comp. Physiol.*, vol. 38, suppl. 1, 245 pp.

Umbreit, W. W., R. H. Burris, and J. F. Stauffer. 1949. "Manometric Techniques and Tissue Metabolism," 2d ed., 227 pp. Burgess Publishing Co., Minneapolis, Minn.

———, and I. C. Gunsalus. 1945. The function of pyridoxine derivatives: arginine and glutamic acid decarboxylase. *J. Biol. Chem.*, **159**, 333–341.

Warburg, O., and W. Christian. 1941. Isolierung und Krystallisation des Gärungsferments Enolase. *Biochem. Z.*, **310**, 384–421.

Werkman, C. H., and P. W. Wilson. 1951. "Bacterial Physiology," 707 pp. Academic Press, Inc., New York.

Wiame, J. M., and R. Storck. 1953. Metabolisme de l'acide glutamique chez *Bacillus subtilis*. *Biochim. et Biophys. Acta.*, **10**, 268–279.

Wiggert, W. P., M. Silverman, M. F. Utter, and C. H. Werkman. 1940. Preparation of an active juice from bacteria. *Iowa State Coll. J. Sci.*, **14**, 179–186.

Williams, R. T., and R. L. Synge. 1950. Partition chromatography, "Biochemical Symposia 3," 103 pp. Cambridge University Press, New York.

Wilson, P. W. 1948. "A Symposium on the Use of Isotopes in Biology and Medicine," 445 pp. University of Wisconsin Press, Madison, Wis.

———, and S. G. Knight. 1949. "Experiments in Bacterial Physiology," 55 pp. Burgess Publishing Co., Minneapolis, Minn.

CHAPTER IX

Serological Methods

E. EDWARD EVANS

INTRODUCTORY

The *in vitro* reactions between antigens and their homologous antibodies have been widely used in the classification and identification of microorganisms. Serological reactions may disclose marked differences among cultures that appear to be similar on the basis of morphology and physiology, and conversely, serological classification may bring a certain order and unity to a biochemically heterogeneous collection like the *Klebsiella-Aerobacter* group (Edwards and Ewing, 1955).

Not only have serological methods, refined by the immunochemist, been of value for extending our knowledge of microbial antigenic structure; they are of potential value as research tools in other areas of microbiology like genetics, metabolism, and cytology. A stimulating application of immunological methods to the study of protein biosynthesis is presented in Cohn (1952).

The present chapter has been written primarily for those trained in areas other than immunology in the hope that original applications for serological methods may be found in other fields. No attempt has been made to present all the fundamentals of immunology or the more advanced aspects of immunochemistry. Those interested may consult several good textbooks covering these disciplines.

PREPARATION OF ANTISERUM

Antiserum for use in the study of microbial antigens is usually prepared by immunizing one of the common laboratory animals with a suspension of microorganisms or with a solution of antigenic material derived from them. Suspensions of intact organisms are called "vaccines," "bacterins," or simply "antigens."

Antigens. In the case of nonpathogens, immunization can be accomplished by injecting suspensions of living or killed organisms into a suitable animal. The dosage will vary depending on the organism under study. While viable cells may be used also in studying pathogenic bacteria, particularly where one wishes to study antibody response during infection, it is more common to employ killed vaccines. Killing is accomplished by heating the microbial suspension at a temperature high enough to kill yet not so high as to cause drastic changes in the antigens, usually 60°C for 1 hr for nonsporulating organisms. An alternative and sometimes preferable choice is to add 0.3–1.0 per cent formalin to the culture and allow it to stand at 37°C for 12 hr.

The medium used for growth of the antigen will vary with the organism and in complexity from the simple synthetic media entirely suitable for certain bacteria, yeasts, and molds to the living host tissues necessary for viruses. The more complex media contain antigenic materials which may interfere with the system under study. This is particularly true with soluble extracellular antigens. Where possible, nitrogen should be supplied by inorganic salts or amino acids. In other instances it may be possible to replace a complex medium with casein hydrolysate or an ultrafiltrate of peptone plus accessory growth factors which are not antigenic. It should be noted that with some pathogenic bacteria, the antigenic structure may be atypical unless the strain has recently been passed through a suitable animal.

After the antigen has been grown, harvested, and killed, it is centrifuged and washed several times with 0.9 per cent sodium chloride solution ("saline"). Following this, the antigen is resuspended in saline and standardized so that the dose used for injection will be contained in 1 or 2 ml. Standardization may be accomplished by direct microscopic count, nephelometry, or Kjeldahl nitrogen determination. If the antigen is toxic, this will be a factor in determining the density; however, for many bacteria and yeasts, the arbitrary selection of 10^8 to 10^9 cells per milliliter will yield satisfactory results. After the vaccine has been standardized, it should be checked for sterility by inoculating a tube of thioglycollate broth and a blood agar slant with 0.1 ml of the vaccine. A preservative like phenol, 0.5 per cent, or merthiolate, 0.01 per cent final concentration, should then be added to the vaccine.

Soluble antigens such as toxins, toxoids, or various protein solutions may often be sterilized by filtration through a Seitz or Berkefeld filter. If the antigen fails to pass through such a filter, sterilization with ethylene oxide may be tried (Reddish, 1954). Sterility tests should be conducted as described for particulate antigens. Soluble antigens may be standardized on a weight basis, 1–2 mg of protein per milliliter of saline usually being satisfactory. In other instances, it may be desirable to standardize by nitrogen content using a micro-Kjeldahl procedure.

Antigens, whether particulate or soluble, should be dispensed in sterile vials with self-sealing rubber caps. The vials of antigen are kept under refrigeration and should be warmed to 37°C immediately before use. Before filling a syringe for immunization, the rubber cap is swabbed with 70 per cent ethanol.

To fill a syringe, the plunger is drawn back so that an amount of air equivalent to the volume desired can be forced into the vial. While the bottle is held inverted, the needle is inserted through the cap, the plunger depressed and then withdrawn, pulling fluid from the vial. The plunger is manipulated back and forth until all air bubbles have been worked out, and then the desired volume is withdrawn.

Adjuvants. Antibody formation may be enhanced by the addition of various substances to the antigen. These substances, called adjuvants, include alum, tapioca, certain bacteria, and paraffin oil.

The water-oil emulsion adjuvants (Freund, 1947) may be used with either particulate or soluble antigens. Water and paraffin oil are emulsified by the addition of lanolin or a derivative of lanolin. The addition of killed *Mycobacterium tuberculosis* or *M. butyricum* also contributes to the adjuvant action. While various ratios of these ingredients have been tried in different laboratories, the following proportions are recommended: 1 part of saline solution containing the antigen, 1 part of anhydrous lanolin, U.S.P., 1–2 parts of paraffin oil and 0.5 mg of killed dry tubercle bacilli per milliliter of the emulsion. The tubercle bacilli should be ground and incorporated into the paraffin oil before the other ingredients are added. Emulsification can be hastened by alternately drawing the mixture into a large syringe and expelling it. Adjuvants of this type are injected intramuscularly or subcutaneously.

When soluble protein antigens are used, adjuvant action may be secured by precipitating the protein with alum. The protein solution is made up at a concentration of 1–2 mg per ml, and 5 ml of sterile 1 per cent alum solution is added to each 100 ml of protein solution. The mixture is adjusted to pH 7.0 with NaOH.

Immunization. Because of its convenient size and superior antibody production, the rabbit is employed more frequently than other laboratory animals for immunization. Antigen may be introduced intravenously, intraperitoneally, or subcutaneously. The following procedures are essentially the same as those recommended for inoculation in Chap. X:

Intravenous injections are made into the marginal ear vein of the rabbit. A state of hyperemia is induced by rubbing the ear with cotton soaked in xylene until the veins become prominent.[1] The xylene is then washed off with 70 per cent ethanol, and the antigen introduced through a 1-, 5-, or 10-ml syringe and a 25- or 26-gauge needle. The needle is held with the beveled edge upward and inserted into the vein. It is

[1] It may not be necessary to use xylene if sufficient hyperemia can be induced by rubbing the ear.

best to start near the tip of the ear so that if the vein is missed, another insertion can be made nearer the base of the ear. During this process, the rabbit can be held by an assistant or placed in an animal box which allows the neck and head to protrude. After the needle is removed, bleeding is stopped by pressing a piece of cotton against the site of injection.

For intraperitoneal injections, the animal is held by an assistant. The fore paws are grasped in one hand, and the hind paws in the other. The animal is held with the head down so that the abdominal viscera are pushed against the diaphragm. The area over the lower part of the abdomen is clipped and washed with 70 per cent ethanol. The skin over the abdomen is held between the fingers, and the needle (21- to 23-gauge) inserted through the skin and peritoneum into the peritoneal cavity. Care should be taken not to perforate the intestine.

Subcutaneous injections are made by pulling up the skin of the back or thighs with one hand and inserting a 21-gauge needle into the raised skin.

If hypersensitivity is suspected, it may be wise to administer a small dose of antigen intracutaneously before these other routes are used. A portion of the back, sides, or belly is shaved, and the antigen injected through a 24- to 25-gauge needle into the epidermis. The needle should be held nearly parallel to the surface of the skin with beveled edge upward. A volume of 0.1 or 0.2 ml is injected, and if done properly, a bleb should be raised.

When antigen is administered to the guinea pig, intraperitoneal, subcutaneous, and intracutaneous injections are made in a manner similar to that described for the rabbit. Intravenous inoculations can also be given to the guinea pig, using the vein on the dorsal inner region of the hind leg; however, this procedure requires some practice and should be demonstrated by an experienced person.

Immunization schedules. There are a number of satisfactory schedules for immunization. One good method is to give daily injections of antigen for the first 3 days of each week, allowing the animal to rest during the intervening 4-day period. This cycle is repeated for 3 to 6 weeks, and a test bleeding is made. If the titer is sufficiently high, the animal is bled from the heart; otherwise, the series is repeated. It is advisable to give the first injection each week by the intraperitoneal or subcutaneous route and the other two by the intravenous route. In the absence of more definite information on dosage, soluble proteins can be injected in amounts of 10–100 mg contained in 1 or 2 ml of saline. With bacteria, 10^8 or 10^9 cells per injection may be used unless the cells are too toxic.

Harvesting of antiserum. Rabbits should be bled from the marginal ear vein before immunization is begun in order to secure a few milliliters of reference serum. The animal is placed in a box with the head protruding, and the ear rubbed with xylene and alcohol as described under "Immunization." When the vein is congested, it is nicked with a razor blade and

5–10 ml of blood allowed to drop into a large tube; the bleeding is stopped by pressure from a cotton pledget. The blood is allowed to clot, the clot is separated from the wall of the tube with an applicator stick, and the serum is centrifuged until clear.

The same technic is used for trial bleedings after a series of injections has been concluded. Six to eight days should elapse between the last injection and the trial bleeding. Serum from the trial bleeding is titrated for antibody content, and if a satisfactory level of antibody has been reached, the rabbit is bled from the heart.

Cardiac bleeding is not a difficult procedure but should be learned under the guidance of an experienced person. The rabbit should not be fed for 24 hr before bleeding to avoid excess lipid in the serum. The animal is tied securely to an animal board, and the hair shaved from the chest. The ribs are palpated until the area of maximal pulsation is located, and the site is painted with tincture of iodine. For quantities of blood up to 50 ml, one may use a syringe fitted with a 17-, 18-, or 19-gauge needle. If a rabbit is to be exsanguinated, the needle may be attached to a rubber tube leading into a flask fitted for suction. All materials should be sterilized in the autoclave.

The animal may be anesthetized with ether or nembutal, but after experience is gained, it is better to use no anesthesia.

The needle is inserted at the site of maximal pulsation, and gentle suction is applied. If the heart is not located on the first attempt, the needle should be completely withdrawn before reinsertion.

The blood is allowed to clot in centrifuge tubes or an Erlenmeyer flask. Clotting will usually occur within a few minutes at room temperature, and the clot can then be freed from the walls of the vessel with a sterile glass rod or applicator stick. Storage overnight in the refrigerator will cause the clot to contract, and the serum can be decanted off and freed from cells by centrifugation. Merthiolate may be added to a final concentration of 0.01 per cent as a preservative; however, glycerol added to serum in a 1:1 ratio is more effective in preventing contamination if its inclusion is not objectionable. If no preservative is desired, the serum may be stored in the deep freeze.

Antibody purification. Many of the reactions between microbial antigens and their homologous antibodies can be studied with crude antiserum. For some operations, however, it may be advantageous to effect at least a partial purification of antibody globulin.

A degree of purification can be accomplished by precipitating the globulins, including antibody, with ammonium sulfate. Each 100 ml of serum is diluted with 100 ml of water and mixed with 200 ml of saturated ammonium sulfate. The globulin precipitate is removed by centrifugation, mixed with water, and dialyzed against cold running water for 2 days. Dialysis can be hastened by placing the cellophane dialysis casing containing the globulin into a large carboy partially filled with saline and allowing the carboy to rotate in a horizontal position on a ball-mill assembly. This operation should be conducted in the cold, and the saline should be changed frequently.

Highly purified antibody for pneumococcus polysaccharide has been prepared by Heidelberger and his coworkers by dissociating the antibody from specific precipitates or agglutinates with 15 per cent sodium chloride solution or with barium hydroxide and barium chloride (see Kabat and Mayer, 1948).

The low-temperature ethanol-precipitation methods devised by Cohn and his coworkers are also useful for isolating gamma globulin and thereby freeing antibody of much extraneous material present in crude serum (Deutsch, 1952).

Various forms of electrophoresis apparatus can also be used for isolating gamma globulins or other globulins with antibody activity. Perhaps the best suited to serum fractionation is the electrophoresis-convection apparatus of Kirkwood (Cann *et al.*, 1949). An application of this procedure to fractionation of rabbit antibody may be found in the paper of Cann *et al.* (1951). Zone electrophoresis (Flodin and Porath, 1954) and continuous-flow paper electrophoresis (Durrum, 1951) would also seem to have merit for some applications.

Antibodies against protein antigens have been purified by a novel procedure (Sternberger and Pressman, 1950). The protein antigen was first coupled to *p*-aminobenzenearsonic acid or *o*-aminobenzoic acid through a diazo linkage. Following this, the coupled antigen was precipitated with antibody. The precipitate was washed and dissolved in saturated calcium hydroxide solution. Calcium aluminate was used to precipitate out the antigen, leaving the pure antibody in the supernate.

Isliker (1953) has purified isohemagglutinin by coupling the blood-group antigen to an ion-exchange resin. The antibody was allowed to bind to the antigen-resin complex and subsequently eluted by a change in pH or by simple sugars.

Antibody absorption. Antibody prepared against one microorganism often cross-reacts with other microorganisms that are taxonomically related, and less commonly the antibody may cross-react with distantly related organisms or even higher plants and animals. Since microbial cells are rather complex mixtures of antigenic substances, a cross reaction may be due to the sharing of one or more common antigens. On the other hand, it sometimes happens that cross reactions occur between two purified antigens that are known to be chemically homogeneous. In this case the cross reaction cannot be due to shared antigens but is due to similarity in structure of the two molecules. In either case, it is usually possible to remove cross-reacting antibodies by absorption. The heterologous antigen is mixed with the serum, allowed to react, and removed by centrifugation. Usually, considerable antibody for the homologous strain remains, whereas antibody for the absorbing strain can no longer be demonstrated.

The technic has been widely employed for establishing the antigenic structure of bacteria, and a classic example is found in the Kauffmann-White antigenic classification of the genus *Salmonella* (Edwards and Ewing, 1955).

To secure bacteria for agglutinin absorption, cultures are grown on a suitable agar medium. Since large quantities of cells are needed, the agar is dispensed in Roux bottles or petri dishes. The cells are washed from the agar with saline and killed by adding phenol to a concentration of 1 per cent. Heat killing may be used instead if the antigen under study is not heat labile or if it is necessary to destroy a heat-labile antigen.

The cells are centrifuged from the saline and washed twice with additional saline in graduated centrifuge tubes. The time and speed of centrifugation should be standardized to permit uniform packing. To 1 ml of packed cells is added 9 ml of antiserum. High-titered sera are diluted 1:5 or 1:10. The cells and serum are mixed and incubated at 37°C for 2 hr with intermittent shaking. Following this, the cells are removed by centrifugation and the serum tested for agglutinins against the absorbing strain. If agglutinins remain, the serum is absorbed again, using a second 1-ml portion of packed cells and the same incubation. When absorption is complete, the serum should fail to react with the absorbing strain but still react with the homologous strain unless the two cultures were antigenically identical. In setting up an absorption, one should include a control serum without cells to be incubated under the same conditions as the test. An absorption of the serum with the homologous strain may be included also.

Absorptions are not always reciprocal, and if two cultures are to be compared, the specific antiserum for each should be absorbed with the heterologous strain (Krumwiede *et al.*, 1925).

Absorption can also be conducted with soluble antigens. One method is to add small amounts of antigen to an antiserum on successive days until a precipitate no longer occurs with that antigen. Another method is to locate the region of slight antigen excess, using the supernate test described under "precipitation." A volume of antiserum and antigen calculated to give slight antigen excess is then mixed and incubated at 0–5°C until precipitation stops. The precipitate is removed by centrifugation, and the supernate is tested with the absorbing antigen and the homologous antigen.

AGGLUTINATION

When bacteria or other particulate antigens are mixed with specific antibody (agglutinin), clumping or agglutination of the cells takes place. The agglutination reaction may be performed in several ways. Slide tests in which a drop of antigen suspension is mixed with a drop of antiserum on a glass slide offer the advantages of speed and simplicity with some sacrifice in accuracy. Such slide tests may be read either microscopically or macroscopically. Agglutinins may also be titrated in a tube test, and this is preferable to the slide tests for some uses, since it affords greater accuracy. Where even greater precision is required, the quantitative determination of agglutinin may be used. In this procedure, the amount of antibody bound to the antigen is determined by micro-Kjeldahl nitrogen analysis.

The preparation of antigen. The microorganisms are cultured on a suitable agar or fluid medium. If grown on agar, the cells are washed off with saline solution, filtered through cotton, and killed by the addition of formalin to a final concentration of 0.3–1.0 per cent or by heating at 60°C for 1 hr. In preparing antigens of the *Salmonella* "O" variety it may be desirable to heat at 100°C for 30 min to destroy the less heat-stable "H" and "Vi" antigens. Broth cultures may be killed by heat or by the addition of 0.3–1.0 per cent formalin directly to the culture and incubation at 37°C for 12 hr. After the cell suspensions are killed, the organisms are centrifuged out and resuspended in saline. This washing process is repeated three to five times, and the final saline suspension is standardized.

Standardization may be accomplished by nitrogen analysis, by direct count using a hemocytometer or by measurement of turbidity with a photoelectric nephelometer or colorimeter or for less precise work by visual comparison with barium sulfate standards (McFarland, 1907). The correct density of the antigen suspension is best determined by preliminary titration; however, for many systems, the arbitrary selection of a density corresponding to McFarland tubes 1, 2, or 3 will yield satisfactory results. After the initial antigen concentration is chosen, it may be duplicated by comparison in a photoelectric nephelometer or colorimeter or by determination of the nitrogen content.

Certain antigen suspensions have a tendency to clump spontaneously. This problem may be overcome in some cases by brief exposure to sonic oscillation; however, the treatment should not be so drastic as to disrupt the cells. Other measures which may be effective include adjustment of the pH and ionic strength of the suspending fluid or the addition of an antigenically unrelated colloid like gelatin (0.01–0.5 per cent) to the system.

Tube agglutination test. Serum is diluted in a series of 10- by 75-mm test tubes using a twofold dilution series as illustrated in Table 21. Into the first tube 0.8 ml of saline solution and 0.2 ml of antiserum are

TABLE 21. THE TUBE AGGLUTINATION TEST

	Tube No.									Antigen control
	1	2	3	4	5	6	7	8	9	
Saline, ml	0.8	0.5	0.5	0.5	0.5	0.5	0.5	0.5	0.5	0.5
Serum, ml	0.2									
Transfer to next tube, ml	0.5	0.5	0.5	0.5	0.5	0.5	0.5	0.5	*	
Antigen, ml	0.5	0.5	0.5	0.5	0.5	0.5	0.5	0.5	0.5	0.5
Reciprocal of final serum dilution	10	20	40	80	160	320	640	1280	2560	0

* Discard 0.5 ml.

pipetted. These are mixed by taking a clean pipet and alternately sucking up about 0.5 ml into the pipet and blowing it back into the tube several times. Following this, 0.5 ml of the 1:5 dilution in tube 1 is transferred into tube 2. This process of mixing and then transferring 0.5 ml into the next tube is followed with each tube in the series, using a clean pipet for each transfer. When the last tube is reached, 0.5 ml is discarded. An additional tube containing only saline solution is included for a control on the antigen suspension. Next, 0.5 ml of antigen suspension is placed in each tube, and the tubes mixed by shaking. The test is placed in a constant-temperature water bath for 2–4 hr. The temperature at which the water bath is set may be 37–55°C, or any constant temperature within this range, depending on the system being studied. As a rule, the arbitrary selection of 37° will yield excellent results with bacterial antigens. The clumping may be read after this preliminary incubation, but as a general rule, more pronounced reactions are secured following storage in the refrigerator for 12–18 hr. Clumping may also be accelerated by centrifuging the tubes.

When a tube agglutination is read, each tube starting with the antigen control is tapped 4–5 times with the finger tip until the cells are resuspended. The antigen control should display uniform turbidity in contrast to the clumping seen in a positive reaction. The size of the clumps will depend on the potency of the antiserum and the physical nature of the antigen. Cells like the "O" antigen suspensions of *Salmonella* usually produce a granular agglutinate, while flagellated or "H" cells yield a more "fluffy" type of clump, and capsulated organisms often produce a solid button which cannot readily be broken up. As the serum becomes more dilute, the agglutination becomes weaker. The "titer" of antiserum is the highest dilution at which clumping can be readily detected. Occasionally, antiserums will be found which fail to agglutinate until the higher dilutions are reached. This range of more concentrated serum where agglutination fails to occur is called the "prozone."

In instances where it is desirable to compare several antiserums or several antigens, all determinations should be made at the same time. Even in the hands of skilled individuals, the agglutination test may yield variations of one tube among tests done on different days.

Microscopic slide test. The microscopic method offers the advantage of requiring only small amounts of serum, but at the sacrifice of some precision.

Antiserum dilutions and antigen are prepared as described for the macroscopic tube test. A loopful of antiserum at a given dilution is mixed with a loopful of antigen on a cover glass rimmed with petroleum jelly, and over this is inverted a hollow ground ("hanging-drop") slide. Controls of antigen plus saline and antigen plus preimmunization serum are also included. The hanging-drop preparations are allowed to incubate for 15–30 min and examined with low-power objective of the microscope. With some antigens there is a tendency toward spontaneous clumping; however, this should be detected in one or both of the control slides.

Macroscopic slide test. This procedure is most useful for the serologic identification of cultures and is widely used for enteric bacteria (Edwards and Ewing, 1955). When used for this purpose, it is essentially a qualitative test.

The antigen is prepared by emulsifying growth from a slant culture in a droplet of saline on a glass slide. The suspension should be moderately dense, but precise standardization is not necessary. A droplet of prediluted antiserum is mixed with the antigen, and the slide is rocked back and forth. With a good antiserum, agglutination should occur rapidly and the clumps should be clearly discernible to the naked eye.

The quantitative measurement of agglutinin. While the methods described previously are entirely adequate for many studies of microbial antigens, it is also possible to measure agglutinin with greater precision. This is done by adding thoroughly washed bacterial suspension to measured volumes of antiserum and allowing the agglutinin to combine with the bacteria. After an equilibrium is reached, the cells are washed free of extraneous protein, analyzed by a micro-Kjeldahl procedure, and the weight of agglutinin nitrogen calculated.

For certain applications, such a method is greatly superior to the less quantitative technics; however, the proper evaluation of all factors involved may well require more information than it is possible to impart in a chapter of this length. Since the procedure is well presented in Kabat and Mayer (1948), further description has not been included here.

PRECIPITATION

When a clear solution of a protein or polysaccharide antigen is mixed with the appropriate antibody or precipitin, the mixture turns cloudy and then precipitates. This reaction differs in principle from agglutination chiefly in the size of the particles involved. In agglutination, the particles are usually cells and in any event are large enough to be seen under the microscope, while in the precipitin reaction, the particles are of molecular dimensions.

Preparation of reagents. It is essential that both antigen and antibody solution be perfectly clear and free from lipid or insoluble material. Lipid in serum can usually be avoided by withholding food from the animal for 24 hr before bleeding. If this precaution has not been taken, lipid can be removed by storing the serum at 0°C for 3–5 days and then centrifuging at 2,000–5,000 × gravity to bring lipid to the surface. Any particulate matter other than lipid will be thrown to the bottom of the centrifuge tube, and the clear serum may be withdrawn with a capillary pipet. Clarification can also be accomplished by passing the chilled serum through a precooled Seitz filter containing a clarifying pad (average pore diameter 5 μ) or a coarse sterilizing pad (average pore diameter 1 μ).

Antigens to be used for qualitative precipitin tests may consist of exudate from an infected animal or cell-free fluid from a broth culture. In other instances, bacteria may be disrupted by mechanical grinding or by sonic oscillation and the soluble products used as antigen.

For precise work, it will be necessary to purify the antigen under study by the physical and chemical procedures customarily employed for proteins, polysaccharides, and other high-molecular-weight compounds.

Concentration of antigen is usually expressed in terms of weight in milligrams or micrograms per milliliter of solvent when dealing with purified antigens. In the case of protein or other nitrogen-containing antigens, concentration may be expressed as milligrams or micrograms of nitrogen based on a micro-Kjeldahl analysis. For crude antigens like cultural supernates or lysates, concentration is ordinarily expressed in terms of dilution of the original cultural fluid. This, of course, is less desirable than expression on a weight basis but is satisfactory for applications where strictly quantitative results are unnecessary.

The electrolyte concentration should be maintained within certain limits. With most systems, a small amount of electrolyte is necessary for precipitation, and if a great excess is present, precipitation will be inhibited. Generally speaking, a satisfactory electrolyte level can be assured by preparing antigen solutions in 0.9 per cent sodium chloride solution ("saline"). This is approximately equivalent to 0.15*M* sodium chloride. If the electrolyte strength is unknown, the sample should be dialyzed against saline solution of known strength. The pH should be maintained near neutrality, at least not lower than 6.5 or higher than 8.0.

The antigen dilution method. Prepare a series of tenfold dilutions of the antigen ranging from 1:10 through $1:10^7$. Place 0.2 ml of each dilution in small serological tubes (10 by 75 mm), then add 0.2 ml of antiserum to each tube. The proper dilution of antiserum may be established by preliminary trial, or undiluted serum may be used arbitrarily. The necessary control tubes consist of antigen plus preimmunization or "normal" serum and saline plus antiserum. All the tubes are mixed by shaking and placed at 37°C for 2 hr. The strength of the reaction may be indicated by plus signs, faint cloudiness being recorded as $\pm$. The tubes are placed at 0–5°C for 24 hr and read again. With weak antiserum, the test may be left in the cold up to 7 days; however, a preservative such as 0.01 per cent merthiolate should be present to prevent contamination and "precipitates" should be checked microscopically to ensure that they are not bacterial growth. When desirable, the test may be repeated, using closer intervals between dilutions of the antigen.

Since the antibody is held constant, this test is actually a titration or measurement of the amount of antigen present and should not be used to indicate the antibody content of a serum. For that purpose, the serum dilution procedure is preferable, or if a higher degree of accuracy is desired, the quantitative determination of antibody nitrogen is used.

The serum dilution method. In a series of 10- by 75-mm test tubes place serial twofold dilutions of serum ranging from 1:2, 1:4, 1:8 through 1:4096. A volume of 0.2 ml is satisfactory for most tests. Add 0.2 ml of antigen solution to each tube. The antigen dilution is held constant, usually at a level found to be optimal in a preliminary antigen dilution titration. Controls of antigen plus preimmunization serum and saline plus antiserum (1:2) are included. The tubes are shaken and incubated as in the antigen-dilution procedure.

Since precipitation is inhibited by an excess of antigen, if the antigen concentration chosen is too great, the serum will appear to have a low antibody level. Martin (1943) suggested a preliminary antigen titration at a constant serum level. The highest dilution of antigen which yields a precipitate is then used as an antigen dose in titrating the serum. In the hypothetical example presented in Table 22, the antigen dilution chosen would be $1:10^5$ and the titer of the serum with this dilution would be 1:160. Had an antigen dilution of $1:10^2$ been chosen, it can be seen that the serum would not have precipitated beyond a dilution of 1:20.

The ring test. The antigen is diluted serially as in the antigen dilution test, and 0.1 ml layered over 0.1 ml of antiserum in a 6- by 50-mm tube. The serum is usually added first, and the antigen must then be added slowly, holding the tube nearly horizontal so that a sharp interface forms between the two solutions. The tubes are placed in a rack at room temperature and observed at 15-min intervals. In a positive test a precipitate forms at the interface. It is essential that the serum have a greater density than the antigen, and the serum is often used undiluted, or in the case of more potent serums, the appropriate concentration may be reached by using normal serum as a diluent. A serum-saline control should be included. If economy of reagents is necessary, the ring test may be performed in small capillary tubing.

TABLE 22. COMBINATION OF SERUM DILUTION AND ANTIGEN DILUTION PROCEDURE

Serum dilution	Antigen Dilutions					
	1:10	$1:10^2$	$1:10^3$	$1:10^4$	$1:10^5$	$1:10^6$
1:20	−	+	+	+	+	−
1:40	−	−	+	+	+	−
1:80	−	−	−	+	+	−
1:160	−	−	−	−	+	−
1:320	−	−	−	−	−	−

The ring test is a qualitative or roughly quantitative procedure and thus has limited value for precise work. It does have the advantages that smaller amounts of material are required and that inhibition due to excess antigen is less likely, since the reactants can diffuse into each other. It has considerable value for the detection of qualitative (type or group) differences among different strains of the same species and has been used for this purpose with organisms like the streptococci and pneumococci.

Optimal proportions. If a precipitin test is set up as described for "Antigen dilution" and the tubes examined at frequent intervals, one tube will usually be observed to precipitate before the others. This is called the point of optimal proportions. The ratio (dilution of antigen/dilution of antiserum) is called the optimal ratio, and should the antiserum dilution be changed, it usually requires a proportional dilution of the antigen to give the most rapid precipitation.

The optimal-proportions method, then, is a way of measuring the velocity of the precipitin reaction. When antigen is diluted and serum held constant, the test is referred to as the "alpha procedure"; conversely, in the "beta procedure," the serum is diluted while antigen is held constant.

A practical application of the optimal-proportions method and one which will serve to illustrate general procedure for optimal-proportions determinations is the Ramon flocculation test (beta method) used for toxins and toxoids.

To a series of test tubes (12 by 75 mm) add increasing amounts of antitoxin as shown in Table 23. This is ordinarily done with a 0.2-ml pipet calibrated in 0.001 ml. Then to each tube is added 2 ml of the toxin or toxoid. The tubes are shaken, placed in a water bath at 50°C, and observed continuously. The tubes in which flocculation is going to appear will first become cloudy. Record the time at which discrete floccules are first seen in one of the tubes. This tube represents the optimal ratio between the particular batches of antitoxin and toxin used. In Table 23, this ratio is represented by tube 6. If a more precise end point is desired, the titration may be repeated, using smaller increments

TABLE 23. BETA PROCEDURE FOR OPTIMAL PROPORTIONS USING DIPHTHERIA TOXIN AND ANTITOXIN (RAMON FLOCCULATION)

Tube No.	Antitoxin, ml	Toxin, ml	Time of flocculation, min
1	0.030	2	
2	0.035	2	
3	0.040	2	
4	0.045	2	60
5	0.050	2	40
6	0.055	2	35
7	0.060	2	40
8	0.065	2	55

of toxin between 0.050 and 0.060 ml. In the case of diphtheria toxin and antitoxin as used here, the L_f value may be calculated. This is defined as the amount of toxin or toxoid which yields optimal flocculation with one standard unit of antitoxin. Let us assume that the antitoxin in this case had 425 units per milliliter, or 0.00235 ml = 1 unit. Since 0.055 ml of antitoxin yielded optimal flocculation with 2.0 ml of toxin,

$$\frac{L_f}{0.00235} = \frac{2}{0.055}$$

$$L_f = 0.086$$

The toxin thus has 11.6 flocculating units per milliliter. The L_f value for the toxin could then be used to standardize an unknown antitoxin.

Those interested in standardization of antitoxin by flocculation or by *in vivo* methods should consult Wadsworth (1947). Quantitative aspects of flocculation are discussed in Boyd (1947), Kabat and Mayer (1948), and Cohn (1952).

It should be pointed out that *flocculation* is characterized by solubility of the antigen-antibody complex in both antigen and antibody excess and is usually found with horse antibody like diphtheria antitoxin. The *precipitin reaction*, on the other hand, displays solubility of the precipitate only in an excess of antigen. This type of reaction is characteristic of most rabbit antisera. The alpha procedure is usually employed for antibodies of the latter type.

The supernate test. In conjunction with the precipitin test, particularly the quantitative method, it may be useful to test the supernate for an excess of antibody or antigen. If the antigen is homogeneous, usually one or more tubes will be found to contain neither excess antibody nor excess antigen. This is called the "equivalence zone." It might be expected that the region of optimal proportions would coincide with the equivalence zone, and in many systems this is so; however, there are some systems where the optimal ratio is outside the equivalence zone.

Failure to find a definite equivalence zone indicates that two or more overlapping precipitin systems are present, and this, in turn, may be used as an indication that the antigen is not pure, a situation which is not unlikely if crude bacterial extracts are being used as antigens. The supernate test, then, may be used to indicate *impurity* of an antigen. It does not necessarily follow from this, however, that if a good equivalence zone is found, the antigen is *pure*. It may be, but additional evidence for homogeneity should be furnished.

To perform a supernate test, a precipitin test is set up in the ordinary manner using the quantitative technique or the antigen dilution method. After the tubes have been incubated a sufficient length of time, the precipitates are removed by centrifugation and the supernatant fluids are poured into clean test tubes. Into a small test tube 0.2 ml of supernate and 0.2 ml of fresh antiserum are placed. Into another tube 0.2 ml of supernate and 0.2 ml of antigen are placed, using antigen that has been diluted considerably to avoid inhibition due to antigen excess. A precipitate in the first tube shows that an excess of antigen was present in the supernate, a precipitate in the second tube shows that excess of antibody was present, and no precipitation in either tube indicates either the presence of an equivalence zone or faulty technique.

Precipitation in gels. The precipitation of antigen and antibody in a gel has been used for the analysis of antigen mixtures by Oudin, Ouchterlony, and others (see Oudin, 1952). In the one-dimensional simple diffusion procedure, one reagent, usually antibody, is mixed with agar and placed in a small test tube. The other reagent (antigen) is layered over the agar, and the test allowed to incubate 4 to 7 days. If only one antigen is present, there will be only one zone of precipitation, but if additional antigens are present, additional rings will be seen, assuming, of course, that the serum contained antibodies directed against the other antigens and also assuming that two rings do not coincide because of diffusion rates and concentration.

In the double-diffusion procedure, agar which contains neither reactant is poured into a petri dish. The antigen and antibody are placed in wells and allowed to diffuse toward each other with the resultant formation of lines of precipitation. Since a complete discussion of these procedures is available in the review by Oudin (1952), the details have been omitted here.

The quantitative measurement of precipitin. The procedures described elsewhere in this section are entirely adequate for some applications involving microbial antigens. In other instances, it may be more profitable to employ a technic that permits the precise estimation of antibody contained in a precipitate. This may be done with a variation of the quantitative precipitin test as developed by Heidelberger and Kendall (1929, 1933, 1935).

When microbial antigens are being studied, the quantitative precipitin test is useful for closely related (cross reacting) antigens, for studying the contamination of one antigen by another, for estimating concentration of an antigen present in a solution, or for any other application where the less quantitative technics are insufficiently sensitive.

In the usual form of the quantitative test, increasing amounts of antigen are added to constant volumes of antiserum and the tubes stored at some constant temperature (0–4°C) for several days until equilibrium has been reached. The precipitates are then carefully removed by centrifugation, and extraneous serum proteins are removed by rinsing the precipitates with ice-cold saline solution. The precipitates are analyzed for nitrogen by a micro-Kjeldahl procedure. In the hands of a careful investigator this method permits accuracy that is comparable to that obtained in quantitative analytical chemistry. In addition to nitrogen analysis, antibody has been measured quantitatively in other ways. A procedure that is useful for serums with low concentrations of antibody is the Folin Ciocalteau analysis for tyrosine (Heidelberger and MacPherson, 1943).

While the quantitative estimation of precipitins is a relatively simple operation, its proper evaluation may require more information than it is possible to impart in a chapter of this magnitude. Since other sources of material (Kabat and Mayer, 1948; Cohn, 1952) are available, further description of the procedures has not been included here.

COMPLEMENT-FIXATION TEST FOR MICROBIAL ANTIGENS

In the normal serum of numerous animal species there occurs a substance known as alexin or complement. It is a mixture of at least four fractions, chiefly globulin in nature. Although it does not increase in amount following immunization, it has a role in serological reactions, since it is able to combine with many antigen-antibody complexes. This combination may be spoken of as the "fixation" of complement.

When erythrocytes are mixed with antierythrocyte serum and complement is added, lysis of the cells will occur. Such lysis requires both antibody and complement. Practical application is made of this lytic phenomenon in the complement-fixation test where sheep erythrocytes and their homologous antibody (hemolysin) are used as an indicator system.

In the complement-fixation test, two separate antigen-antibody systems are allowed to compete for complement. When the test is performed the antigen and antiserum under study are mixed and a carefully measured amount of complement is added. If the serum contains antibodies that react with the antigen, complement is absorbed by the antigen-antibody complex and no free complement remains. If, at this point, sheep erythrocytes and their homologous antibody (hemolysin) are added, the sheep cells will not lyse, since there is no free complement. If, on the other hand, the first antigen-antibody reaction had failed to occur owing to deficiency of either antibody or antigen, free complement would be available and the sheep cell indicator system would lyse.

The complement-fixation test is not necessary for some systems where other antigen-antibody reactions are available; however, it sometimes yields information that cannot be obtained by other procedures. The test described here is a modification of the Kolmer method for syphilis (Kolmer *et al.*, 1951).

Reagents needed for the test include 0.9 per cent sodium chloride solution ("saline"), a suspension of sheep erythrocytes, hemolysin, complement, antigen, and antiserum.

Preparation of sheep erythrocytes. Blood is collected from the jugular vein of the sheep. The erythrocytes may be preserved for several weeks by mixing the blood immediately with an equal volume of sterile modified Alsever solution (Bukantz *et al.*, 1946) of the following composition:

Glucose	20.5	g
Sodium citrate	8.0	g
Sodium chloride	4.2	g
Citric acid	0.55	g
Distilled water	1	liter

If a sheep is not available, sheep blood may be obtained commercially; however, the erythrocytes are likely not to survive shipment through the mail. Freshly collected blood should be stored 48 hr before use.

In the preparation of erythrocyte suspensions for testing, the blood in Alsever's solution is diluted with 2 vol of 0.9 per cent saline solution placed in a 50-ml centrifuge tube and centrifuged at approximately 500 × gravity for 5–10 min. The supernate is siphoned off and discarded, and the cells resuspended in 35–40 ml of saline solution. The cells are washed in this manner two additional times. If the last supernate is not colorless, the cells are too fragile and the blood should be discarded. If the erythrocytes are satisfactory, dilute 1 vol of erythrocytes with 49 vol of saline to produce a 2 per cent suspension. This suspension should be filtered through a small wad of absorbent cotton to remove any clumps. The suspension should be kept in the refrigerator when not in use. Fresh suspension should be prepared each day.

The red-cell counts of different batches prepared in this way may vary somewhat, and therefore for work of precise nature it is preferable to standardize by colorimetric determination of hemoglobin.

For the initial determination, 5 per cent red-cell suspension may be diluted to yield several concentrations ranging from 0.5 through 3 per cent. One milliliter of each suspension is lysed with 9 ml of distilled water, and the optical densities of the hemoglobin solutions read at 5,500 A in a spectrophotometer. If a spectrophotometer is not available, a Klett-Summerson colorimeter with a No. 54 filter may be used. With the values obtained, a calibration curve is plotted, and future 2 per cent suspensions can be standardized against the curve. The same instrument and cuvette should be used for all determinations.

Sensitization of erythrocytes. The treatment of sheep erythrocytes with their homologous antibody is spoken of as sensitization. The antibody, called hemolysin or amboceptor, may be prepared by immunizing rabbits with sheep red cells; however, it is usually preferable to purchase commercial hemolysin.

Before the proper concentration of hemolysin to use can be known, it is necessary to perform a titration of the hemolysin. From a stock 1:100 dilution, prepare dilutions of 1:1,000 and 1:1,500. Arrange eight 15- by 85-mm test tubes in a wire rack and pipet 0.5 ml of 1:1,000 into the first tube and 0.5 ml of 1:1,500 into the second. Into the remaining tubes, place 0.5 ml of saline and prepare serial twofold dilutions of the 1:1,000 and 1:1,500 hemolysin to yield dilutions of 1:2,000, 1:3,000, 1:4,000, 1:6,000, 1:8,000, and 1:12,000. After 0.5 ml is discarded from the last tube, each tube should contain 0.5 ml of diluted hemolysin. Since the complement has not yet been titrated, a constant but excess amount is added to each tube. In this case, add 0.3 ml of 1:30 normal guinea-pig serum to each tube. Then place 1.7 ml of saline solution and 0.5 ml of 2 per cent erythrocyte suspension into each tube. The contents of the tubes are mixed by shaking, and the rack incubated in a 37°C water bath for 1 hr. The *unit* of hemolysin is taken as the highest dilution that yields *complete* hemolysis. A satisfactory hemolysin should yield a fairly high titer. The unit should be 1:4,000 or higher.

To prepare sensitized erythrocytes, place the desired volume of 2 per cent erythrocytes in a flask and add an equal volume of hemolysin. Incubate at room temperature for 30 min. The dilution of hemolysin used should contain two units as calculated above; therefore, if the unit were found to be 1:4,000, one would select the 1:2,000 dilution for sensitization. Sensitized erythrocytes should not be kept from one day to the next.

Complement. The serum from nonimmunized, healthy guinea pigs is used as a source of complement. Since the serum from some animals may be low in complement activity, it is necessary to pool the serum from at least five animals. Commercially prepared lyophilized complement may be used; however, occasional lots are unsatisfactory.

The guinea pigs are bled from the heart or anesthetized and exsanguinated by cutting the large blood vessels on both sides of the neck. The animals should not be fed for at least 12 hr prior to bleeding.

The blood is allowed to clot in centrifuge tubes, and the serum is centrifuged until free from cells. Serum which appears slightly turbid may contain a few cells and should be recentrifuged. Since complement is destroyed rapidly by higher temperatures,[1] it is best stored in the frozen or dried (lyophilized) state. A convenient method is to place 1.0- to 5.0-ml aliquots of the pooled serum in tightly stoppered vials. These are then stored in the deep freeze until needed. For the complement titration, a vial is removed from the deep freeze, thawed in cold water, and diluted to 1:30 with *cold* saline. The diluted serum should be kept in the cold until needed and should be discarded if not used within a few hours.

In dealing with systems where the correct antigen "dose" has been established, it is customary to add the optimal amount of antigen (see next section) to each tube in the complement titration. This will offset any slight anticomplementary effect of the antigen.

Using a 0.2-ml pipet graduated in 0.001 ml, place the following amounts of 1:30 guinea-pig serum in a series of 15- by 85-mm test tubes: 0.20, 0.25, 0.30, 0.35, 0.40, 0.45, 0.50 ml. Add sufficient saline to each tube to bring the volume to 1.5 ml (1.3 ml in the first tube), and then place 0.5 ml of the proper antigen dilution into each tube, shake the rack, and incubate in a 37°C water bath for 1 hr. This incubation period will allow time for the

[1] Complement is also inactivated by passage through Berkefeld and other types of sterilizing filters.

antigen to exert any possible anticomplementary effect. Following the initial incubation, place 1.0 ml of sensitized erythrocytes in each tube, mix, and incubate in a 37°C water bath for 30 min.

Since the optimal antigen dilution will not be known at the outset, the titration may be performed as above, substituting an additional 0.5 ml of saline for the antigen. In either procedure, a control consisting of 1.0 ml of sensitized erythrocytes and 2.0 ml of saline should be included. The tubes are read by comparison with this control, which, of course, should display no lysis. Complete lysis is reported only in those tubes with no trace of turbidity. The smallest amount of complement yielding *complete* hemolysis is called the *exact unit;* however, the unit actually employed is called the *full unit* which is contained in the next higher tube, or the tube with 0.05 ml more complement. In the actual test, two full units are added to each tube. Thus, if the exact unit were contained in 0.35 ml, the full unit would be 0.40 ml and 0.8 ml would be the volume employed in the complement-fixation test. It is usually more convenient to calculate a complement dilution such that two full units are contained in 1 ml. This is done by dividing the reciprocal of the dilution by the volume containing two full units. In the example given,

$$\frac{30}{0.8} = 37.5 \text{ or } 1 \text{ ml of } 1{:}37$$

The complement units as determined by the foregoing (Kolmer) procedure are based on lysis of 100 per cent of the cells and are sufficiently precise for most practical applications. Some workers prefer to use a 50 per cent end point in calculating the unit, since the 50 per cent unit may be determined more accurately. A description of this procedure may be found in Kabat and Mayer (1948).

Antigens. Microbial antigens used for complement-fixation tests are often suspensions of washed, whole cells prepared like antigens for the agglutination test. It is not uncommon to find that such antigens possess some anticomplementary activity; however, this can usually be removed by dilution. Special procedures have been devised for preparing certain microbial antigens used in serodiagnosis (Kolmer *et al.*, 1951). In addition to whole cells, one may employ a cell lysate or mechanically disrupted cells as antigen, and where possible, the use of a purified antigen should be considered. Certain antigens for the complement-fixation test, mostly viral and rickettsial, can be purchased from biological houses.

Antigens are standardized and preserved by procedures described under "Agglutination" and "Precipitation." Special procedures for viral antigens will be found in the chapter on viruses.

Antigens suitable for use in the complement-fixation test must not be hemolytic or anticomplementary within the range where they have sufficient combining power, and these properties must be measured when a new antigen is prepared.

Anticomplementary activity of antigen. The stock antigen is diluted in tenfold (1:10, 1:100, 1:1,000, etc.) steps for a rough titration or in twofold increments after the approximate range has been established. Then to each of a series of 15- by 85-mm tubes are added 1.0 ml of complement (two full units), 0.5 ml of saline solution, and 0.5 ml of a given antigen dilution. These are shaken and then placed at 37°C for 1 hr. At this point 1.0 ml of sensitized erythrocytes is added and the tubes are shaken and placed once again in the 37° water bath for 30 min. Hemolysis should occur, and any evidence of inhibition shows anticomplementary activity of that particular antigen dilution.

Hemolytic activity of antigen. Into a series of test tubes, place 0.5-ml portions of the antigen dilutions used in the foregoing section. Add 2.0 ml of saline and 0.5 ml of nonsensitized 2 per cent erythrocyte suspension to each tube. Mix, and incubate in a 37° bath for 1 hr. If hemolysis is observed, the antigen concentration used in the actual test must fall below the hemolytic level.

Antigen titration. In arriving at the correct amount of antigen to use in the complement-fixation test, one may choose a fixed concentration of antibody and titrate various dilutions of antigen against it. It occasionally happens, however, that such a procedure yields misleading results, and it is generally safer to titrate *both* dilutions of antibody and dilutions of antigen in a "checkerboard" protocol of the type shown under "Complement Fixation" in the chapter on viruses or Table 22 in this chapter.

One should choose an antiserum known to contain a reasonably high titer of antibody. Serial twofold dilutions of the serum (1:2, 1:4, 1:8, etc.) are placed in 15- by 85-mm test tubes, using 0.5 ml per tube. At the same time, serial twofold dilutions of antigen are prepared and added to the serum dilutions; thus, if five dilutions of serum and eight dilutions of antigen are being used, 40 tubes are required. It is also customary to include controls as shown in Table 24. The concentrations of antigen used must be less than the hemolytic and anticomplementary levels.

Saline and complement are added to the tubes (Table 24), and the primary incubation is made at 37°C for 1 hr.

After the primary incubation, sensitized erythrocytes are added and a secondary incubation is conducted in the 37° bath for 30 min. The tubes are read as described for the complete test. The optimal amount of antigen is the dilution that gives the most sensitive reaction with the smallest concentration (highest dilution) of antibody.

Antiserum. The antiserum is prepared by immunizing rabbits or other experimental animals as described earlier. The serum should be free from erythrocytes and lipid. Although a slight amount of hemolysis does not harm the serum, excessive hemolysis should be avoided, since this may make the serum anticomplementary. Contaminating bacteria may also render the serum anticomplementary.

It is necessary to inactivate complement present in the serum by heating in a water bath at 56°C for 30 min. Some serums may possess lytic

TABLE 24. ANTIGEN TITRATION

	Antigen, 0.5 ml	Antiserum, 0.5 ml	Saline, ml	Complement 2 full units, ml	Primary incubation	Sensitized erythrocytes, ml	Secondary incubation
Antigen titration...	*	*	...	1.0		1.0	
Antigen control.....	1:2	...	0.5	1.0		1.0	
Antiserum control..	...	1:2	0.5	1.0		1.0	
Hemolytic control..	...	...	1.0	1.0		1.0	
Erythrocyte control.	...	...	2.0	...		1.0	

* Various dilutions (see text).

The dilutions shown for the antigen control and antiserum control may differ from the values indicated, but each should be the lowest dilution in the series used.

activity because of their content of natural antisheep hemolysin. If this difficulty is encountered, the natural antibody may be absorbed out by adding one drop of washed packed sheep erythrocytes per milliliter of serum. The mixture is stirred with an applicator stick and allowed to stand at room temperature for 30 min. The cells are then removed by centrifugation.

The complete test. If several tests are to be made with each antiserum, greater accuracy can be obtained by preparing the serum dilutions in small flasks. Using separate pipets for each dilution, 0.5-ml quantities are added to a series of test tubes as shown in Table 25. The dilutions shown are arbitrary and may be changed if antiserum of greater or lesser potency is being titrated.

The optimal concentration of antigen, contained in 0.5 ml, is then added to each tube, followed by the saline and complement as in Table 25. The reagents are mixed by shaking, and the primary incubation is carried out in the 37°C bath for 0.5–2 hours or at 0–6°C for 4–24 hr. The time and temperature of the primary incubation may be varied to find the conditions best suited to the system under study. In the serodiagnosis of syphilis, the primary incubation is conducted in the refrigerator for a period of 15–18 hr followed by 0.5 hr at 37°C. When using 37°C incubation, a fixation time of 1 hr is usually satisfactory and in any case should not exceed 2 hr, since complement deteriorates much faster at this temperature than in the refrigerator.

After the primary incubation is complete, 1.0 ml of sensitized erythrocytes is added to each tube. The secondary incubation is conducted at 37°C for 15–30 min.

When the results are read, absence of hemolysis indicates fixation of complement due to union between the antigen and antiserum. Partial hemolysis represents a weaker reaction, and complete hemolysis is a negative reaction. The control tubes 8, 9, and 10 should display complete hemolysis, while the erythrocyte control, tube 11, should not be lysed. Failure to observe lysis in tube 8 indicates that the antiserum was anticomplementary at this dilution. If this occurs, another test should be run, using all dilutions of antiserum but substituting saline for the antigen. This will show if the anticomplementary range extends far enough to obscure the titer of the serum. Tube 9 is a control on the antigen, and this control is usually satisfactory if the preliminary titrations on the antigen have been performed correctly; however, some antigens may become anticomplementary after storage for some time.

TABLE 25. PROTOCOL FOR COMPLEMENT-FIXATION TEST

Tube No.	Antiserum, 0.5 ml	Antigen, ml	Saline, ml	Complement 2 full units, ml		Sensitized erythrocytes, ml	
1	1:10	0.5	...	1.0	Primary incubation	1.0	Secondary incubation
2	1:20	0.5	...	1.0		1.0	
3	1:40	0.5	...	1.0		1.0	
4	1:80	0.5	...	1.0		1.0	
5	1:160	0.5	...	1.0		1.0	
6	1:320	0.5	...	1.0		1.0	
7	1:640	0.5	...	1.0		1.0	
8	1:10	...	0.5	1.0		1.0	
9	.....	0.5	0.5	1.0		1.0	
10	.....	...	1.0	1.0		1.0	
11	.....	...	2.0	..		1.0	

In addition to the controls described, it is advisable to include a titration with known positive and known negative serum when a new system is being studied. The negative (preimmunization) serum titration is particularly important, since some antigens may react nonspecifically with normal serum.

BIBLIOGRAPHY

Boyd, W. C. 1947. "Fundamentals of Immunology." Interscience Publishers, Inc., New York.

Bukantz, S. C., C. R. Rein, and J. F. Kent. 1946. Studies in complement fixation. II. Preservation of sheep's blood in citrate dextrose mixtures (modified Alsever's solution) for use in the complement fixation reaction. *J. Lab. Clin. Med.*, **31**, 394–399.

Cann, J. R., J. G. Kirkwood, R. A. Brown, and O. J. Plescia. 1949. The fractionation of proteins by electrophoresis convection. An improved apparatus and its use in fractionating diphtheria antitoxin. *J. Am. Chem. Soc.*, **71**, 1603.

———, D. H. Campbell, R. A. Brown, and J. G. Kirkwood. 1951. Fractionation of rabbit antiserum (antiphenylarsonicazo-bovine globulins) by electrophoresis convection. *J. Am. Chem. Soc.*, **73**, 4611–4615.

Cohn, M. 1952. Techniques and analysis of the quantitative precipitin reaction, (A) Reaction in liquid media. In "Methods in Medical Research," vol. 5, pp. 301–335. Year Book Publishers, Inc., Chicago.

Deutsch, H. F. 1952. Separation of antibody-active proteins from various animal sera by ethanol fractionation techniques. In "Methods in Medical Research," vol. 5, pp. 284–300. Year Book Publishers, Inc., Chicago.

Durrum, E. L. 1951. Continuous electrophoresis and ionophoresis on filter paper. *J. Am. Chem. Soc.*, **73**, 4875–4880.

Edwards, P. R., and W. H. Ewing. 1955. "Identification of Enterobacteriaceae." Burgess Publishing Co., Minneapolis, Minn.

Flodin, P., and J. Porath. 1954. Zone electrophoresis in starch columns. *Biochim et Biophys. Acta.*, **13**, 175–182.

Freund, J. 1947. Some aspects of active immunization. *Ann. Rev. Microbiol.*, vol. 1. Annual Reviews, Inc., Stanford.

Heidelberger, M., and F. E. Kendall. 1929. A quantitative study of the precipitin reaction between type III pneumococcus polysaccharide and purified homologous antibody. *J. Exptl. Med.*, **50**, 809–823.

———, ———, and C. M. Soo-Hoo. 1933. Quantitative studies on the precipitin reaction. Antibody production in rabbits injected with azo protein. *J. Exptl. Med.*, **58**, 137–152.

———, and ———. 1935. A quantitative theory of the precipitin reaction. III. The reaction between crystalline egg albumin and its homologous antibody. *J. Exptl. Med.*, **62**, 697–702.

———, and C. F. C. MacPherson. 1943. Quantitative micro-estimation of antibodies in the sera of man and other animals. *Science*, **97**, 405–406; **98**, 63.

Isliker, H. C. 1953. Purification of antibodies by means of antigens linked to ion exchange resins. *Ann. N.Y. Acad. Sci.*, **57**(3), 225–238.

Kabat, E. A., and M. M. Mayer. 1948. "Experimental Immunochemistry." Charles C Thomas, Publishers, Springfield, Ill.

Kolmer, J. A., E. H. Spaulding, and H. W. Robinson. 1951. "Approved Laboratory Technic," 5th ed. Appleton-Century-Crofts, Inc., New York.

Krumwiede, C., G. Cooper, and D. J. Provost. 1925. Agglutinin absorption. *J. Immunol.*, **10**, 55–239.

McFarland, J. 1907. The nephelometer: An instrument for estimating the numbers of bacteria in suspensions used for calculating the opsonic index and for vaccines. *J. Am. Med. Assoc.*, **49**, 1176–1178.

Martin, D. S. 1943. A simplified serum dilution method for the quantitative titration of precipitins in a pure antigen-antibody system. *J. Lab. Clin. Med.*, **28**, 1477–1482.

Oudin, J. 1952. Techniques and analysis of the quantitative precipitin reaction. (B) Specific precipitation in gels. In "Methods in Medical Research," vol. 5, pp. 335–378. Year Book Publishers, Inc., Chicago.

Raffel, S. 1953. "Immunity-Hypersensitivity-Serology." Appleton-Century-Crofts, Inc., New York.

Reddish, G. F. 1954. "Antiseptics, Disinfectants, Fungicides, and Chemical and Physical Sterilization." Lea & Febiger, Philadelphia.

Sternberger, L. A., and D. Pressman. 1950. A general method for the specific purification of antiprotein antibodies. *J. Immunol.*, **65**, 67–73.

Wadsworth, A. B. 1947. "Standard Methods." The Williams & Wilkins Company, Baltimore.

CHAPTER X

The Detection of Bacterial Pathogenicity

Joel Warren

INTRODUCTION

The determination of bacterial pathogenicity for animals requires not only a knowledge of those factors which contribute to microbial virulence but skill in the technical procedures of inoculation, autopsy, and bacteriological examination. In recent years there has been a diminished interest in pathogenic studies in animals, and with a few exceptions, they are no longer required in routine diagnostic bacteriology. The use of selective and enriched media and the advent of antibiotics have made nearly extinct such formerly routine procedures as pneumococcus typing by mouse intraperitoneal inoculation or guinea-pig toxicity tests for diphtheria. On the other hand, certain diagnostic procedures still depend upon pathogenicity determinations, e.g., in tuberculosis, and will probably continue in use. In research bacteriology, especially medical and veterinary, the mouse, rabbit, and guinea pig are widely employed for studies of natural or induced bacterial infection, and these reports comprise a large segment of the literature.

The section which follows is designed to provide only an introduction to the study of bacterial pathogenicity. For more specific information the reader should consult the literature on a particular microorganism or one of the standard bacteriological texts such as Topley and Wilson (1946) or Dubos (1948). For good diagrams of animal anatomy and inoculation sites, see Haudauroy (1948).

In a general sense, pathogenic bacteria are those capable of causing the death of an animal when they reach a site which permits their multiplication. Bacteria which may multiply but do not cause fatal disease (as, for example, *Leuconostoc mesenteroides*) are eliminated by this definition, since illness alone is generally difficult to establish in the smaller animals. Thus, from a pragmatic point of view this chapter is primarily concerned with those organisms which, when introduced into a host, will cause death or readily recognizable symptoms. The determination of pathogenicity of such undoubted pathogens as anthrax usually presents no special problem. However, a number of the bacteria which may be pathogenic for man, such as the hemolytic Group A *Streptococci*, frequently lose virulence when cultivated on artificial media or even on initial isolation. Such organisms may require repeated animal passage, etc., before their lethal properties are demonstrable (see below).

GENERAL FACTORS UNDERLYING VIRULENCE

The pathogenicity of microorganisms represents the resultant of a number of single components shared by both the bacterium and its host. Thus, specific toxicity, speed of multiplication, tissue tropisms, strain, and age of the cell all contribute to the virulence of a microorganism and the outcome of infection. In addition, an array of host factors, as, for example, portal of entry, species or age resistance, avitaminosis, partially determines whether the test animal will live or die. In certain instances those factors which determine virulence for animals are the same as for man; in others they are not.

Some of the most prominent components of pathogenicity are the following:

1. Septicemia is an important factor in fatal infections caused by most of the gram-positive cocci and certain bacilli, such as *Pasteurella pestis*. Massive hemal multiplication of the invading bacterium results in an overwhelming infection, and in general, experimental animals developing septicemia succumb after a short illness. In contrast, a bacteremia per se is not indicative of pathogenicity, and bacteria can be readily isolated from the circulation of many healthy animals or those inoculated with an attenuated or nonpathogenic strain of certain microorganisms. In localized infection showers of organisms may appear in the blood stream at intervals. This is usually, but not always, associated with a rise in body temperature.

2. Toxin production is an important feature of bacterial virulence and sometimes one of the most difficult to establish. A toxic death may be considered when animals succumb shortly after infection and often without evidence of extensive bacterial multiplication. The inoculation of culture filtrates will then be required to characterize the agent further. In this connection one should remember that there need not be a correlation between pathogenicity and high toxin production and, as in the case of diphtheria or plague, strains isolated from mild cases may be very toxigenic when grown on artificial media.

The exotoxins of gram-positive bacteria are ready demonstrable in the filtrates of autolysis-free cultures. Among the major toxins are those of

Corynebacterium diphtheriae, *Clostridium botulinum*, and *Clostridium tetani*. The endotoxins, sometimes designated as "gram-negative endotoxins," or O antigens, are bacterial complexes which are easily released into the culture on autolysis. They are not so readily detectable as the exotoxins and frequently require considerable concentration by chemical methods before they can be demonstrated.

In establishing the toxicity or toxigenicity of a bacterial species, the selection of the test animal is of vital importance. Whereas the mouse and rat are generally less susceptible than the guinea pig to exotoxins, the mouse is particularly susceptible to the endotoxins of *Pasteurella pestis*, and the rabbit to the *Salmonella* or *Escherichia* endotoxins (especially *E. coli* strains O-111 and B-55).

3. Important components of bacterial pathogenicity are a number of extracellular enzymes and metabolites. Some of these, such as the hemolysins, are possessed by many species. Others, for example, lecithinase (or alpha-toxin of *Clostridium perfringens*), are found only in certain organisms. Fibrinolysins, collagenases, coagulases, hyaluronidase, and a variety of still uncharacterized cytolysins, such as streptococcal leucocidins, all contribute to the virulence of bacteria. However, it should be noted that these toxic substances can rarely be directly demonstrated in the animal dead of bacterial infection. For this purpose, rather, it is usually necessary to perform specific *in vitro* tests with the bacterial culture. Finally, the presence of extracellular toxicity in a species does not of itself indicate that it is a pathogen. For example, autolysates of some saprophytic bacteria, such as *Leuconostoc mesenteroides*, can be shown to be cytotoxic when injected into the skin of rabbits and other animals.

4. As stated above, bacterial infection is the resultant of a complex of opposing variables on the part of the host and the invading cell. Any determination of pathogenicity, therefore, must be made in healthy animals whose resistance can be assumed to represent the norm for that species. If one injects a bacterial culture into animals which are fatigued, overcrowded, dehydrated, or suffering from an avitaminosis, etc., death of the animal, even with an overwhelming bacteremia, does not necessarily signify that the organism is a virulent one. The possibility should always be considered that the resistance of this particular host was so lowered by unfavorable living conditions or some other organic disease that it did not respond in the normal fashion to bacterial invasion. With regard to the strain of animal, it should be borne in mind that certain substrains of rabbits or of mice will vary in their susceptibility to a particular pathogen. It is not profitable to list each instance here, for this information can be found in appropriate reference texts.

5. One must maintain a certain sense of proportion in selecting the size

of the bacterial inoculum. For example, if an otherwise resistant host were to be injected with a massive dose, something of the order of 500 or 600 mg dry weight of a particular organism, the animal might die of a massive artificial toxemia or, if it survives, might carry viable bacteria for long periods of time. In neither instance is this a satisfactory demonstration of pathogenicity. Conversely, inoculation of a few cells of a moderately pathogenic agent, as *Clostridium perfringens*, may be insufficient to establish either infection or a lethal toxemia. For these reasons, when attempting to measure pathogenicity, it is well to inject a high (several million) and low (200–300) number of bacteria in separate sets of test animals. In this manner the potential invasiveness of the organism is more likely to be demonstrable and the host's phagocytic resistance mechanisms will not be overwhelmed through the sheer mass of foreign inoculum.

6. Pathogenic bacterial populations are a collection of dishomogeneous microorganisms whose total invasive capacity is labile and subject to great variation. Both the number of cells and their virulence are markedly affected by the external environment. Prolonged cultivation may, and usually does, lead to the development of a strain whose mutant composition differs markedly from the original isolate. The relationship of pathogenicity to M, S, R, or G colony types and to the morphology, size or age of the cell may have to be determined for each culture. With some organisms, e.g., *Salmonella typhosa* and *Corynebacterium diphtheriae*, the smooth colony type is the most pathogenic, whereas the mucoid phase of *Diplococcus pneumoniae* and the rough phase of other organisms, such as *Bacillus anthracis*, are the most pathogenic. Unfortunately the colonial morphology of cell mutants is not always associated with virulence in the same fashion. Some species, such as *Malleomyces*, may be lethal in the rough, smooth, or mucoid stages, or virulence can change markedly without any alteration in colonial or cell morphology. For each specific microorganism the worker should familiarize himself with the range of variation and with associated states of pathogenicity. This is absolutely essential for intelligent pathogenicity studies.

In certain instances, as in the case of the meningococci, an organism which is pathogenic for man is only weakly so or avirulent for animals. By inoculating the bacteria in combination with certain adjuvants, as starch or mucin, it is possible to block phagocytosis in the host temporarily, thus permitting the original inoculum to multiply and eventually overwhelm the animal's resistance. Such artificially enhanced virulence is, strictly speaking, not within the province of this section, since it does not necessarily reflect an inherent pathogenicity of the organism but rather the ability of the adjuvant to blockade host defenses. On the other hand it may occasionally be necessary to use adjuvants with old cultures of true pathogens when serial passages are employed to regain the virulence of a degraded type.

Although no longer regarded as the *sine qua non* standard of pathogenicity, it is well to conclude this section with a brief recapitulation of Koch's postulates. Fulfillment of the postulates is often difficult, even with accepted pathogens, as with the gonococcus. In certain instances, such as synergistic infection, they are not applicable. Nevertheless, the basic intent of Koch is valid and directed against premature or unwarranted assumptions of pathogenicity. The postulates are as follows: (1) The organism must always be present where the disease occurs, (2) the organism must be obtained in pure culture from the infected tissue, (3) this culture must cause the disease when injected into a favorable region or tissue of a normal susceptible animal, and (4) the organism must be recovered from the latter.

PREPARATION OF INOCULA

It is always desirable to prepare bacterial inocula in a consistent and standardized manner so that experiments can be readily duplicated. Commonly used diluents for the preparation of suspensions are any of several nutrient broths, physiological saline solution, distilled water, or an isotonic buffer solution. (Methods followed in preparing tissue suspensions will be found in Chap. XI.) Suspensions may be obtained from washings or scrapings of growth on solid media or from fluid cultures. Old cultures are generally less pathogenic than young growths. In each case the largest cell aggregates are removed by light centrifugation, and when necessary, the final suspension is standardized for density with a colorimeter or densitometer.

As a check of purity approximately 0.1 ml of the final inoculum should be cultured on a suitable agar medium. Bacterial counts of the inoculum can be performed either turbidimetrically or by culture; the latter is often the preferable procedure, as it counts viable bacteria only. When the toxicity of a preparation is being determined, the cells or cell debris are usually removed by centrifugation or filtration. It should be borne in mind that certain toxins, e.g., the murine toxin of *Pasteurella pestis*, may be inactivated by filtration, and this procedure is generally not recommended for toxin preparation. Sterility tests of toxins are also desirable to ensure that the death of an experimental animal is not due to a contaminated preparation.

The storage of bacterial inocula for periods of a few days is best done at 4–8°C, whereas for longer periods freezing, at −20 or −70°C, is often resorted to. However, considerable killing may occur with even a single freeze and thaw, and this method is not recommended for the storage of inocula containing relatively few bacteria. Toxins generally retain their potency exceedingly well when frozen at −20°C. Inocula, such as seed cultures, are sometimes lyophilized, and the recommended procedure for this can be found in Chap. V. Particular care should be taken in the labeling of cultures which are to be stored at subzero temperatures, since freezing will often make carelessly written or crayon labels unintelligible. It has been found good practice to typewrite labels onto a piece of adhesive tape, and these, when attached to a glass container, will usually adhere for considerable periods of time. If a crayon pencil is used, it is a good idea to put a piece of transparent plastic tape over the crayoned label. Cultures which are to be stored in the frozen state for any length of time should never be left plugged with cotton alone. A sterile rubber stopper or glass sealing of the tube is necessary to prevent evaporation of of frozen fluid during prolonged periods of storage. When autopsy tissues must be stored pending bacterial examination, it is preferable to retain the tissue in its original structure, since whole tissue fragments retain their infectivity better than the same tissue ground into a suspension. For this reason many workers take double samples of tissue from an autopsy, one portion of which is immediately frozen and the other portion used for workup. Although it is a point well known to the experienced technician, the amateur should be reminded that if a blood sample is intended for serological studies, it should never be frozen before separation of the cells from the serum; otherwise the specimen will be badly hemolyzed.

SELECTION AND PREPARATION OF LABORATORY ANIMALS FOR PATHOGENICITY STUDIES

Animals not only are necessary for determining the etiology of specific infectious diseases and the pathogenicity of particular cultures of bacteria but are also utilized as a means of isolation, to determine in more detail certain specific pathogenic mechanisms, to maintain species that grow best *in vivo*, to increase pathogenicity, and to produce antibodies.

The bacterial species and the property to be studied determine the choice of the experimental animals. They should be kept in a well-ventilated and -lighted room with properly supervised animal care. Animals inoculated with suspect or known pathogens should, whenever possible, be caged apart from the common stock, and the cages marked in a suitable fashion. Where insects, such as flies, mosquitoes, or ticks, are troublesome, the caging area should be thoroughly sprayed with DDT or a similar insecticide. Do not cage large numbers of animals in a single enclosure. Whenever possible, the larger species, such as rabbits, monkeys, or dogs, should be kept in individual cages. Marking of laboratory animals may be done by tattooing, particularly on the ear for rabbits; by the use of numbered metal ear tags; or by some suitable stain. For albino mice and rats a saturated alcoholic solution of picric acid is often employed. Ample supplies of feed and water are absolutely essential. In the case of guinea pigs one should guard against an excess fluid intake, as these animals easily develop severe diarrhea when given too much water. Usually a daily feeding of greens, supplemented with ascorbic acid, is adequate for guinea pigs.

Infected animals should be inspected daily, not only to determine death but also to ascertain the first appearance of any clinical symptoms. It is not good practice to bring normal animals into the same laboratory where one is performing bacteriological work. Whenever possible it is a better policy to take the material to be injected from the laboratory into the animal area, where a separate section can be equipped for inoculations and autopsy.

Animals which appear to be sick or mangy or have snuffles, diarrhea, marked loss of weight, middle-ear (rolling) disease, or obvious tumors should not be used for pathogenicity studies. New shipments of animals, especially when they are received in a crowded condition, are best utilized after at least a 24- to 48-hr period of rest in clean cages.

METHODS OF INOCULATION

Because most pathogens are able to produce disease in the experimental host only when inoculated via a suitable portal of entry, it is essential that the operator be familiar with the technical aspects of animal injection. There are innumerable methods of inoculating animals, and most laboratories will be found to have modified the standard procedures in some fashion that suits them best. However, the following general routes of inoculation and procedures are discussed to provide the beginning worker with a basis for acquiring his own training.

Cutaneous. This route of inoculation is rarely used for bacterial infection. It is more common in the testing of certain viruses, such as those of the herpes group. The term is a rather loose one and includes scratching the skin or covering the deposited inoculum with an adhesive patch. In some instances the inoculum is placed directly upon the adhesive patch, which is then attached to the skin. The precise method is determined by the object to be attained. The skin should be cleansed and sterilized with an antiseptic that has brief action, as 70 per cent ethyl alcohol. The hair may be removed by plucking or shaving. If it is desired to determine whether or not an organism can penetrate the normal skin, the material should be spread over a nonirritated area; therefore shaving or depilation should be avoided. Following inoculation the site may be covered with sterile cloths or coated with collodion to exclude air. However, the latter procedure may create a degree of anaerobiosis, and this makes the conditions abnormal. It is sometimes desirable in cutaneous inoculation to abrade the epidermis by scratching or scraping with a sharp instrument. A small piece of tin similar to that used when vulcanizing a rubber tire tube can be employed for this purpose. Scraping aids penetration by removing the outer defensive epithelium and is similar to intracutaneous injection. When very small amounts of inoculum are available, it is sometimes necessary to resort to the latter type of inoculation.

Intracutaneous. By intracutaneous injection is meant the introduction of material into the epithelial layers. The formation of a bleb during inoculation indicates successful injection. It is advisable to use animals with a nonpigmented skin, and rabbits should not be in molt. Shaving and the application of antiseptics may interfere with making a test and should be used judiciously. A 25- to 27-gauge needle is used and, with the bevel uppermost, is inserted at a slight angle to a depth sufficient only to cover the bevel. Pressure should be maintained on the needle during inoculation, and the maximum permissible quantity is about 0.1–0.2 ml.

Subcutaneous. This is one of the most common routes of inoculation and adaptable to a wide range of pathogens. The skin may be shaved or the hair clipped or plucked without interfering with the procedure. Material should be injected into the subcutaneous tissue and, when performed on the abdomen, with care not to puncture the peritoneal wall. The area may be disinfected with a nonirritating disinfectant, such as tincture of zephiran chloride, alcohol, merthiolate, or green soap. The area of injection may be marked with an indelible pencil. The size of the needle and the amount inoculated will depend upon the material and the animal. A 21- to 25-gauge needle and volumes up to 2 ml are customary.

If the material will not pass through the needle, as in the case of thick bacterial suspensions, adjuvant preparations, or certain toxoids, the skin may be sterilized, after removal of the hair, and a V-shape opening cut into the skin with sterile scissors or a sharp scalpel. The flap is then lifted up and loosened until a pocket is formed into which the material to be injected is inserted. After the flap is replaced, the area is sterilized and covered with collodion or aseptically sutured.

Intramuscular. The skin is treated as for subcutaneous injection, but the inoculum injected deep into the muscle. Be certain to use a needle of adequate length. With larger animals it is good practice to withdraw the syringe plunger before injection to make sure that the point of the needle has not accidentally entered a large vessel.

Intravenous. This is a commonly used site for deposition of pathogenic bacteria, toxins, and other virulent exudates. The choice of veins is largely a matter of convenience and will vary with the animal. Rabbits are usually injected into the marginal ear vein; mice and rats in a tail vein; guinea pigs in the ear or jugular vein; horses, cows, and sheep in the jugular vein; swine in the ear vein; dogs and cats in the jugular or the vein crossing the inner surface of the thigh. Fowl may be injected, after a little practice, in the radial vein that crosses the elbow joint of the wing. Veins may be dilated by rubbing the surface of the skin with xylene-soaked gauze or cotton or by warming the area with warm water. In the case of rats or mice, it is simplest to immerse the tail for a few seconds in a container of water adjusted to a temperature of about 45°C. If the material to be inoculated is considerably acid or alkaline, it is adjusted to p.H 7.3–7.4. Any coarse particles are removed by centrifugation. The area is cleansed in the same fashion as for subcutaneous injection. The previously warmed material, *free from air bubbles*, is then slowly injected. As the needle is withdrawn, alcohol-soaked cotton should be pressed over the puncture until the bleeding stops. The following needle sizes are suggested for general use: mice, 27-gauge; rats, 25-gauge; guinea pigs, 25-gauge; rabbits, 23-gauge; larger species, depending upon the size.

Intraperitoneal. This route permits of simple inoculation of relatively large volumes of material, and because of the rapid absorption of substances from the peritoneum, it provides tissue saturations only slightly slower than in the case of the intravenous. The area is disinfected as before, and the needle passed through the skin and then through the abdominal wall with a short stabbing motion. Care should be taken to avoid puncturing the viscera, and this danger can be minimized by injecting into one of the lower abdominal quadrants.

Intrapleural. This procedure is the same as for intraperitoneal injection except that one injects into the pleural cavity anterior to the diaphragm. The actual point of inoculation depends upon the experimental animal. The intrapleural route is usually employed for the purpose of obtaining exudates, and additional details of the procedure can be found in the standard bacteriological texts.

Per os. Introduction of the material into the stomach or intestines may be readily accomplished by means of a catheter or capsules or by mixing the inoculum with food or drink. To avoid exposure to the gastric juices the material may be closed in enteric-coated capsules. Liquids may be mixed with starch and made into pills which are readily digestible. In the case of mice and rats it is simplest to blunt the end of a 19- or 21-gauge needle by grinding it down and then passing it gently into the esophagus while holding the rodent vertical and retracting the head to provide a relatively straight passage.

Intranasal. In most instances the material is dropped into the nostrils and the animal is allowed to insufflate or the material is sprayed onto the membranes of the nose and throat. Light anesthesia is generally necessary and desirable to quiet the animal before intranasal inoculation is attempted. The operator should take care that he himself does not inhale any of the material, and it is absolutely essential that intranasal inoculation never be performed unless the operator is suitably masked. In certain instances it may be desirable to use a small chamber, such as a metal mouse box covered with a glass plate, in which the animal can be exposed to an infective aerosol. The latter may be introduced through an opening in the side of the chamber with an atomizer or nebulizer. The aerosol chamber minimizes the exposure to the operator. However, it is essential that care be taken to prevent leakage from the system, and the animals should be removed from the chamber with proper precautions.

Intratracheal. The material may be introduced into the trachea through a tube placed in the larynx or by means of a syringe through the side of the neck. In the latter method the skin may be incised after shaving and sterilization.

Ophthalmic. Material is dropped into one eye, the other left untreated, as a control. It may also be inoculated upon the scarified conjunctiva.

Intracerebral. The method of injection will vary with the different species of animals and the desired location for the inoculum. In most instances it is deposited into one of the frontal lobes. Care should be taken not to inject on the midline of the skull, since a large sinus lies below and fatal hemorrhage may result. A trephine or punch is necessary to open a hole in the case of guinea pigs and larger animals which have relatively thick skulls. A trephine is not necessary in the case of mice, hamsters, or smaller rats. Do not use a volume which is great enough to cause pressure. For larger animals one can inoculate 0.5–1.0 ml; for medium-sized animals, as guinea pigs, 0.1–0.25 ml; and for small animals, as mice and hamsters, 0.03–0.05 ml.

Intracardial. Initial attempts at intracardial injection may fail, but numerous attempts are inadvisable because the animal usually develops a hemothorax from punctures of the heart. This method is most easily performed if the animal is anesthetized, the skin disinfected, and then the area of maximum beat of the apex of the heart is palpated by means of the forefinger. When the heart has been located, a sharp needle (the size depending upon the animal) should be inserted between the ribs and pushed in the direction of maximum palpation. One should then be able to feel the heartbeat pulsating against needle and syringe. When this is established, the needle can be readily plunged into the heart, but care should be taken not to push too deeply or it will pass through the chamber and become embedded in the opposite wall. The presence of the needle point in a cardiac chamber is usually indicated by the appearance of blood in the syringe. The inoculum must be injected slowly, or the animal may go into syncope.

THE CARE OF EXPERIMENTAL ANIMALS

It must always be borne in mind that unknown agents and many known infectious materials may be dangerous. It is the responsibility of the person who inoculates animals to ensure that everyone coming into contact with these animals does not inadvertently run a risk of infection. Unless there is a valid and specific reason to do so, it is ill-advised to inoculate animals unnecessarily with certain agents, e.g., anthrax or tularemia. A large forceps and container of suitable germicide should be kept in the animal area, and sick or dead animals handled with these forceps, especially when being removed from the cage. In many laboratories gowns or laboratory coats and a suitable type of face mask must be donned before entering the area of infected animals. This is particularly important in the case of tuberculosis and the mycotic diseases. When sick animals are to be disposed of or it is desired to sacrifice them for autopsy, they may be killed by dropping them into a suitable chloroform-filled container. This same container should not be used for the anesthetization of healthy animals. The disposal of infected carcasses is best accomplished by incineration, and waterproof disposal bags are now commercially available which serve as excellent containers for transportation of animals to the incinerator. All animal quarters should be kept scrupulously free of arthropods, especially when such agents as *Pasteurella pestis* are being studied.

CLINICAL SIGNS IN ANIMALS

Certain types of clinical signs are readily recognizable and, with increasing experience, will be found to provide a valuable guide to the health of the animal. However, all too frequently a sick animal will be found to be showing symptoms which are not caused by the injected material but rather by such unforeseen circumstances as starvation, thirst, spontaneous tumors, etc. Furthermore it should be remembered that the same microorganism may cause a different disease picture, depending upon the organs which are involved.

In mice a hunched back and quiet position in the cage may signify gastrointestinal infection or a systemic disease. Central nervous-system lesions may cause tremors, rolling, ataxic gait, or a tendency to jump wildly at loud noises or sudden motion. Constant pawing of the nose is often indicative of infection of the upper respiratory system. A gasping, spasmodic type of respiration can be seen in mice which are suffering from pulmonary infection or irritation.

In rabbits and guinea pigs the most helpful clinical sign is an elevation of the temperature. It is always well to obtain a preinoculation temperature on an experimental animal in order to establish his normal level as well as to make certain that the animal is not suffering from some intercurrent infection at the time it is first used. Animals which are prostrated and about to die may very often have a subnormal temperature. An animal which is cold, either because it has a fever or because of low temperature in its surroundings, may shiver. In rabbits and guinea pigs diarrhea, snuffles, discharging eyes, and particularly encrusted conjunctivae are all indicative of ill health.

When examining inoculated animals it is not sufficient merely to open the cage and glance at the animals. Move the animal around in the cage, or in the case of mice, pick them up with a forceps and permit them to move about on the wire-screen top of the box. Frequently an animal which appears to be perfectly well when left alone will be found to be lethargic, paralyzed, or trembling when disturbed.

RECOVERY OF BACTERIA FROM INFECTED ANIMALS

It not infrequently happens that an entirely different organism is isolated from an animal which has been previously inoculated with some known strain. For example, *Pasteurella lepiseptica* can readily be isolated from rabbits injected with suspect acid-fast or mycotic agents. Thus, before proceeding with either an animal autopsy or removal of serum or exudates from an infected species, it is always well to consider beforehand the precise procedures which will be taken, to collect all the necessary materials and instruments, and, when the operation is completed, to prepare proper and adequate notes. Certain laboratory equipment should be maintained in readiness and preferably in one particular portion of the laboratory for animal-autopsy work. The bacteriologist should bear in mind that the recovery of microorganisms from experimentally infected animals can be one of the most hazardous operations in bacteriology.

Procedures will vary in different laboratories, but in general it is well to have a large covered container of $HgCl_2$ or similar germicidal solution, large towels which can be placed in the germicide before and after completion of the procedures, a set of sterile instruments, bunsen burner, sterilizers, and sterile glass pipets, cotton, and similar materials for removal of exudates. It is a good general policy to employ an assistant, if available, for the collection of specimens from larger animals, such as rabbits, dogs, and monkeys.

The aseptic collection of blood for purposes of culture is a relatively simple procedure, but certain essential precautions must be observed. Particularly important is adequate sterilization of the injection site, and for this purpose it is advisable to use one of the stronger germicides, as tincture of iodine. A suitable location for collection of blood for culture is directly from the heart, and this may be done in a manner analogous to that described for intracardiac inoculation. Animals may be anesthetized with ether to ensure their remaining quiet during the procedure. Anticoagulants, as heparin or citrate, are toxic to certain organisms, and wherever possible blood should be collected without preservative and immediately inoculated into culture media. For this reason it is often desirable to bring the culture materials directly to the site of bleeding.

For routine cultures, liquid media, such as a tryptophane broth, should be previously sterilized in 60- to 80-ml volumes in 100-ml glass serum bottles which are stoppered with a rubber vaccine cap. The use of such a container makes it relatively simple to inject the blood directly through the closure into the broth medium upon removing the hypodermic needle from the animal. An important consideration is the ratio of the volume of blood to that of the culture fluid. In general, one should try to maintain a ratio of 1 part of blood to 15 or 20 parts of medium. It is well to prepare multiple cultures so that anaerobic, aerobic, or other special procedures can be performed. It is sometimes also desirable to place 0.1–0.5 ml of the blood specimen on the surface of a blood agar or similar plate and observe for surface growth. It is not good procedure to delay blood cultures until after death, since a massive bacteremia can occur shortly before or immediately after the death of an animal. This may even develop with some saprophytic strains, such as *Pseudomonas,* thus giving a false result.

The collection of exudates, especially peritoneal, is frequently desired to obtain material for microscopic examination. The recovery of peritoneal fluid is readily performed in the living animal by aspirating a small amount of fluid with a syringe and hypodermic needle while the animal is held with its ventral abdominal wall parallel with the working surface. When the needle has been inserted at an angle tangential to the surface, the point of the needle is then depressed with the bevel upward and the peritoneal fluid will run into the pocket thus formed and can be withdrawn. If the material is viscous or small in quantity, it may be necessary first to inject 3–5 ml of sterile saline solution to flush out the peritoneal cavity.

Animal-autopsy procedures constitute one of the backbones of bacteriological technique, and it is essential that basic techniques be properly learned and that the autopsy procedure be treated with considerable respect. It should always be remembered that examination of a dead or dying animal may be a hazardous operation because the cause of death remains to be established. For this reason a series of standardized and properly thought-out safety and aseptic precautions should always be followed even when autopsying animals which have died following inoculation of a known pathogen. As stated above, it not infrequently happens that following injection of known cultures, the animal may die from some supravening agent and one which may be pathogenic for man.

The following information is sought at autopsy and in subsequent workup of the materials: the cause of death, the type of distribution and histopathology of the lesions, distribution of the infected organism, changes that may have taken place in the microorganism itself, and finally, if antibodies have developed.

If an autopsy cannot be made immediately after death, the carcass should be refrigerated. It is well to cover the operating surface with a germicide-soaked paper or cloth towel, and before the animal is opened, the entire carcass should be washed down with disinfectant. In the case of small animals the entire carcass may be immersed into a jar of some disinfectant, such as lysol. Carefully examine the area of the site of inoculation, if any. Then with a sharp scissors make a small keyhole opening through the skin but not penetrating the muscle wall over the lower peritoneum. By blunt dissection separate the skin from the underlying muscle. When this is completed, it is a relatively simple matter to open a median ventral incision, using the scissors and pulling the skin back on each side. It is not advisable to use a scalpel for the autopsy of small animals because of the inherent danger to the operator and because it is often too easy to cut deeper than is intended and to contaminate an area. After the peritoneal muscularis has been exposed, the surface should be sterilized by swabbing with tincture of iodine, following which, a fresh set of instruments is used to continue the operation. Open the peritoneal cavity, leaving the pleural cavity intact. Examine the peritoneal contents, paying particular attention to the surface of the large organs, such as the liver and spleen. Make note of the size of the spleen, since splenomegaly often accompanies acute infections. Using separate instruments, remove portions of the tissues for cytological and cultural study to sterile petri plates, which should be kept at hand for this purpose. Do not make any smears or cultures of the tissues until the autopsy is completed, as this can usually be done at leisure. If it is desired to remove exudates, this can be done by means of a glass pasteur pipet or syringe. When the examination of the peritoneum is completed, the thoracic cavity is opened, care being taken not to cut the diaphragm or pierce the lungs. After intracerebral inoculation it is sometimes desirable

TABLE 26. ACCELERATED IDENTIFICATION PROCEDURES FOR PATHOGENIC BACTERIA

I Microorganism	II Type specimen	III Presumptive identification	IV Definitive identification	V Time required		VI Serologic methods	VII Test animal	VIII Occurrence
				Presumptive	Definitive			
Bacillaceae:								
Bacillus anthracis (anthrax)	Animal tissue, blood, sputum	1. Direct smears: gram stain	1. Inoculation of mouse (IP) with suspected material	1. 10–15 min	1. 24–48 hr	None	1. Mouse	Epidemic or sporadic
		2. Thermo precipitin test	2. Direct culture: blood agar	2. 24 hr	2. 48 hr		2. Guinea pig	
Clostridium botulinum (botulism)	Food (esp. canned), water, stomach contents	1. Animal inoculation for toxin	1. Antitoxin protection test[1]	1. 12–14 hr	1. 1–3 days	See IV	1. Mouse	Sporadic
		2. Direct smear: gram stain	2. Direct anaerobic culture (organism usually not isolated)	2. 10–15 min	2. 7 days	Complement-fixation test with food filtrate antigen[2]		
Clostridium tetani (lockjaw, gangrene)	Tissue, pus from cutaneous lesions	1. Direct smear: gram stain, spore stain	1. Direct anaerobic culture: enriched media	1. 10–15 min	1. 5–7 days	See IV, 2	1. Guinea pig	Sporadic
			2. Antitoxin protection test[3]		2. 24–96 hr			
Gas gangrene group:								
Clostridium: *perfringens* (Welchii) *septicum* *sordelli* (Bifermentans) *novyi* (Oedematiens) *histolyticum*	Tissue	1. Direct smear: gram stain	1. Biochemical studies	1. 10–15 min	1. 5–7days	See IV, 2	1. Guinea pig	Sporadic
		2. Acrolein test[4,5]	2. Animal inoculation	2. 4 hr	2. 12–72 hr		2. Rabbit	
Corynebacteriaceae:								
Corynebacterium diphtheriae	Nose and throat swabs, skin	1. Direct smear: gram and	1. Direct culture: yeast-enriched	1. 10–15 min	1. 18–48 hr	1. Antitoxin protection tests:	1. Guinea pig	Epidemic or

	lesions, genito-urinary swabs	methylene blue stain 2. Leoffler's culture for morphology	blood agar and potassium tellurite agar 2. Biochemical studies	2. 8 hr	2. 48 hr	subcutaneous and intracutaneous (Fraser-Weld[6]) 2. *In vitro* toxin detection (Elek[7])	2. Rabbit	sporadic
Micrococcaceae: *Micrococcus pyrogenes* var. *aureus* (enteritis, pyogenic infections)	Food, stomach contents (if early)	1. Impression smear: gram stain	1. Direct culture: blood agar 2. Salt sensitivity 3. Mannite fermentation 4. Toxin production (Dolman's medium[8])	1. 10–15 min	1. 18–24 hr 2. 18–24 hr 3. 18–24 hr 4. 40 hr	None	1. Kitten (6–8 wk old) 2. Monkey	Epidemic or sporadic
Lactobacteriaceae: *Streptococcus* species (enteritis, pyogenic infections)	Blood, sputum, nasopharyngeal and throat swabs, pus, exudates	1. Direct smear: gram stain	1. Direct culture: yeast-enriched broth and blood agar for hemolytic activity 2. Bile solubility	1. 10–15 min	1. 24 hr 2. 24 hr	1. Grouping with Maxted's carbohydrate extract	1. Mouse	Epidemic or sporadic
Diplococcus pneumoniae (pneumonia)	Sputum, blood, spinal fluid, pus	1. Direct smear: gram stain 2. Quellung reaction: peritoneal exudate from injected mouse	1. Direct culture: blood agar 2. Bile solubility	1. 10–15 min 2. 4 hr	1. 24 hr 2. 24 hr	1. Type with specific antisera	1. Mouse	Epidemic or sporadic
Pseudomonadaceae: *Pseudomonas aeruginosa* (pyogenic infections)	Pus, sputum, urine	1. Direct smear: gram stain 2. Presence of green pigment in specimen	1. Direct culture: blood agar 2. Pigment production, nutrient agar 3. Biochemical studies	1. 10–15 min 2. 12–18 hr	1. 12–18 hr 2. 12–18 hr 3. 24 hr	None	1. Rabbit 2. Guinea pig	Sporadic

TABLE 26. ACCELERATED IDENTIFICATION PROCEDURES FOR PATHOGENIC BACTERIA (*Continued*)

I Microorganism	II Type specimen	III Presumptive identification	IV Definitive identification	V Time required		VI Serologic methods	VII Test animal	VIII Occurrence
				Presumptive	Definitive			
***Pseudomonadaceae:* (*Cont.*)**								
Vibrio cholera (cholera)	Feces, water, food	1. Direct smear of mucus flake, gram stain 2. Growth in alkaline peptone broth	Direct culture: alkaline peptone broth, Aronson's medium	1. 10–15 min 2. 6 hr	1. 24 hr	1. Type with specific antisera 2. Greig hemolysin test[94] 3. Pfeiffer test 4. Cholera red test	1. Guinea pig (IP only)	Epidemic
Enterobacteriaceae:*								
Shigella species (dysentery)	Feces, food, water	1. Growth on *Salmonella-Shigella* and MacConkey's agar plates 2. Growth in microculture on carbohydrate media[10]	1. Biochemical studies	1. 16–24 hr 2. 7 hr after isolation from initial plating	1. 24 hr 2. 7 hr	1. Agglutination with specific antisera	None	Epidemic and sporadic
Salmonella species (enteritis, typhoid, paratyphoid)	Feces, food, water, pus, blood	1. Growth on *Salmonella-Shigella* and MacConkey's agar plates 2. Growth in microculture on carbohydrate media[10]	1. Biochemical studies			1. Agglutination with specific antisera	None	Epidemic and sporadic
Parvobacteriaceae:								
Malleomyces mallei (melioidosis)	Blood, pus, sputum, exudate, water	1. Direct smear: gram stain (for evidence of mycelial and branching mi-	1. Direct culture: glycerol agar (+) 2. Strauss test	1. 10–15 min 2. 15 min	1. 18 hr 2. 48–72 hr	1. Complement-fixation	1. Equines 2. Cat	Epidemic

		croorganisms) 2. Acid-fast stain (−)	3. MacConkey's agar (no growth) 4. Motility (−)		3. 24 hr 4. 12 hr		3. Guinea pig	
Malleomyces pseudomallei	Blood, pus, sputum, exudate, water	1. Direct smears: gram stain (for evidence of mycelial and branching microorganisms) 2. Acid-fast stain (−)	1. Direct culture: glycerol agar (−) 2. MacConkey's agar (growth) 3. Motility (+)	1. 10–15 min 2. 15 min	1. 18 hr 2. 24 hr 3. 12 hr	1. Complement-fixation	1. Rat 2. Guinea pig	
Pasteurella pestis (plague)	Sputum, blood, bubo pus, vesicular fluid, tissue from suspected rodents	1. Direct smear: gram stain 2. Thermo precipitin reaction[11]	1. Direct culture: blood agar 2. Staining characteristics 3. Motility test 4. Biochemical studies	1. 10–15 min 2. 24 hr	1. 18–24 hr 2. 18–24 hr 3. 12 hr 4. 7–14 days	1. See III, 2 2. Agglutination	1. Mouse 2. Guinea pig 3. Rat	Epidemic and sporadic
Pasteurella tularense (tularemia)	Blood, biopsy pus, lesion scrapings, water	1. Direct smear: gram stain 2. Thermo precipitin test[12] 3. Skin test (Foshay[13])	1. Direct culture: Francis cystine agar 2. Chick embryo 3. Biochemical studies	1. 10–15 min 2. 24 hr 3. 48–72 hr	1. 24–48 hr 2. 4–5 days 3. 24–48 hr	1. See III, 2 2. Agglutination	1. Mouse 2. Rabbit 3. Guinea pig	Sporadic
Pasteurella pseudotuberculosis	Sputum, blood, pus	1. Direct smear: gram stain 2. Direct culture: bile broth	1. Direct culture: nutrient agar 2. Motility test at 20°C			1. Agglutination	1. Rat 2. Mice 3. Guinea pig	Sporadic
Brucella: *abortus* *suis* *melitensis* (brucellosis)	Blood, tissue, milk, sputum	48–96 hr Direct culture: 1. Rabbit-blood meat-infusion broth 2. Liver-infusion broth 3. Blood agar pour plates 4. Strain differentiation by selective inhibition of dyes 5. Guinea-pig passage 6. Biochemical studies				1. Agglutination	1. Rabbit 2. Guinea pig	Sporadic

* Other Enterobacteriaceae are identified in a similar manner.

TABLE 26. ACCELERATED IDENTIFICATION PROCEDURES FOR PATHOGENIC BACTERIA (*Continued*)

I Microorganism	II Type specimen	III Presumptive identification	IV Definitive identification	V Time required		VI Serologic methods	VII Test animal	VIII Occurrence
				Presumptive	Definitive			
Neisseriaceae:								
Neisseria meningitidis	Spinal fluid, blood, nasopharyngeal swabs, autopsy tissue	1. Direct smear: gram stain 2. Quellung reaction following mouse passage (IP)	1. Direct culture: yeast-enriched blood agar 2. Biochemical studies	1. 10–14 min 2. 4–6 hr	1. 18–48 hr 2. 36–48 hr	1. Agglutination	1. Mouse	Epidemic and sporadic
Neisseria gonorrhoeae	Urethral and vaginal discharge	1. Direct smear: gram stain	1. Direct culture: yeast-enriched agar 2. Biochemical studies	1. 10–15 min	1. 18–48 hr 2. 36–48 hr	None	None	Sporadic
Bacterium anitratum (*Mimeae, novum genus*)	Urine, sputum, nasopharyngeal swabs, spinal fluid, pus	1. Direct smear: gram stain	1. Direct culture: plain blood agar 2. Biochemical studies (inactive in carbohydrates)	1. 10–15 min	1. 18–24 hr 2. 7 days	1. Group with specific antisera	1. Mouse 2. Guinea pig	
Hemophileae:								
Hemophilus influenzae (bacterial influenza, meningitis)	Nasopharyngeal swabs, sputum, spinal fluid, blood	1. Direct smear: gram stain	1. Direct culture: yeast- and cystine-enriched agar 2. Biochemical studies	1. 10–15 min	1. 18–24 hr 2. 36–48 hr	1. Type with specific antisera	1. Mouse	Epidemic
Hemophilus pertussis (whooping cough)	Nasopharyngeal swabs, sputum	1. Direct smear: gram stain	1. Direct culture: Bordet-Gengou medium 2. Cough plate, Bordet-Gengou medium	1. 10–15 min	1. 48–72 hr 2. 48–72 hr	1. Type with specific antisera (agglutinins do not appear early nor in sufficient concentrations to be of diagnostic value)	1. Mouse 2. Chick embryos	Epidemic

Hemophilus ducreyi (chancroid)	Urethral and vaginal discharge	1. Direct smear: gram stain	1. Direct culture: broth enriched with fresh inactivated rabbit blood	1. 10–15 min	1. 24–72 hr	1. None	1. None	Sporadic
Moraxella lacunata	Discharge from eye	1. Direct smear: gram stain	1. Direct culture: alkaline blood agar 2. Loeffler's blood serum	1. 10–15 min	1. 24 hr 2. 24–36 hr	None	None	Epidemic
Mycobacteriaceae:								
Mycobacterium tuberculosis (tuberculosis)	Sputum, gastric washings, urine, spinal fluid, exudates, food (milk)	1. Concentrated smear: Ziehl-Neelsen or Kinyoun stain 2. Direct smear: Kinyoun stain	1. Direct culture of concentrate to: Dubos' liquid medium, Lowenstein's egg medium	1. 2 hr 2. 15 min	1. 7–21 days	1. Cytochemical reaction[14]	1. Guinea pig	Epidemic and sporadic
Mycobacterium leprae	Swabs from nasal mucosa, lepromatous skin lesions	1. Direct smear: Ziehl-Neelsen stain	None	1. 15 min	None	(High percentage of biologically false positive Wassermann and Kahn reactions)	None susceptible	Sporadic

[1] Max S. Marshall. 1947. "Applied Medical Bacteriology," p. 162. Lea & Febiger, Philadelphia.
[2] R. Kelser and H. Schoening. 1943. "Manual of Veterinary Bacteriology," p. 658. The Williams & Wilkins Company, Baltimore.
[3] Marshall, *op. cit.*, p. 281.
[4] *Ibid.*, p. 198.
[5] Frederick B. Humphries. 1924. *Am. J. Infectious Diseases*, **35**, 282–290.
[6] American Public Health Association, Inc. 1950. "Diagnostic Procedures and Reagents," p. 168. Publication Office, New York.
[7] E. O. King, M. Frobisher, Jr., and E. D. Parsons. 1949. *Am. J. Public Health*, **39** (10), 1314–1320.
[8] C. E. Dolman *et al.* 1936. *Can. Public Health J.*, **27**, 489–493.
[9] E. D. W. Greig. 1915. *Indian J. Med. Research*, **2**, 623.
[10] R. A. McCready and M. B. Holmes. 1953. *Am. J. Public Health*, **10**, 285.
[11] C. L. Larson, C. B. Philip, W. C. Wright, and L. E. Hughes. 1951. *J. Immunol.*, **67**, 289.
[12] C. L. Larson. 1951. *J. Immunol.*, **66**, 249.
[13] David T. Smith and Normal F. Conant. 1952. "Zinnser's Textbook of Bacteriology," 10th ed., p. 506. Appleton-Century-Crofts, Inc., New York
[14] W. C. Morse, M. C. Dail, and I. Olitsky. 1953. *Am. J. Public Health*, **43**(1), 36–39.

to remove brain tissue, and the technique of this procedure is relatively simple. The skin over the skull is dissected and reflected back, the surface of the skull is disinfected as before, and with an appropriate pair of small sharp scissors a series of small cuts is made around the temporal and occipital edges of the skull. The skull cap can then be removed. In the case of large animals, such as monkeys, cats, or dogs, it may be necessary to use bone forceps for this purpose. After the roof of the skull has been removed and the larger cranial nerves cut, the entire brain is lifted out and placed in a sterile petri dish. Upon completion of the autopsy the animal should be carefully placed in a paper bag or similar container and the carcass either sterilized or incinerated; the latter procedure is preferable. Instruments, etc., are sterilized and then washed and dried.

In an ordinary standardized autopsy it is customary to remove portions of the liver, spleen, kidney, lungs, and occasionally the lymph nodes; other tissues, such as brain and heart, or exudates are taken as indicated. Small portions of the tissues, removed with sterile instruments, may be dropped into tubes of culture media or rolled across the surface of solid media in plates.

A simple and often profitable procedure is the preparation of impression films from tissues. It is sometimes possible to save a considerable amount of time by an examination of impression films prepared immediately after autopsy. The proper manner of preparing films is to lift a small piece of the tissue with a pair of sterilized forceps and carefully blot the cut surface on sterile gauze. It is then momentarily pressed, gently but firmly, against the surface of a glass slide which, after fixation and staining, provides an excellent field for detection of bacteria and occasionally even typing of the cellular exudate or histopathology. When tissues are to be fixed for sectioning and microscopic examination, portions of the tissues can be placed in the fixative at the same time that cultures are prepared. If it is desirable to retain the original tissues, the material should be placed in sterile wide-mouthed bottles fitted with a plastic screw lid, and these can then be further sealed with liquid paraffin at the junction of the bottle and the lid. Material can be either stored in a 4–8°C refrigerator or frozen, depending upon the conditions.

THE DETECTION OF ANTIGEN IN TISSUES

In contrast to the virus diseases it is not usually necessary to resort to antigenic extraction from tissues for making a definitive diagnosis. It is simpler and more reliable to demonstrate microorganisms in infected tissues by cultural methods or by direct microscopic examination. However, in recent years, certain procedures have been developed whereby extraction of infected tissue and subsequent use of the extract as an antigen will provide a more rapid method. Perhaps the leading example is that described by Larson *et al.* (1951) in which the tissue of plague-infected rodents is extracted and the extract used as precipitin or complement-fixing antigen. Another, and older, example is the so-called thermal precipitin or Ascoli test for anthrax. In heavily infected animals pneumococcal polysaccharide has been extracted from visceral tissues and the extract used as antigen with antipneumococcal rabbit serum.

CONCLUSION

In the foregoing sections no attempt has been made to provide the reader with an exhaustive description of the various techniques and reactions required in the determination of bacterial pathogenicity. Each organism requires special conditions, and the available bacteriologic literature contains numerous and excellent descriptions of the specific diagnostic requirements in pathogenicity of each microorganism. For purposes of a brief summary Table 26 has been prepared to provide an over-all view of the bacteriologic tests used for the more common pathogens. In almost every instance it is necessary to supplement the animal tests with cultural or serological procedures as well. It is obvious that good techniques and skill in their application are not enough to interpret the results of animal inoculation intelligently. The operator must at all times be aware of the pitfalls of these procedures and should have a good background in pathogenic bacteriology.

REFERENCES

Dubos, R. J. 1948. "Bacterial and Mycotic Infections of Man." J. B. Lippincott Company, Philadelphia.

Haudauroy, P. 1948. "Traite de bacteriologie," Masson et Cie, Paris.

Larson, C. L., C. B. Philip, W. C. Wright, and L. E. Hughes. 1951. *J. Immunol.*, **66,** 249.

Topley, W. W. C., and G. S. Wilson. 1946. "Principles of Bacteriology and Immunity," 3d ed. The Williams & Wilkins Company, Baltimore.

CHAPTER XI

Virological Methods

ALBERT P. MCKEE

INTRODUCTORY

While the reading knowledge of the virologist is usually sufficiently broad to include the viruses attacking higher plants, animals, and man, the bacterial viruses or bacteriophages, and the rickettsiae, his area of laboratory concentration is usually more definitive. Certain fringe areas cannot be ignored by the virologist. Some similarities between enzymes and viruses exist, though the enzymes seem unable to reproduce themselves. On the other hand, pleuropneumonia organisms may be very small and filterable and possess a growth cycle, as do some of the viruses, though they can be grown apart from living cells.

To the extent that viruses and their behavior are similar to other microorganisms and their behavior, we may draw upon knowledge about a vast group of organisms to help us understand the viruses. In such instances as they differ, we must, if we are to succeed, improvise and develop new methods to serve the peculiarities of virology.

Viruses and rickettsiae may be studied along similar lines to the bacteria, though the areas of emphasis and knowledge are in some respects quite different. The conventional approaches of studying morphology, cultural characteristics, physiology, serology, and pathogenicity used in bacteriology can be applied only partially to virology. Since intensive study of virology is relatively recent, technical knowledge changes almost daily. No attempt is made in this discussion to report the very latest technics, nor will all available technics or modifications be treated.

Since viruses vary tremendously in the host cells they attack (from bacteria to man), in size (from near molecular to bacterial), and in many other respects, generalizations and analogies must be used with caution. The technics that may work well with one virus may fail miserably with another. Standardization of methodology has some advantages and some drawbacks. Standard methods not uncommonly discourage improvement, stunt the adaptation necessary when the test is applied to other than its original purpose, and encourage avidity for preciseness over principle. In virology, as in other studies, standard technics make a valuable contribution if used with awareness of the purpose for which they were designed and of their inherent limitations and as a point of departure or reference when the need arises.

It is hoped rather that beginners may find herein some basic tools with which to start. Application of each technic to specific diseases requires space provided for in books but not in leaflets or sections. No apology is offered for what must be redundant emphasis on caution and control. Viology is a science based almost predominantly on indirect methodology. One may work weeks with a virus without seeing it. Hence, caution and control are extremely important.

PRECAUTIONS

Laboratory infections are not unknown among virologists. Many of the viruses and rickettsiae isolated from man or animals are capable of infecting man even after having been cultivated in the laboratory for a long time. Perhaps the most reliable method for the novice is to train under an experienced virologist. Nevertheless there are areas encompassing greater dangers than others. Generally speaking less hazard is encountered when working with tissue cultures and embryonated eggs than with laboratory animals. The use of adequate isolation quarters for the animals and means for the worker to handle the animals without coming in direct contact with them or their products (excrement, expired air, etc.) is often mandatory while certain highly pathogenic viruses are being investigated.

Any material containing viruses or rickettsiae pathogenic for man must be treated with respect. Any maneuver which tends to create an aerosol presents a hazard. Grinding infected tissue with a mortar and pestle, disintegration of infected tissue with a tissue blender or sonic vibrator, or the simple acts of transfer or mixing with the pipet may produce dangerous aerosols. The use of special hoods and ventilating systems, where possible aerosols may be moved away from the operator and subsequently sterilized or diluted beyond danger, is important to safeguard the worker.

Large droplets caused by grinding or manipulating infected tissue may be dealt with by placing a towel soaked in 1 per cent phenol on the table top and placing receptacles and various pieces of equipment over it.

Working in a gown or laboratory coat which can be readily removed and sterilized is a worth-while practice. Washing the hands with soap and water immediately after working with infectious material or at any time the hands become contaminated aids in preventing one from contaminating himself or his environment further. A sink and liquid-soap container operated by foot pedals have much to recommend them for this purpose. Cloth face masks may contribute to negligence by inspiring a false sense of security.

Probably the greatest assets to protect one are patience and forethought. The time is well spent if the procedures to be employed are gone over mentally in an attempt to recognize danger areas and make plans to minimize or obviate them. Artificial active immunization when practical is strongly advised. It should not be used as a substitute for vigilance, however.

STAINS

Many viruses are submicroscopic in size when the light microscope is used. These viruses can be viewed by the electron microscope. Considerable knowledge and training are required to make proper preparations for the electron microscope and for the operation of this instrument. It is believed that such procedure is beyond the scope of this chapter and should be learned under the experienced investigator. Large viruses and rickettsiae as well as certain cellular inclusions caused by viruses can be observed by the light microscope when properly stained.

Preparations. In many instances success or failure of a staining procedure depends on the initial preparation of the material to be stained. Generally speaking, best results are obtained by fixing the tissue, sectioning and staining it. By the use of this method the position of the infectious agent or cellular inclusion is not disturbed in relation to the host cell. The position of the infected cells in relation to adjacent cells or tissue is also maintained. Fixation, sectioning, and staining more properly belong in the field of histology or pathology. Special equipment and experience are usually requisite to success.

Nevertheless acceptable preparations of infectious agents and cellular inclusions can be demonstrated by smears. The smear technic is quick and does not require special equipment. Tissue should be spread in a thin film over the slide, or impression smears may be made. The latter are made by sectioning an involved organ or tissue and pressing the cut surface against a slide. Only scrupulously clean slides should be used for any preparation to be stained. It is of paramount importance that the tissue is harvested at a time when the agent or inclusions should be present.

Yolk-sac smears require special preparation. A small portion of the membrane is detached, rinsed in saline, and placed on a paper towel. With a pair of fine forceps the membrane is teased across the towel to remove excess yolk. The membrane is then removed to be macerated between two slides. Making two preparations at once allows for two stains to be used, e.g., Giemsa's and Mcachiavello's. Other embryonic membranes from the egg require blotting only before making the smear.

While many stains and modifications could be recorded, it would be better if the novice learns to use these two basic ones well. Two stains which usually are found in almost every virologist's laboratory are the two mentioned above, Giemsa's and Macchiavello's.

Staining. Giemsa's stain and method are well known. The stain is available commercially, and if the manufacturer's directions are followed, good results can be obtained. This stain is used for both viruses and rickettsiae. In addition to the infectious agent being demonstrated, the host cell can be differentiated. Inasmuch as it is a polychrome stain, inclusions may be classified with regard to the reaction, i.e., basophilic, neutrophilic, or acidophilic. The position of the inclusion, whether intranuclear or intracytoplasmic, should be noted. Macchiavello's method is given in Chap. II, page 31. The stain is useful for locating rickettsiae, which appear as red rods on a blue background.

Staining for Negri bodies involves examining the proper area of the brain as well as suitable staining technic. For this reason the reader is referree to more extensive treatises dealing specifically with this problem (Francis, 1948).

Since viruses and rickettsiae are small and because they are observed against a background of complex host material, one must always be aware of artifacts. Examination of numerous preparations from normal material followed by examination of numerous preparations known to contain infectious agents is time well spent. Only in this fashion can one become somewhat proficient in recognizing significant factors in a stained preparation.

Examination of tissue containing viruses by phase microscopy holds some promise but will not be discussed here.

CULTIVATION

Viruses may be cultivated by a variety of methods all of which most probably utilize living cells. Not uncommonly only certain cells will support the increase of these rather fastidious organisms. Usually one or more of three methods is employed: susceptible animals, embryonated eggs, and tissue cultures.

Cultivation in Animals

Normal animals vary considerably as to susceptibility to various viruses and rickettsiae, depending primarily on species and age. Cortisone, X ray, vitamin-deficiency states, artificial selection, and other methods may be used to alter the natural susceptibility. It seems advisable to confine this treatise to normal animals and to suggest that those interested in special methods of altering resistance consult the original articles.

The choice of the experimental animal depends on many factors. Usually mice, guinea pigs, rabbits, monkeys, and chickens offer a sufficient species spread for propagating many viruses and rickettsiae. Generally speaking, young animals are more susceptible than older animals. Suckling animals are necessary for propagating some viral agents, and pregnancy may alter resistance to certain viral infections. The expense in procuring the animals and in feeding and caring for them as well as the available space for their housing may limit the proposed small list even further. Since a discussion on the use of laboratory animals appears in Chap. X, the material covered there will not be repeated. Working with viruses and rickettsiae, however, does require some considerations that, for obvious reasons, do not appear there.

Natural inapparent infections. The necessity of obtaining animals which are not carrying certain viruses in their tissues cannot be overemphasized. Animals plagued with any other disease are undesirable, but bacterial, parasitic, and fungal diseases among animals are usually quite apparent if one looks for them. A herd of mice carrying lymphocytic choriomeningitis or mouse pneumonitis viruses as well as other viral agents may appear healthy. It becomes very important therefore to attempt to provoke the inapparent state to become the obvious one and to check periodically for neutralizing antibodies against viruses causing any natural infections in a colony. Inapparent infections sometimes may be made apparent by serial passage of tissue from a suspected site of such an infection. For example, brains removed from several sacrificed mice are pooled and ground to a 10 per cent suspension. This suspension is inoculated intracranially into several mice. After an appropriate incubation period has been allowed, which will vary with the disease suspected, the inoculated mice are sacrificed and a suspension of their brains passed on to additional mice. Some inapparent infections become overt only after six or more passages. The same general procedure is followed in passing lung tissue, except the inoculations are made intranasally. The regular inoculation of control animals under identical conditions of the experimental ones, minus the infectious agent in the inoculum, is a procedure which must be followed religiously.

Market vs. local animals. If the investigator is procuring his mice from one breeder, the state of health may be fairly constant. If a jobber is involved who secures mice from a number of breeders, the chance of running into trouble is proportionately greater. If it is feasible, maintaining one's own colony has much to recommend it. Even when a colony is maintained locally, certain pitfalls must be guarded against. It should be remembered that wild rodents, usually mice, if given the opportunity may carry diseases to your stock. Feeds not properly stored can become contaminated with excretions from wild mice. Carrying cages returned to the colony room without first having been sterilized may bring infection to the stock animals. It becomes apparent that maintaining a colony of animals gives control over many factors, but good judgment must still be exercised in their handling.

Inoculum. In general the routes and methods of inoculation described in Chap. X suffice where viruses and rickettsiae are concerned. There are, however, some exceptions and additions that should be noted.

The investigator of viral and rickettsial diseases cannot be too careful about the inoculum to be used for the animals. Several precautions are necessary in this regard. While laboratory animals under certain conditions can resist the introduction of a few bacteria of mediocre pathogenicity or a considerable number of nonpathogenic bacteria, usually it is wise to introduce no bacteria along with the viral or rickettsial agents. Under certain conditions the animals are inoculated with an exudate containing bacteria as well as viruses, and the animals succeed in separating the virus by overcoming the bacteria. It can be readily understood, however, that bacteria may overgrow the virus, causing failure in the attempt to propagate the virus. It should become almost second nature to the investigator to culture the inoculum aerobically and anaerobically to check its bacterial sterility. Bacterially contaminated viral preparations inoculated intracranially will almost invariably result in the bacterial infection predominating.

Various exudates containing bacteria may be received in the virological laboratory, and occasionally pure cultures of viruses or rickettsiae become bacterially contaminated. There are several methods for removing the bacterial contaminants. Differential centrifugation at speeds that will sediment them but not the virus will permit the removal of many bacteria. Almost never will this method render the material free from bacteria. Filtering the specimen through a filter known to be fine enough to remove the bacteria but not so fine as to remove the viruses or rickettsiae may be the only successful method available. Where larger viruses and rickettsiae are concerned, this method may fail. Not uncommonly one sacrifices half the virus content or more because virus adheres to the filter. Filtration will not remove bacterial protoplasm which has gone into solution. For example, a bacteriologically sterile filtrate may produce toxic manifestations when introduced intranasally into mice because of soluble bacterial products rather than the accompanying virus or its products.

Antibiotics as well as other chemicals have been used to inactivate or to inhibit bacteria where these organisms contaminated a viral or rickettsial suspension. Needless to say the chemical must be one which affects only the bacteria adversely. Such a method will prepare the mixture satisfactorily for some routes of inoculation such as into the respiratory tract or intraabdominally. However, the use of an inoculum containing an appreciable quantity of some antibiotics, such as penicillin or streptomycin, may produce neurological signs and even death when inoculated intracranially. The amount and choice of the antibiotic will depend on the task it has to perform. Ordinarily 100–500 units of the antibiotic per milliliter of inoculum are employed. Usually a short incubation period is advisable to allow the antibiotic time to deal with the bacteria before inoculation. The antibiotic may be injected into the abdominal cavity of the animal prior to the intracranial inoculation and obviate toxic reactions due to the antibiotic.

While bacterial contamination may represent one of the more common hazards to the virologist, viral contamination can be more insidious and therefore more disastrous. As an example it has been demonstrated that two different types of influenza virus can be propagated together over a considerable period of time without either overcoming the other (Sugg, 1951). It takes little imagination to surmise the amount of trouble inherent in such an experience as far as research or diagnostic work is concerned. Viruses that grow under widely divergent conditions, differ in susceptibility to antibiotics, or are not related antigenically usually are separated with relative ease. Strains within a type could be impossible to separate. A situation as awkward as the latter one should serve to make the virologist "virus-contamination-conscious." In addition to all the usual procedures to avoid contamination it is advisable to work with only one strain at a time and between times to institute decontamination procedures. It takes little imagination to project mixed-strain antigen to mixed-strain antibody and utter confusion. It is very necessary to have pure virus antigen and antibody available to obviate the strain-contamination dilemma.

While the quantities of inoculum suggested in Chap. X usually suffice for the virologist, the question of concentration of virus in the inoculum requires special attention. The concentration must vary considerably depending upon the results desired. For antibody production a maximum antigenic stimulus is commonly desired. If inactivated virus is used or if the virus is inoculated by a route wherein propagation of the virus is not anticipated, a high concentration of the virus antigen, but short of a toxic dose, is desirable. This is usually necessary because of the relatively small mass of viral antigen present in the inoculum. Considering the other extreme, a relatively small concentration of virus is desirable for maximum production of viable virus. Not uncommonly one may sacrifice high-infectivity titers in his harvests if a high concentration of seed virus is used. This is particularly true if strictly fresh seed virus is not used. It is said to be the result of inactive virus interfering with the growth of active virus. Inocula of high viral concentration are more likely to produce the course of disease and pathology of toxicity rather than that of infection. It should be remembered that high dilutions of virus are prone to become inactive more readily than low dilutions. Obviously, the investigator must vary the concentration of the inoculum to suit his special purposes.

Inoculation. A few additional suggestions might be in order concerning the actual inoculation of animals. The mortality rate of intranasal inoculation attributable to the inoculation procedure, per se, increases under certain conditions. Hot humid weather, deep instead of light ether anesthesia, and an inoculum of high viscosity may all increase the mortality rate. It is especially desirable to anesthetize the animals lightly for intracranial inoculation and to inoculate as quickly as possible.

Intracranial. Before starting intracranial inoculations it is advantageous to open the skull of a deceased animal and study the dimensions and relationships of the brain to the cranial vault. Placing the inoculum in the center of the brain substance is usually desirable. The mouse is probably used for intracranial inoculation in virology more often than other animals. A 0.25-ml syringe fitted with a sharp 27-gauge needle aids in injecting and controlling the small volumes required, 0.01–0.03 ml.

While the operator is seated at a desk, the mouse is grasped by the dorsal skin behind the head, using the thumb and forefinger, and pressed gently against a firm surface. The hand holding the syringe is steadied on the desk and rocked slowly forward to make the inoculation. It is wise to hold the needle *in situ* for a few seconds. Usually less inoculum follows the needle when you withdraw it if this procedure is followed. Larger animals may require trephining as described in Chap. X.

Intravenous. Several factors aid in making successful intravenous inoculations in mice. Three- to four-week-old mice are easier to inoculate than older mice. If the mice are placed in a 37°C incubator ½ hr before inoculation, the tail-vein dilation makes the task easier. A 0.25-ml syringe fitted with a sharp 27-gauge needle is ideally suited for the procedure. The distal one-half of the tail vein is more readily entered than the proximal one-half. The tail, needle, and syringe should be held so the approach is one of putting a stilette in a needle. Trying to enter the vein at an angle will usually fail. If the distal end of the tail is bent over the forefinger and secured with the thumb, the remainder of the tail vein can be approached as described above.

Intratesticular. Intratesticular inoculation is used to propagate some viruses. Since this method is not covered in Chap. X, it will be described briefly. In order that the progress of infection may be adequately observed, it is desirable to remove any hair covering the scrotum. This may be accomplished by clipping or by the use of a depilatory. The scrotum of the animal should be cleansed and treated with a topical disinfectant. The inoculation is made by securing a testis between the thumb and forefinger and introducing the needle to approximately the center of the testis before expelling the inoculum. For small animals 0.05 ml and for larger ones up to 0.2 ml is customarily used. A 27-gauge needle is used for small animals, and a 25-gauge for larger ones.

Postinoculation observation. Precautions must be taken to prevent cross infection among experimental animals. Such an occurrence is more likely with some viral diseases than others. Suffice it to say that the hands should be washed when going from one experiment to another and proper precautions in regard to feed, water, and ventilation must be observed to obviate the problem. Controlling aerosol hazards requires considerable vigilance (Smadel, 1951; Reitman *et al.*, 1954).

After having inoculated the animals by whatever method, you should observe them daily under good light after they have settled down from the initial disturbance of your first presence in the room. Nocturnal animals tend to appear drowsy during the daytime and by the novice may be pronounced ill. Respiration may appear labored during this time. Only by observing normal experimental animals over a period of time can one readily select the initially ill from the healthy. Daily observation is required for some viral and rickettsial infections in laboratory animals because the signs are transient and would be overlooked otherwise. In this regard it is particularly important to take daily rectal temperatures on some animals, e.g., guinea pigs used for rickettsial study. Guinea-pig temperatures vary considerably, but a fever may be considered present if the temperature definitely exceeds 39.6°C.

A certain amount of differentiation among certain rickettsial diseases is possible by careful examination of the testes after intraabdominal inoculation. The testes and scrotum are examined daily for evidences of edema, redness, necrosis, and the reducibility of the testes into the abdomen. The time these changes make their appearance should be recorded.

Some viruses can produce encephalitis or paralysis following intracranial inoculation. Not uncommonly a hyperkinetic state will be observed preceding lethargy or paralysis. Convulsions can be provoked by rotating the mouse to and fro by twisting his tail between thumb and forefinger while he is suspended in air. Tics may also develop in the animal. Since mice are commonly infested with mites, they scratch often. A tic is a definite repeated action pattern of involuntary nature and is best determined by comparing the experimental animals with appropriate controls. Quite commonly experimental animals die within 24 hr after exhibiting the first neurological signs.

Intracutaneous inoculations produce a variety of lesions. These should be described accurately as to character and time of appearance. Erythema, edema, size, configuration, and differentiation should be determined. Is the lesion macular, papular, vesicular, pustular, or ulcerative? What is the duration of the lesion and each successive change? Does it heal with or without scarring? It is well to measure the lesion. Histological study of the involved areas at various time intervals is advisable in some instances.

A given property of a virus does not necessarily correlate with another property of that virus. To assume that a certain quantitative increase in any one property, for example, hemagglutinin potency, is a direct indication of other functions, such as infectivity titer, is not only unwarranted but may lead to considerable error. Under certain standardized conditions direct relationships may hold, but too much faith should not be extended in this direction. The time the virus is harvested may influence the various propensities of the virus, for it is known that some functions may increase before others.

It is not particularly difficult to follow the increase of a well-adapted virus in animals. Following the progress of a freshly isolated strain or determining if a virus has been isolated may be very difficult. An appropriate number of days may elapse after the inoculation of a laboratory animal with material suspected of containing a virus without any evidence of infection. The experienced investigator does not accept this as a failure but harvests the tissue that should be involved and prepares it for passage to new animals. This procedure has been referred to variously as "passage by faith" and "blind passage." The object is to cause an inapparent infection to become an apparent one. After several passages one concludes that he has failed or is rewarded by noting some evidence of infection. Serial passage of some tissues, for example lungs, at too short an interval encourages bacterial pneumonia. If a 4-day interval is used, this complication usually does not occur. Too long an interval between passages encourages antibody production and results in defeat for this reason.

Inapparent infections may be made obvious sometimes by other methods. Although an infection may not have progressed to the state of recognition, it may have induced antibody formation. The tissue or organ used for serial passage may also be examined for the presence of specific antigen. The appearance of this antigen after several passages should encourage continued passage.

The repeated passage of viruses or rickettsiae through experimental animals usually leads to some degree of adaptation. This adaptation may be of little consequence or may be significant, depending upon the object of the investigator or the conclusions he draws. One should assume the possibility of antigenic change in a well-adapted virus and check it against the same strain stored shortly after its isolation. Such antigenic changes have been demonstrated (Hartman *et al.*, 1954), and viruses have been deliberately adapted to altered environmental conditions (Jones, 1945). Since viruses and rickettsiae are quite labile in many of their characteristics, carefully controlled experiments cannot be overemphasized.

Animal autopsy. Usually it is advisable to autopsy the animal when the first definite signs of an infection are present. If this procedure is neglected, the animal may die after working hours and the tissues become unsatisfactory for pathological study. The bacterial content of the tissue harvested for virus may rise sharply if the animals are not autopsied shortly before death.

All the usual precautions are observed when autopsying the laboratory animals. It is a good practice to wet down the fur or hair of the animal with a disinfectant such as a solution containing 1 per cent phenol. The object of the autopsy may vary considerably—from harvesting a specific organ for obtaining a quantity of virus to performing a rather extensive examination to learn something of the nature of the infection. It is a common practice to determine the mass of the tissue excised either gravimetrically or volumetrically so that some reference point for quantitative work pertaining to the virus can be established. In practice, if one uses the same kind of animal of the same age, an average weight or volume is accepted to save time. For example, the average weight of the lungs from 4-week-old mice will be very close to 0.25 g. An extensive and solidifying infection, of course, may vary the weight somewhat.

In addition to observing the tissue for signs of gross pathology the investigator may be interested in searching for cellular inclusions. The tissue may be sectioned, and the freshly cut surface smeared on a slide and stained, or the more precise histological methods of fixation, embedding, and sectioning may be used. The smear method is more likely to rupture cells and distort relationships. Since many of the viruses are not sufficiently large to be seen by the ordinary microscope and fail to form cellular inclusions, other methods are necessary to determine the presence and extent of viral propagation. When a well-adapted virus is employed to induce an infection, the involved tissue can be serially diluted to determine its potency in establishing an infection, in causing agglutination of a constant concentration and volume of appropriate red cells, or in functioning as a specific antigen in a satisfactory serological test. The production of pathology in a given tissue need not necessarily indicate the increase or growth of a virus in that tissue because some viruses, in sufficient concentration, are toxic even though there is little or no evidence of increase.

Harvesting tissue. Chapter X discusses general procedures for autopsy. Some departures from the general methods are peculiarly adapted to the virologist. Some viruses are propagated in the lungs for a supply of antigen. A considerable number of lungs are desired at times if mice have been used. In the interest of speed considerable time may be saved if the lungs are removed from the back as follows. The sacrificed mouse is pinned to the autopsy board in a prone position. After the dorsal fur has been wet with 70 per cent ethanol, the skin over the backbone is pulled up with forceps. A V-shaped cut is made with scissors running from the middle of the backbone up to the neck. The backbone is secured with forceps just distal to the bases of the lungs. Three more cuts with the scissors expose the lungs: one to sever the vertebral column, a second anterolaterally from the severed column to the left axilla, a third to the right axilla.

Obtaining spinal cord from mice may be tedious or easy depending on the method employed. The mouse is secured as above. A skin flap is removed from over the vertebral column. The entire column is removed by severing it from ribs and muscle attachments. With the column held vertically, a stiff wire of such size as to just slip through the canal is used to rod out the cord.

If the tissues are to be used for cutting sections, care must be exercised to avoid damaging the tissues. Two common errors are (1) to submit tissue crushed by forceps and (2) failure to fix the tissue before autolysis has progressed to a significant degree.

If a suspension or extract of the tissues is desired, they can usually be disintegrated by grinding with mortar and pestle. Sterile powdered glass or other suitable abrasive aids the grinding process where fibrous tissues are encountered. The abrasive should be removed by low-speed centrifugation before the suspension for inoculum is used.

Cultivation in Chicken Embryos

Supply Source. Embryonated eggs have become one of the most useful culture media for virologists. While eggs from various fowl have been used, chicken eggs, because of the availability and relatively low cost, are most popular. As with animals the source of supply is important. Usually the embryonated eggs can be supplied by a local hatchery or by contract with owners of farm flocks. Clean, white-shelled eggs of high fertility are desirable. Usually the investigator has little control over the feed used and cleanliness of quarters where the eggs are produced. There is some reason to believe that the more or less regular use of antibiotics in many chicken feeds may interfere with the growth of rickettsiae and some viruses (Greiff and Pinkerton, 1951). Obtaining embryonated eggs from a flock on an antibiotic-free diet may be difficult but necessary for some types of experimental and diagnostic work.

Preinoculation incubation. Fertile eggs set at 37.5°C develop satisfactorily for most viral and rickettsial work. The setting date should be recorded on the eggs or elsewhere so the embryo age can be determined as required. While it is preferable to set the eggs soon after they are received, this may be delayed for a day or two. If they are held, they should be stored at a low temperature, 60–65°F, so as to minimize the drop in fertility. Hatching eggs properly handled should yield 70 per cent fertility or better. Proper attention should be paid to humidity and air circulation. Regular hatching incubators with forced-air circulation and dual thermostats usually are satisfactory. Dual thermostats so arranged that the spare is cut in when the other fails and a warning light or buzzer comes on will save several days of lost time, not to mention the investment in eggs. Moisture content of the air should be maintained at 90 per cent saturation.

Candling. After the eggs have been incubated for the required length of time, which ordinarily will vary from 5–10 days, they are candled. Five- to 7-day-old embryos are frequently employed when yolk-sac inoculations are used, whereas when other areas are injected older embryos, up to 10 days, are selected. An inexpensive commercial candler or any apparatus causing a strong light to go through the egg without scattering to blind the operator is satisfactory. If candling is carried on with the proper equipment in a dark room, the air sac, embryo, and some of the blood vessels can be recognized easily, especially when white-shelled eggs are used. Usable eggs are chosen on the basis of the distinct blood vessels and active embryos. For the beginner it is well to check his discards and some of those he considers acceptable by breaking open the shell and observing the actual situation which he has characterized indirectly.

Marking. Time can be saved by marking the eggs as they are candled. Rotating the egg on the candler with the pencil following the air-sac periphery outlines this area. A mark may be placed over the spot where the embryo shadow is most visible. An additional mark can be made at the intended point of inoculation for the allantoic sac.

Windowing and Inoculation. If it is desired to inoculate the amniotic sac or embryo under direct observation, or if the inoculum is to be applied directly to the chorioallantoic membrane, the embryos should be windowed. Any small motor tool that can be fitted with an emery disk may be used for preparing the window in the shell. An area of about 1.5 cm square is marked off approximately equidistant from each end of the egg and on the embryo side. The square and slightly beyond is disinfected by painting it with tincture of iodine. With the use of the grinding tool the shell is ground through along the pencil line, care being taken not to penetrate to the chorioallantois. A small slit is ground over the air sac so the air can escape when the membrane is to be dropped.

After the slit and square have been cut, the remainder of the work may be done under a glass hood to reduce contamination from air-borne bacteria. First the shell is removed with sterile forceps. Then the shell membrane is wet with a 70 per cent ethanol solution containing 0.1 per cent thymol. This membrane is pierced carefully with a sterile lance so as not to injure the underlying chorioallantois. With the vacuum broken, the contents of the egg settle by gravity, displacing the air from the normal air sac. With sterile forceps the shell membrane is then removed, leaving a square aperture through which the embryo and embryonic membranes can be observed. The shell surrounding the hole is again disinfected by carefully painting it with a swab partially soaked with tincture of iodine. One must be cautious not to allow the iodine solution to enter the embryo. Finally a small square of transparent gummed tape is stuck on the egg to cover the hole.

Dropping the chorioallantois successfully requires considerable practice. If the membranes are torn, the embryo usually has lost its usefulness. Any unnecessary manipulation is likely to cause hemorrhage of or nonspecific lesions on the membrane. Windowed eggs require considerable time for preparation but allow the worker subsequently to inspect the results of his inoculations under direct observation and permit definitive inoculation of specific sites.

Amniotic sac. Using a 27-gauge 12-in. side-delivery needle, it is possible to inoculate the amniotic sac without windowing the egg. The embryo is prepared as for allantoic sac inoculation with one hole made directly over the embryo. Introducing the needle directly toward the embryo, which is visualized by the aid of a candler, one can be successful in about 90 per cent of the attempts. It is advisable to practice several times, using a methylene blue solution to determine one's aptness. One-tenth-milliliter inoculum may be used on the membrane, and 0.05 ml in the amniotic sac. More or less may be used as the experiment dictates.

Yolk sac. Yolk-sac inoculation is commonly used for the propagation of rickettsiae and certain viruses, e.g., those of the meningopneumonitis group. A single hole drilled over the air sac is adequate. Usually a large-bore needle is used for the injection. A 1½-in. 20-gauge needle is suitable. With the egg in a vertical position, large end up, the needle is pushed through the hole and straight downward. To check the accuracy of position a slight pull on the syringe plunger causes yolk to appear in the syringe. The inoculum can then be discharged.

Allantoic sac. When considerable numbers of embryos are to be inoculated, e.g., in determining infectivity titers, the allantoic sac may be inoculated quickly and simply. Two holes are punched in the egg, one over the air sac and one slightly eccentric from the embryo and toward the blunt end. A ½-in. 27-gauge needle is introduced at an acute angle from the surface of the shell. This procedure avoids injuring the embryo and usually places the inoculum in the allantoic sac. Practice using methylene blue will permit you to appraise your technic. Drilling a small hole with a 27-gauge needle and drilling the hole over the air sac both contribute to prevent leak-back of the inoculum. One-tenth milliliter of inoculum is commonly used for the allantoic sac. With proper care large amounts up to 2.0 ml may be used.

Inoculum. The preparation of inoculum for embryonated eggs is important and frequently requires considerable judgment. Inoculating too much virus may encourage lower titers than anticipated. This is especially true if the virus is not from a fresh harvest. It is held that inactive viral particles compete for cell receptors but do not grow. If this is true, one can see why it would be wiser to dilute the inoculum considerably. Too great a dilution may take the virus concentration beyond a minimal infective dose. Until the peculiarities of a particular strain are known, it is probably best to employ at least two dilutions, e.g., 10^{-1} and 10^{-3}.

Diluent. Dilute virus may be quite unstable. For this reason it is usually desirable to use a menstruum protective for the virus such as 10 per cent serum in saline. One must never cease to consider the menstruum as often as he considers the virus suspended therein. For example, antibodies and inhibitors may be present in any serum or body fluid. Buffered salt solutions are usually preferable to ordinary saline. Some viruses are precipitated by phosphates. Calcium ions are necessary for the receptor-destroying enzyme to function properly; citrate, however, interferes with the complement-fixation test. Equal parts of hormone broth and saline have served well as diluting fluids in some virological laboratories, although this diluent should not be used in the hemagglutination test. No matter which diluting fluid is employed in infectivity titrations, the inoculations had best be done quickly starting with the most dilute viral preparation.

Tissue Preparation. Material obtained from infectious diseases frequently presents several problems when it is prepared for inoculation into embryos. Since viruses grow in intimate contact with body cells, it may be advisable to disintegrate the host cells by grinding the exudate or tissue with an abrasive. Freezing and thawing, sonic vibration, or disintegration in an homogenizer or blender may free a greater number of viral particles. The use of antibiotics may circumvent the problem of bacterial contamination in the embryonated egg. It must be remembered, however, that certain viruses are also susceptible to antibiotics. Filtration through filters of proper porosity to withhold bacteria but not viruses would seem ideal. Some viruses are held back by bacterial filters, and organisms of the pleuropneumonia group pass these filters quite readily. Some of the latter organisms grow quite well in the chick embryo. Not uncommonly it is wise to treat material containing an unknown virus by more than one method prior to inoculation. Penicillin and streptomycin in concentrations of 100–500 units per milliliter are frequently employed antibiotics used to discourage bacterial growth.

Postinoculation incubation. After the inocula have been appropriately prepared and the eggs inoculated, they are ready for sealing, marking, and incubation. Small apertures such as punch holes may be sealed by swabbing with sterile melted paraffin or nail polish. A small piece of gummed tape may be used for this or for closing windowed eggs. Windowed eggs are best incubated in a rack which does not allow the window to deviate from its upright position lest the contents be spilled or contaminated. Punched eggs may be incubated in almost any position. A convenient rack for handling punched eggs can be made simply and for low cost. A double row of 1½-in. holes drilled 1¼ in. deep and 2 in. on center in standard 2- by 4-in. lumber prepares such a rack. Material adaptable to repeated autoclaving is best where highly infectious material is used. Eggs may be placed in the holes small end down. Scrubbing with iodine, punching, inoculating, sealing, labeling, and incubating may all be done without removing the embryos from such a rack.

Incubation temperature. The incubation temperature employed will vary depending upon the virus to be propagated. Most commonly this range is small, 35–37°C. For well-adapted viruses altering the temperature from an optimum of 37 to 35°C results in less growth but not a complete failure. Temperature for isolation may be more critical wherein one may fail sometimes at 37°C and succeed at 35°C. It is not uncommon to find experienced virologists employing two temperatures for initial isolation of a virus. Some incubators vary in temperature from one shelf to another. It is well to keep a thermometer on each shelf and examine them daily. Temperatures may vary with line load, with a full or an empty incubator, and for other reasons. Generally speaking, incubation temperatures are more critical for viruses than for bacteria and must be controlled carefully.

Periodic inspection. The time of incubation will vary with the agent under investigation. Previously described well-adapted viruses have the time of incubation well worked out. This may vary from 1 to 7 days for different viruses and will vary also depending on what the investigator desires. Unidentified viruses must be handled differently. If the inoculation is on the chorioallantoic membrane of a windowed egg, the membrane can be examined daily for lesions by carefully peeling back the gummed tape. An ophthalmoscope may aid in differentiating shell dust from opaque lesions. Punched eggs can be candled daily or twice daily to learn the viability of the embryo. Some viruses kill the embryo, while others do not. Control eggs of the same age (uninoculated or inoculated with the same diluent used for the experimental eggs) should be included. Examination of these controls will help interpret findings among the experimental eggs. When possible a sufficient number of eggs should be inoculated to allow one to sample and pass a certain number each day. Under such circumstances it is possible to determine the gross state of the embryo and membranes as well as test for hemagglutinin and complement-fixing antigen. Determining the appropriate time interval and the material to be passed on to new embryos requires experience and judgment. Any significant positive finding should suggest passing the material involved. Pooling membranes and embryo may be tried in the absence of a significant finding. Each fluid passage should be accompanied by culturing for bacteria and, when indicated, pleuropneumonia organisms. Smears should be made of the membranes, stained with Giemsa's and Macchiavello's stains, and a search made for larger viruses, rickettsiae, and inclusion bodies. It is also advisable to use a saline extract of the pooled materials as antigen in the complement-fixation test against known antibodies and patients' acute and convalescent sera. Where sufficient antibody and antigen are available, a checkerboard complement-fixation test is advisable to avoid the Neisser-Wechsberg phenomenon (see Chap. IX).

Harvesting. The procedure for harvesting the fluids and membranes from embryonated eggs will vary with the method of inoculation and the information the investigator wishes to obtain. It is easy to rupture a vessel during harvest; therefore, it is a common procedure to expose embryos to 4°C overnight to congeal the blood, thus avoiding hemorrhage into the fluid one expects to harvest. Allantoic fluid harvest from punched eggs can be accomplished quite easily. The shell over the air sac is swabbed with tincture of iodine. After the solution has dried, the shell is broken and removed, exposing the sclerotic membrane under the air sac. The membrane may be clarified with a few drops of 95 per cent ethanol or stripped away with sterile forceps prior to harvesting the fluid. It is well to examine each egg before harvesting. The odor and appearance of the eggs contents have definite meaning to the experienced investigator. A number of bacteria and molds growing in the egg give off characteristic odors. *Pseudomonas aeruginosa* commonly produces its soluble green pigment to alert one to its presence. Death of the embryo and disintegration of the membranes where they are not anticipated results should be viewed with suspicion. Sometimes only a few eggs out of several inoculated will show the undesirable signs and can be discarded, thus saving the bulk of the harvest. A 20-gauge needle fitted to a 5- to 10-ml syringe works well for removing the fluid. The needle is inserted into one of the larger pockets of fluid close to the shell, and with a gentle suction the fluid is removed. If suction is applied too rapidly, the membranes are pulled to the needle opening and block the aspiration procedure. Depressing the membranes with a sterile spatula helps obviate the latter problem.

Allantoic fluid can be aspirated with ease from the windowed eggs because a considerable area of the membranes and embryos is under direct vision. A good light source appropriately directed and close to one's work is invaluable for any type of harvest. From 5 to 10 ml of allantoic fluid can be drawn from the allantoic sac, depending on the age of the embryos and the agent under propagation.

Amniotic fluid. Withdrawing amniotic fluid requires more technical skill than other methods of harvest, especially if it is desired to keep it free from the other fluids. Perhaps one of the more certain methods involves preliminary removal of as much allantoic fluid as possible, followed by careful evacuation of the eggs' contents into a sterile petri dish. Care is exercised not to break the allantoic and yolk sacs and to have the embryo end up on the top side. While one is seated at a table to steady one's self, a 22-gauge 1½-in. needle, fitted to a 1.0-ml tuberculin syringe, is used to withdraw the fluid. By careful observation a pocket of fluid somewhere around the embryo is located. With the hand holding the syringe steadied on the table, the needle is carefully inserted into the pool of fluid. The other hand is held in readiness to draw back on the plunger of the syringe. Success comes to those with sharp eyes and steady hands. About 0.1 ml of air can be injected to verify the position if desired once the operator feels he has the needle in the amniotic sac. The bubble will be trapped in the sac surrounding the embryo and not move out into the large allantoic sac. From 0.1 to 2.0 ml may be drawn from the amniotic sac, depending on many variables.

Yolk fluid. Yolk fluid is sometimes harvested, although the membrane is probably more commonly collected. Fluid can be withdrawn if a large-bore needle, e.g., 18–20-gauge, is used. Following some infections the yolk granules are quite fluid. Other infections do not produce this effect, and the yolk, particularly in older embryos, is thick and viscid and therefore difficult to withdraw.

Membranes. If viruses are propagated on the membrane, the egg has probably been windowed. When the chorioallantoic membrane is harvested, the gummed tape is removed and the shell broken away down to the membrane. Care is used to avoid shell fragments falling on the membrane. Under good light the membrane is grasped in an area away from the better lesions by sterile forceps. With sterile small scissors (curved nail scissors are good) the membrane is cut loose at the periphery. If the operator is steady and gentle, the membrane disk can be removed intact. The membrane can then be floated in sterile saline in a petri dish. If the dish is held over a dark background and at the edge of a desk-lamp shade, lesions are usually visible. A better method is to view the membrane under a dissecting microscope by both transmitted and reflected light. It is well to be alert for several changes in the membrane. The amount of edema, presence of lesions, perivascular changes, condition of the blood vessels, hemorrhage, and membrane luster are some of the things one should look for. The beginner could profitably spend considerable time examining normal membranes and membranes inoculated with various menstrua containing antibiotics, sterile tissue, and other substances in order to become well acquainted with a range of nonspecific changes.

Following allantoic, amniotic, and yolk-sac inoculations, it may be necessary to harvest these membranes as well as the fluids. If the contents of the egg are evacuated into a sterile petri dish, individual membranes may be pulled away from the remaining material. With a little practice in learning where to grasp the membranes, whole membranes can be disengaged readily rather than picking the membrane out by pieces. If one wishes to remove as much contaminating fluid as possible, the membranes can be rinsed quickly in several changes of cold sterile saline. Any of the above membranes may be ground with an abrasive and saline to prepare a suspension for passing. The suspension should be cultured for bacterial sterility. Stained smears may be made from them, or they may be fixed, sectioned, and stained for observation.

Embryo. If the embryo is to be collected, it may be removed after opening the amniotic sac. If the embryo is to be removed intact, care must be exercised, for it is very fragile.

Determining infectivity titer. Certain investigative problems and diagnostic procedures require the worker to have some concept as to the potency of his viral preparation. Infectivity titers of some viruses can be determined in embryonated eggs. The viral preparation is serially diluted in a protective diluent as mentioned above. Using a separate pipet for each of the preliminary dilutions 10-fold decreases are usually employed, since finer increments not uncommonly result in a scattered end point. Several factors may influence the end point. As mentioned earlier, freshly harvested virus gives a more reliable end point. Preparations with very high titers require careful treatment. If the anticipated titer is high, it is advisable to dilute into the proper range by larger increments such as 100-fold decreases. For example, if the anticipated titer is 10^{-9}, one could proceed as follows: Dilute the preparation 10^{-2}, 10^{-4}, and 10^{-6} using three 100-fold serial decreases and a separate pipet for each dilution. The same pipet and 10-fold serial decreases could be used to prepare 10^{-7}, 10^{-8}, 10^{-9}, 10^{-10}, and 10^{-11} dilutions. A 10-fold increase in titer is not uncommon if either of the above suggestions are not followed. Whatever method is used, it is advisable to use it consistently, especially if one wishes comparable results.

At least 5 and preferably 10 embryos should be inoculated with an equal volume from a single dilution. If the worker starts with the least concentrated sample and progresses toward the most concentrated, a single syringe may be used throughout. In certain instances it may be desirable to use a separate syringe for each dilution in order to avoid contamination. Any investigator does well to analyze his technic in handling syringes, diluting virus, etc., so as to minimize the probability of his introducing viral or bacterial contamination into his experiment.

Embryos receiving the most dilute preparation are best harvested first, progressing to those receiving the least dilute. If one is interested in minimal infective amounts, material from embryos receiving the same dilution may be pooled. If a 50 per cent egg infective dose (E.I.D. 50) is desired, each embryo harvest is tested separately and the 50 per cent end point calculated mathematically. Methods for calculating 50 per cent end points are discussed below. It should be remembered, however, that regardless of the method used, there is no substitute for employing significant numbers in any experiment. Nor does mathematical preciseness overcome an inherent wide biological variation.

Obviously the infectivity end points will be determined by several methods because viruses vary in what they can do. Some will produce lesions, others kill embryos, still others agglutinate red cells, and probably all produce a certain amount of detectable antigen. Any of these methods can be used to determine the results of extinction dilution. In some instances a combination may be employed, such as dilution factor and lesion count.

In any experiment it is well not to lose sight of other properties of the virus when testing its infectivity. Under standard conditions the infectivity might be directly proportional to other functions, yet these proportions are easily upset, and too much should not be left to presumption. For example, under standard conditions a ratio of $6{,}400:10^{-9}$ may exist between hemagglutinin and infectivity. Yet because the stability of hemagglutinin and infectivity may differ considerably, the ratio may be changed to $6{,}400:10^{-7}$ if the specimen is allowed to stand at the proper temperature. It follows that ratios of other functions should be watched with equal care.

Eggs can be deembryonated and used for propagating viruses. For this technic the reader is advised to consult Bernkopf (1950).

***Determination of the* 50 *per cent infectious dose for embryos* (*E.I.D.* 50).** It is generally agreed that the 50 per cent infective dose is less subject to extremes than the minimal infective dose. Reed and Muench (1938) have worked out a method of obtaining such an end point which is in wide use. The method is illustrated in Table 27 on an actual protocol wherein 10 embryos were inoculated in each dilution group.

TABLE 27

Reciprocal of log dilution	Infected embryos	Uninfected embryos	Cumulative infected	Cumulative noninfected	Fraction infected	Per cent infected
6	10	0	34	0	34/34	100.0
7	8	2	24	2	24/26	92.3
8	6	4	16	6	16/22	72.7]
9	5	5	10	11	10/21	47.6]
10	3	7	5	18	5/23	21.7
11	2	8	2	26	2/28	7.1
12	0	9	0	35	0/35	0

The 50 per cent end point lies between 10^{-8} and 10^{-9}. The proposed formula for determining the proportionate distance is: Number above 50 per cent or $72.7 - 50 = 22.7$, divided by the number above 50 per cent or $72.7 -$ the number below 50 per cent or $47.6 = 25.1$. The fraction $22.7/25.1 = 0.904$, hence the 50 per cent end point is $10^{-8.904}$ dilution-wise, or the specimen contains $10^{8.904}$ E.I.D. 50. The cumulative totals are obtained on the presumption that eggs infected with 10^{-11} dilution of virus would also have been infected with 10^{-10}, 10^{-9}, and so on. Uninfected eggs are accumulated in reverse order.

Several comments are in order concerning this method. It is convenient and therefore has become widely used. One obtains a false sense of preciseness with a figure carried out to the third decimal. Although from the cumulative totals it appears that 189 eggs were used in this test, actually only 70 were used. One could question if $10^{-8.901}$ is a more accurate end point than 10^{-9} wherein actually 50 per cent of the embryos were infected. Even using 10 eggs per dilution group it is difficult to reproduce results consistently within 0.5 log units. This in itself is not a criticism of the calculation procedure but a commentary to remind one that mathematical preciseness does not alter an unprecise technic.

Needless to say the same method may be used for estimating a 50 per cent infectious dose in animals and in determining neutralization titers of antisera wherein log dilutions of the serum have been mixed with a constant dose of challenge virus. If the reverse approach is desirable, the serum may be held constant and the virus diluted by increments of 10. The difference in infectivity titer between that in the presence of the antiserum and that of the control can be attributed to the neutralizing properties of the serum.

Other methods for estimating the 50 per cent end point are also available (Finney, 1947; Irwin and Chaseman, 1939).

Cultivation in Tissue Cultures

Uses. Many of the known viruses and rickettsiae have been propagated in tissue cultures. This method of culture has had a rather wide adaptation for the virologist, including isolation, identification, study of reproduction cycles, and cellular pathology. Large quantities of virus have been cultivated to be used for vaccines and for other reasons. Antibody neutralization of virus and screening of various chemicals including antibiotics for their effects on viruses and rickettsiae may be carried out in tissue culture.

Materials. Tissue-culture methods change at such a rapid rate and vary so much in complexity that it would be advisable for the beginner to start with a simple method. After he has acquired some basic experience with this method, various original works and texts can be consulted to extend his proficiency. Basically the ingredients of a tissue culture are tissue and extracellular fluid. Tissues used will vary considerably with the virus to be propagated. Embryonic tissue usually suffices quite well, although testicular tissue and propagated cells of various kinds may also be used.

Chick embryo, because of its availability, is often used. The air-sac ends of 10- to 12-day-old embryonated eggs are painted with tincture of iodine, and the solution allowed to dry. Sterile forceps are used to break and pull away the shell over the air sac. A different pair of forceps is used to strip away the shell membrane, and the contents of the egg are poured out into a sterile petri dish. The embryos are then collected in a sterile receptacle prior to mincing. After the eyes of the embryos are removed, the remainders are placed on a firm cutting surface such as bakelite. With a sharp blade, e.g., razor blade, the embryos are cut into tissue fragments approximately 1 mm square. The fragments are transferred to a sterile receptacle, e.g., petri dish or watch glass, for washing with salt solution containing no sodium bicarbonate. Three washings usually suffice for removal of most of the blood cells. Certain tissue, e.g., embryonic lung, can be selected and used exclusively for the culture.

The size of the flask or container used will determine roughly the number of tissue fragments used. Tissue from each embryo can be suspended in 3.0 ml of salt solution to facilitate handling. If 4.5 ml of balanced salt mixture are deposited in each 50-ml flask, 0.25 ml of embryo suspension will give about the right proportion. If a large-bore pipet is used and the embryo suspension shaken just before use, one can withdraw a representative specimen.

Several salt-solution formulas have been used successfully. The formula for Simms and Sanders' (1942) salt solution follows:

Solution A	*G. per liter*
Sodium chloride (NaCl)	160.00
Potassium chloride (KCl)	4.00
Calcium chloride ($CaCl_2 \cdot 2H_2O$)	2.94
Magnesium chloride ($MgCl_2 \cdot 6H_2O$)	4.06
Solution B	
Sodium bicarbonate ($NaHCO_3$)	20.20
Disodium phosphate (Na_2HPO_4)	4.26
Dextrose	20.00
Phenol red	1.00

Solution A is sterilized by autoclaving; solution B by filtration. Before use 50 ml of solution A and 50 ml of solution B are added to 900 ml of sterile, glass-distilled water. Stock solution B is best refrigerated until needed. Reagent chemicals should be used throughout.

Too much dilution of the culture by the addition of inoculum is undesirable. Approximately a 1:10 dilution is satisfactory. If 0.5 ml of inoculum is added to the culture described above, this will be achieved.

Incubation. Temperature and time of incubation will vary with the virus employed. From 35 to 37°C usually suffices for most viruses. The more rapidly growing viruses will produce satisfactory yields in 2 days. Slower growing viruses require longer.

Transfer. The virus may be serially transferred to new media by pipetting 0.5-ml quantity of the culture. Preparation, inoculations, harvests, and transfers can best be done in a cubicle to avoid air contamination. Antibiotics in subtoxic concentrations may be used judiciously in some cases to help combat bacteria. Bacteria should be excluded, if possible, rather than killed or inhibited after admission.

Practically all the precautions urged under the discussion of embryonated egg propagation of viruses are urged here. Inoculated controls held at 4°C and uninoculated controls held at the experimental temperature should be employed. If there is no propagation in the former and no evidence of virus in the latter, the worker may feel secure. One still has to prove growth in the experimental cultures, however. This proof may be accomplished by checking an increase in some reliable function of the virus or by extinction dilution. In the latter case, if the virus is serially diluted from one culture to the next for a sufficient number of times, without propagation, the virus will disappear. The two methods can be combined by noting the titer of some function upon each serial transfer. If it remains constant or increases, propagation must be taking place. Functional end points vary with the virus under study. In addition to those mentioned under embryonated egg culture, cytopathogenic effect may be followed. This will apply wherein the virus can produce inclusion bodies or cellular changes in the host cells. Needless to say the cells from the control cultures should have their histology scrutinized under the same conditions as the experimental. Every harvest should be checked for bacterial sterility.

HEMAGGLUTINATION

A number of viruses will agglutinate red blood cells. Since a given amount of a given concentration of red cells will be agglutinated under standard conditions by a minimal amount of virus, it is possible to titrate this function. Although the hemagglutination test has many modifications, one of the simplest to perform and most likely to give reproducible results in the hands of the inexperienced worker is the pattern test (Salk, 1944).

Materials.

1. Serology rack capable of holding 20 Wassermann-size tubes. A piece of white paper may be placed on top of the bottom of the rack, or better still, the metal bottom may be replaced by a transparent material such as a ⅛-in.-thick Lucite slab. The latter rack is then set on a white background or on frosted glass with a light source beneath.
2. Wassermann tubes that have been carefully cleaned and rinsed free from any cleaning reagent. They should be clear and dry.
3. Saline solution made from redistilled water and containing 0.85 per cent of sodium chloride may be used as a viral diluent. Under special circumstances a phosphate buffer of 0.1 *M* and pH 7.0 may be used. This is particularly true if the saline solution is allowed to become very acid while in use.
4. Red blood cells from adult roosters or O-type red cells from man serve well, although erythrocytes from some other animals may be used. Where chickens are used, blood may be collected from the wing or jugular veins in citrate solution. After the blood is centrifuged, the plasma and buffy coat are removed and the red cells washed three times. During the last centrifugation at 1,200 rpm for 10 min a calibrated centrifuge tube may be used in order to make the packed cells into a 3 per cent suspension in physiologic saline. The suspension

should be checked for sterility and stored at 4°C. Contaminated batches should be discarded. Before use, the 3 per cent suspension is diluted 1:12, giving a 0.25 per cent suspension.

5. Virus suspension. Sharp end points are most often obtained from virus suspensions that are free from macroscopic particulate matter. Usually a brief period of centrifugation will accomplish this.

Methods. The method of performing the test will vary with the virus tested and the menstruum suspending the virus. For illustrative purposes chorioallantoic fluid containing influenza virus will be discussed.

Fluids having a low hemagglutinin titer or harvested from older embryos may possess sufficient inhibitor to obscure sharp end points. In such instances it may be necessary to treat the preparation with receptor-destroying enzyme or to adsorb the virus on red cells and elute it therefrom before using it. These two methods are discussed elsewhere.

A suitable number of Wassermann tubes, e.g., 8–10, are placed in a single row in the serology rack. One-half milliliter of saline is pipetted into each tube. Virus fluids in which a high concentration of virus is suspected should be diluted considerably before they are serially diluted in the tubes in the rack. A preliminary dilution of 1:100 is advisable. If such is the case, the pipet used to transfer the potent virus preparation to the diluting fluid should be discarded and a new pipet used for serial dilution in the rack. One-half milliliter of the 1:100 virus fluid is added to tube 1 in the rack, and after mixing, 0.5 ml is transferred to tube 2, and so on, as in an ordinary agglutination test. The test fluid is omitted from the last tube so it can serve as a control.

One-half milliliter of 0.25 per cent erythrocytes is then added to each tube, and the rack shaken to ensure mixing of the ingredients. The test should be left undisturbed for 2 hr at either room temperature or 4°C. Some virus suspensions require 4°C to cause hemagglutination. Adequate time has elapsed when the cells in the control tube have settled out, usually 1–2 hr.

The white background for the serology rack allows light to be reflected up through the tubes, and by looking down into them the results may be read. The control tube containing only saline and red cells should have the cells settled out in a small, compact button. In the tubes showing hemagglutination the cells will be spread out in an even film over the whole of the round bottom of the tube. Tubes containing insufficient virus to cause hemagglutination will resemble the control. The so-called "bull's eye" or "doughnut" reaction, a mixture of positive and negative results, will sometimes be found between the last positive and first negative and is to be read as negative. The tube containing the least virus causing a complete positive is considered to be the end point, and the final dilution of the virus fluid in this tube is the hemagglutinin titer.

Since the relationship of hemagglutinin to virus mass or particle number is not accurately known either between viruses of different antigenic types causing the same disease or between viruses causing different diseases, one should be cautious about quantitative conclusions along this line.

SEROLOGICAL PROCEDURES

Antihemagglutination or Hemagglutination Inhibition

As with a number of other viral functions hemagglutination can be prevented by specific antibody. This inhibition is not to be confused with inhibition of hemagglutination caused by substances other than antibody and discussed elsewhere. A certain minimal amount of antibody is required to prevent hemagglutination caused by a specified amount of virus under standard conditions. Therefore, quantitative antibody estimations can be made under appropriate conditions such as the difference in antibody titers of acute and convalescent sera from a given patient against a given sample of virus.

The same materials are used as for the hemagglutination test except, of course, a specimen of serum is also required. The serum may require preliminary centrifugation to render it free from particulate matter. Ten tubes are arranged in the first row, and seven in the back. Nine-tenths milliliter of saline is pipetted into tube 1, row 1; 0.25 ml to the remaining nine tubes in the same row; 0.5 ml to tubes 1–5, inclusive, and tube 7 in row 2; and 0.25 to tube 6. When sera suspected of having a high antibody is used, a preliminary dilution of 1:10 may be necessary. Otherwise 0.1 ml of serum is added to tube 1, row 1. After the contents of this tube are mixed, 0.75 ml is withdrawn, 0.25 ml discarded, 0.25 ml added to tube 6, row 2, and the final 0.25 ml used to continue the serial dilution in tube 2 and remaining tubes in row 1. Discard the final 0.25 from tube 10.

The virus concentration used in the test is held constant. The hemagglutinin titer of a given sample of virus having been determined, the sample is diluted in saline to contain 16 hemagglutination doses per milliliter. Virus so diluted should be used immediately. Add 0.25 ml of the diluted virus to each of the 10 tubes in row 1. Pipet 0.5 ml to tube 1, row 2, and serially dilute through tube 5 by carrying 0.5 ml. Discard the 0.5 ml from tube 5. Shake the rack to mix the ingredients well, and after a brief waiting period (5 min) pipet 0.5 ml of 0.25 per cent red cells to each of the 16 tubes. Shake the rack again, and allow it to stand unmolested for 2 hr at room temperature or refrigerator (4°C) temperature. The results in the antihemagglutination test will be the reverse of those in the hemagglutination test. The tube in row 1 containing the highest dilution of serum that completely prevents hemagglutination is considered the end point and is recorded as the antihemagglutination titer. This figure means little unless it is known against how much hemagglutinin the serum was employed. Tubes 1–5 inclusive in row 2 represent a final check on this matter under the conditions of the experiment. If the hemagglutination end point is in tube 1, you have used 1 hemagglutinin dose in the antihemagglutination test; in tube 2, 2 doses; in tube 3, 4 doses; in tube 4, 8 doses; and in tube 5, 16 doses. Some workers like to correct the antihemagglutination titer by multiplying it by the hemagglutinin doses used. This is permissible only if a small number of hemagglutinin doses are employed. If large doses, e.g., only slightly diluted, high-titered virus fluids, are employed, the results are not comparable to those obtained by more dilute virus even after correction involving the large number of doses.

Tube 6, row 2, controls the greatest concentration of the serum used in the test in regard to its ability to agglutinate the red cells by itself. There should be no agglutination. Tube 7, row 2, containing only saline and red cells, is a control on spontaneous agglutination of the cells.

It is almost always desirable to run normal serum from the same species from which the serum containing antibody was obtained. Since some normal sera possess antihemagglutination properties, the titer of the experimental serum may lose its significance. Heating either normal or immune sera at 56°C for 30 min may increase its antihemagglutination power. The extent to which this occurs varies considerably with the serum obtained from different animals.

The fact that hemagglutination is prevented by specific antiserum permits the antihemagglutination test to be used in identifying the unknown virus that possesses hemagglutinin. By the same token the presence of specific antibody may be noted by using a known virus. Acute and convalescent sera may be compared to detect a rise in antibody titer against a known virus for the serologic diagnosis of the disease. The last technic, because of its importance, necessitates consideration in some detail.

Blood drawn during the acute phase of the disease is allowed to clot; the serum is removed and stored in the frozen state. A second sample of blood is drawn 10–14 days later and treated likewise. In some infections weekly samples drawn for 3 weeks after the acute specimen are advisable. The technic for running the hemagglutination test described above is used, but it is important that the sera under comparison are run at the same time, using identical reagents and conditions. Under such circumstances a fourfold or more rise in antibody titer in the convalescent serum is considered significant. This conclusion assumes that the test was conducted with precision and with adequate controls.

The antihemagglutination test is also used in the antigenic analysis of antigenically related viruses. The principle is not too different from cross agglutination among the bacteria. There is an important difference, however. In viral hemagglutination, using the proper menstruum, the antigen-antibody combination interferes with a result that would otherwise occur. On the surface this may seem inconsequential, but it could obviate the very quantitative end hoped for. If antigenically related viral particles A and B possess hemagglutinin activity X and $2X$, respectively, it may take twice as many viral A particles to equal a hemagglutinin dose as those of virus B. If the antibody-binding ability of hemagglutinin groups are disproportionate in the opposite direction or not directly proportionate in the same direction, direct comparison of the two viruses from an antigenic standpoint as elicited by the antihemagglutination test would be in error.

If the antibody functioned indirectly, i.e., not acting against the hemagglutinin but against other portions of the virus particle and thus preventing hemagglutination by not allowing the hemagglutinin to seat properly with the surface of the red cell, antigenic comparisons by this test could be in even greater error. Since some virus particles are known to fix firmly to red cells in the presence of considerably greater quantity of antibody than is necessary to prevent hemagglutination, the above hypothesis may not be so unlikely as might be presumed (Hultin and McKee, 1951). The fact remains that with three major reacting entities, virus, red cell, and antibody, the direct conclusions drawn from a two-way system, antigen-antibody reaction, should be viewed with caution.

To discover qualitative antigenic differences among related viruses antibody adsorption much as is employed in bacteriology may be used. Advantage being taken of the fact that adequate centrifugal force will sediment the majority of viral particles from a given viral preparation and that a number of different viruses apparently form a firm union with their specific antibody, the antibody-adsorption technic can be carried out. Both the adsorption and centrifugation should be carried out in the cold (4°C) unless the virus and antibody are known to be very stable.

Dilution of the antibody may be essentially avoided if the viral particles are sedimented first and then resuspended in the antibody. The time to be allowed for adsorption will vary, as will the gravitational force necessary to sediment the viruses with the virus and antibody employed. Experimentation and consultation with previous works will have to be made to arrive at this end. Adsorption carried to the end of no function of the antibody against a heterologous virus while retaining the function for the homologous virus is of qualitative significance. It must be remembered, however, that the procedure described above, while effective in removing antibody for the available antigens of the virus particle, may not have removed antibody for soluble antigens, depending on the extent of aggregation of the antigen-antibody molecules and perhaps on other factors. Although such soluble antigens may be presumed to be bound to their antibodies, the particle-adsorbed serum should be used only with the awareness that the soluble antigen-antibody system is probably subject to the same factors affecting other antigen-antibody systems.

Receptor-destroying Enzyme

It would be amiss to attempt a prolonged discussion of inhibitors in this technical section. Some consideration of inhibitors, however, cannot be omitted from even the most basic discussion of technic. One can do almost no virological investigation without using reference antiserum to be certain of his virus. Antiserum is one source of inhibitor. The presence of an inhibitor in antiserum used in hemagglutination-inhibition studies may be the cause of false conclusions. If the inhibitor titer is greater than the antibody titer, one cannot expect to read the antibody function accurately. A case in point would be the parallel titration of acute and convalescent serum. If the convalescent serum possessed a relatively low titer of antibody and the acute serum possessed a relatively high titer of inhibitor, a rise in antibody would be overlooked. Since the inhibitor can cause results comparable to antibody in the hemagglutination-inhibition test, it is best removed.

Certain strains of *Vibrio cholera*, e.g., 4-z, elaborate an enzyme capable of inactivating the inhibitor in question. To prepare cholera receptor-destroying enzyme (RDE) a young agar slant culture of the vibrio is used to inoculate a shallow layer of brain-heart-infusion broth, 1.5–2.0 cm deep, in a large flask. The culture is incubated 24 hr at room temperature (26–28°C), and most of the bacteria are removed by centrifugation. The remainder are removed by filtration through a Seitz pad, and the filtrate is checked for sterility and activity. The activity may be checked against chicken red cells or against serum inhibitor. If it is to be used to remove serum inhibitor, it probably would be better to use the latter substrate. The enzyme apparently loses its activity slowly at 4°C, for it is usable for several weeks if so stored.

To treat a serum with unpurified enzyme, 1 part of serum is added to 1 part of filtrate and the mixture is incubated 18 hr at 37°C. The ratio of serum to enzyme will vary somewhat with the potency of the enzyme.

The enzyme is then inactivated at 62°C for 30 min. The treated sample and a sample of untreated serum are titrated under the same conditions to demonstrate the change in inhibitor titer. The dilution caused by the addition of the enzyme, of course, must not be overlooked. If the enzyme is not inactivated prior to the hemagglutination-inhibition test, it is free to act on the red-cell receptors, thus equivocating the whole test.

Receptor-destroying enzyme requires calcium ion, but its activity is impaired by citrate ion. The formula for buffered calcium saline follows:

	Grams
Calcium chloride ($CaCl_2$)	1.000
Sodium chloride ($NaCl_2$)	8.500
Boric acid (H_3BO_3)	1.203
Sodium tetraborate ($Na_2B_4O_7 \cdot 10H_2O$)	0.052
Distilled water	1,000.000

The enzyme preparation can be serially diluted in the calcium saline when its activity is assayed against red cells. If the concentration of the red cells and the incubation time and temperature are maintained constant, the potency of the enzyme can be determined by extinction dilution. The end point is described as the least amount of enzyme capable of rendering the red cells inagglutinable by the virus in question under standardized conditions. It is advisable that the worker consult original articles to determine further activities of the receptor-destroying enzyme (Burnet *et al.*, 1946).

COMPLEMENT FIXATION

Basic Technic

The basic technics employed in complement-fixation tests have been described in Chap. IX. It would seem advisable to comment only on the aspects of the test more troublesome or peculiar to virology.

Antigens. Since we are unable at this time to propagate large quantities of viruses on artificial culture media, obtaining a strong antigen is one of the more common problems. Freeing viral antigens from host-tissue elements which may be anticomplementary or be otherwise objectionable not uncommonly presents another problem.

The viral antigen functioning in the complement-fixation test may be intimately associated with the viral particle, soluble and separate from it, or may be found under both circumstances. During an active infection of host cells it may increase simultaneously with or independent from other virus propensities.

The preparation of antigen varies considerably with the virus and/or tissue in which the virus was propagated. Antigens from allantoic or amniotic fluids or chorioallantoic or amniotic membranes usually present no major problems if a sufficient concentration of antigen can be obtained. Membranes may be disintegrated to some extent by one of several methods, e.g., grinding with an abrasive or using a mechanical blender or sonic vibrator. After disintegration with or without the addition of an extractive, e.g., saline or phosphate buffer solution, the material is centrifuged at sufficient speed to remove the grosser particulate matter. Antigen from allantoic or amniotic fluids can be concentrated by placing the fluid in a viscose tube and hanging it in the air current from an electric fan. Quite often the fluid should first be dialyzed against 20 vol of distilled water at 4°C overnight to prevent large precipitates of salts from forming as the evaporation proceeds. Antigen may also be concentrated sometimes by high-speed centrifugation and resuspension of the sediment in a fraction of the amount of the original suspending fluid. See discussion on concentration.

Antigen propagated in yolk-sac membrane requires special attention. After the membranes have been disintegrated, extracted, and centrifuged, the lipoidal cake forming at the top of the centrifuge tube is usually discarded. Placing the preparation at 4°C until the lipoidal material has become more solid, circling the cake with a wire, and removing it with a curved spatula may aid in what otherwise can be a troublesome procedure.

Depending on the size of the organism and the speed at which the preparation has been centrifuged, the desired portion of the extract may be in the supernatant fluid or sediment. In some instances the supernatant fluid may be used as a satisfactory antigen. Further centrifugation at higher speeds will concentrate the antigen if it is of sufficient size, and this procedure may be necessary to obtain one of appropriate strength. If the antigen is of adequate size to appear in the sediment at slower centrifuge speeds, then it may be purified by repeated sedimenting and resuspending in fresh menstruum.

Not infrequently it may be necessary to shake the yolk-sac preparations with equal volumes of ethyl ether for the purpose of extracting sufficient lipoid to render the antigen complementary. A separatory funnel aids in separating the aqueous from the ether layer. This same procedure may be necessary when dealing with mouse-brain virus antigen. Lyophilized preparations can be extracted with much more efficiency than wet preparations. It is advisable to determine the effect of the lipid extraction procedure on the stability of the antigen.

Antibodies. While the complement-fixation procedure is invaluable to the virologist, it must be rigidly controlled to prevent embarrassing errors. It is a not uncommon occurrence to encounter human sera with antibody titer against egg products. Hence, one must always include controls to learn if the antibody titer is greater against an egg-propagated antigen than against the egg-products control. If egg-propagated virus has been used as an antigen to produce control antiserum, the problem involved in using this serum as a control against egg products is obvious. If the chicken is the donor of the antiserum, this might be obviated as far as heterologous tissue antigen is concerned, but chicken sera may function poorly in the complement-fixation test for other reasons. Whenever possible, however, it is probably advisable to have reference serum containing antibody against only the virus and/or its products. Sometimes this is possible by the use of homologous tissue virus antigens to produce the antibody, e.g., mouse-lung virus to produce antibody in mice. Otherwise, adsorption with heterologous antigen may be employed.

To avoid excess of either antigen or antibody most workers cross-titrate the serially diluted antigen against a serially diluted antibody as indicated in Table 28.

TABLE 28. CROSS TITRATION OF ANTIGEN AND ANTIBODY*

Dilution of antibody	Dilution of antigen						
	1:2	1:4	1:8	1:16	1:32	1:64	1:128
1:2	4+	4+	4+	4+	4+	–	–
1:4	4+	4+	4+	4+	4+	1+	–
1:8	4+	4+	4+	4+	4+	2+	–
1:16	4+	4+	4+	4+	4+	–	–
1:32	4+	4+	4+	4+	2+	–	–
1:64	4+	4+	4+	4+	2+	–	–
1:128	4+	4+	4+	4+	2+	–	–
1:256	–	–	–	3+	1+	–	–
1:512	–	–	–	–	–	–	–

* For experimental details on this titration, see Chap. IX.

Both the antigen and antibody used in the above test were strong. Either could have been weak, such as an early convalescent serum or an unconcentrated freshly isolated and poorly adapted viral antigen. In such cases, unless the cross-titration method is employed, negative results might be obtained where they should be positive.

Peculiarities. Soluble antigen may be difficult to detect without the complement-fixation technic. Artificial active immunization using inactive virus may fail in some instances to call forth the antibody involved in the complement-fixation test, whereas an active infection may more regularly do so. In a number of instances the complement-fixation reaction has a broader spectrum in recognizing antigenic relationships among viruses than does the hemagglutination-inhibition test.

Thus in the case of the influenza viruses, complement fixation readily succeeds in placing a number of strains within a given type while separating one type from another quite sharply. The antihemagglutination test on the other hand is less sensitive in recognizing widely separated strains as belonging to the same type but quite readily shows differences among strains.

A rise in antibody titer during an infection may be elicited by the complement-fixation technic using paired serum specimens much as in the antihemagglutination test. Positive complement-fixation tests may be obtained earlier or later than positive results using other serological methods. The complement-fixing antibody may be detectable for longer or shorter periods of time than antibody revealed by other methods. These last two peculiarities vary with the disease in question.

Neutralization

General observations. *In vitro* neutralization tests, e.g., hemagglutinin inhibition, may or may not be directly related to *in vivo* neutralization. *In vivo* neutralization tests vary with the virus and the living cells to be protected. Not uncommonly among virus diseases active immunity seems to transcend circulating antibody as an entity. Nevertheless neutralization tests are useful in assaying the protective power of circulating antibody. Generally speaking an *in vivo* neutralization test has as its goal the determination of the amount of specific antibody required to prevent the virus from producing observable changes in host cells or to prevent viral propagation.

Ingredients. It should be remembered that inactive virus can bind antibody, but only active virus enters into establishing infection. Therefore, it is easy to imagine disparities in experimental results (McKee and Hale, 1947). A preparation containing nothing but active virus would be desirable but probably impossible to achieve. Harvesting the virus at an optimum time and using it immediately minimize the ratio of inactive virus to active virus. If nonviral antigen adheres to the virus or is present in the mixture, neutralization tests may be in error. For example, if virus is produced in the egg and rabbit antiserum to egg virus is used, at least two antigen-antibody systems are involved. Undoubtedly this problem can be minimized by using virus and antibody from the species used as the test animal. This is not always convenient or possible.

Dilutions. Either the antibody or the virus may be diluted while the other is held constant. As with other antigen-antibody systems the straight-line function fails when too much or too little of either reagent is employed. Biological variation is great enough that serial dilutions by less than increments of 10 may produce equivocal end points. This is not invariably true, for doubling dilutions sometimes produce acceptable end points. Increments between these two extremes have not been popular, but there is no reason why they could not be used.

If it is anticipated that either antibody or virus is to be used in high dilution, a protective diluent should be used. This precaution is discussed under "Cultivation in Chick Embryos," infectivity titers.

Incubation time. All viruses do not react with their antibodies at the same speed. Whereas 15 min at room temperature may be an acceptable time for the influenza virus and its antibody, vaccinia virus-antibody mixtures require a longer period of time. As the exposure time is extended, temperature should probably be reduced, for some viruses are perishable rather quickly at higher temperature. The time-temperature variable undoubtedly plays some role in altering the preciseness of the test. Considerably more inactivation may occur with the virus than with the antibody. Inactive virus may bind antibody, yet only active virus produces the end point. Preliminary determination of the lability of the virus and the use of adequate controls usually reduce these variables to an insignificant level. To report an exposure time of virus and antibody as 15 min is inaccurate if all dilutions are mixed at once and it takes 1 hr to inoculate the experimental animals. If antibody and virus dilutions, or vice versa, are mixed together at intervals paralleling inoculation time, accuracy is increased. Longer exposure time, e.g., overnight at 4°C, probably requires less or no adjustment of this potential error.

Test system. Susceptible animals, embryos, or tissue cultures are used depending on their suitability for achieving the objective in mind. Certain precautions should be kept in mind when the outcome of the neutralization test is being determined. The end point should be clearly designated. For example, at least four end points could be used in a mouse neutralization test where a mouse-adapted strain of influenza virus is used: death, signs of illness, production of lung lesions, and no infection on subsequent passage of lung tissue. Considerable disparity would occur among these end points. If the production of antigen is used as an indication of propagation, the antigen introduced with the inoculum must be taken into consideration. Neutralized virus may not grow or agglutinate red cells, but it will function in the complement-fixation test. A sufficient amount of neutralized virus could make the tissue into which it had been introduced appear anticomplementary. Sufficient antibody to prevent viral growth need not prevent it from adhering to susceptible cells (Hutlin and McKee, 1951).

Generally speaking one should allow more incubation time on a neutralization test than for determining infectivity. Antibody insufficient in quantity to neutralize a virus frequently will slow down its progress. For example, viruses reaching their propagation peak in

24–48 hr should be allowed to incubate an extra day. End points may be reported on pooled specimens of like dilutions, or the 50 per cent end points may be calculated as discussed elsewhere.

Route of inoculation. The route of inoculation will vary according to the potentiality of the underneutralized virus to produce a recognizable end point. Dermatropic viruses ordinarily would be inoculated intracutaneously in a susceptible animal, although other methods, such as corneal scarification, might be used. Neutralization end points are known to vary, even with the same virus and antibody preparations, when different routes of inoculation are used even in the same species of animal.

Bleeding mice for serum. Mice are not uncommonly used to produce antibody where small amounts of several different kinds are desired. Eight-week-old mice will yield about 0.25 ml of serum each. Since mice are susceptible to several virus diseases, convalescent as well as hyperimmune serum may be obtained. Mice can be bled quickly if a suitable method is used. The mice are anesthetized in an ether jar. They are then held in one hand by the scruff of the neck ventral side up. The thorax is punctured at the xiphoid process with the sharp blade of a pair of scissors. The blades are closed to make a diagonal cut toward one shoulder and withdrawn. The incision is inverted over a test tube having an 18-mm opening. The blood that will flow out quickly is collected, and the mouse is placed on the table, cut side up. A second mouse is treated in the same fashion. In returning to the first mouse more blood can be collected. With a little practice and forethought one operator learns to have three to four mice under the proper depth of anesthesia at all times as he bleeds others more or less continuously. A practiced operator can bleed 100 mice in an hour. While this method of bleeding will cause some hemolysis, the hemolysis can be minimized by centrifuging the blood immediately after it is collected.

To obtain clear serum or serum specimens from individual mice requires more time and patience. The anesthetized mouse is secured, ventral side up, on a mouse board. Spring clamps, which can be easily and quickly released, are used to immobilize his extremities. At least 0.1 ml of blood may be obtained by cardiac puncture, using a 27-gauge needle and 0.25-ml tuberculin syringe. It is advisable for two people to cooperate, one to handle the mice and the other to bleed. Depending on the amount of blood removed, the animals may be allowed to recover from the anesthetic or sacrificed by returning them to the ether jar.

CONCENTRATING VIRUSES

Gravitation. For various reasons the virologist may wish to concentrate viruses or their products or both. To concentrate particles of virus a high-speed centrifuge may serve the purpose adequately. The gravitational force and time employed will vary with the virus, and published work will often serve as a guide for the individual viruses under study. Certain generalities, however, are worth noting. The virus particles within a given suspension may not be uniform in size; therefore, to calculate the amount of virus sedimented from the amount left in the supernatant fluid may or may not be correct, depending on the qualitative identity of the two fractions. Inhibitive products may be left in or removed from the supernatant viral particles, causing them to produce a misleading titer.

The resuspension of virus from high-speed-centrifugation sediments may be more difficult than ordinarily assumed. Usually best results are obtained by repeated and gentle agitation over a period of time. This may be especially difficult in concentration procedures where less than the original volume is used for resuspension. Resuspending virus in a menstruum different from the one from which it was removed may change its stability and the functions by which it is recognized and quantitated. It is well to control such changes carefully. So-called soluble substances as well as small viral particles will not sediment at the same speed as larger viral particles; therefore, this presents a method of separation. It seems preferable for the novice to learn the methodology for operating a high-speed centrifuge from an experienced operator. Therefore, such operation will not be discussed here.

Evaporation. Certain viral preparations, such as an allantoic fluid harvest, can be concentrated by evaporation. If it is desirable to have the whole preparation concentrated, this method offers some advantages. Since some of the salts in allantoic fluid are present almost to the point of saturation, it frequently is desirable to dialyze the preparation against distilled water at 4°C before evaporation. The time of dialysis and the number of water changes depend upon the amount of salt to be removed. Most of the salts can be removed by six exposures to 20 vol of water over a 72-hr period. Approximately 80 per cent of this time can be saved by employing a stirring apparatus during the dialysis. If a precipitate forms in the dialysate, it can be removed by low-speed centrifugation. The precipitate usually contains some of the virus.

The preparation is placed in a viscose tube of suitable diameter and length with a wall thickness of 0.18 mm. The tube should be washed inside and out. After double-knotting one end, fill it with water to test for leaks. If the bag is intact, empty it and tie a funnel in the open end. The apparatus can be sterilized in flowing steam. This bag can be used for dialysis and for evaporation. To evaporate the preparation the tube is hung in front of the air current from an electric fan. The evaporation keeps the preparation reasonably cool, and several viruses will not fall in infectivity during the process. If the column of fluid is measured before and during the evaporation, the process can be halted when the proper end point can be obtained. Carrying the evaporation slightly beyond the desired amount allows one to dilute up to the correct volume with distilled water. The bag is hung up slightly below eye level, and the outside cleansed with cotton saturated with alcohol. A large syringe and 20-gauge needle can be used to withdraw the concentrate. The needle is inserted just above the fluid level with the point down in the fluid. A small amount of distilled water may be used to rinse out the remnant in the bottom of the bag. Some preparations, after dialysis, may be concentrated 8–10 times without a precipitate developing. This same method may be employed for concentrating soluble antigen from virus preparations, i.e., the supernatant fluid following high-speed centrifugation. Whether or not sodium chloride is added to physiological concentration depends on the use to which the preparation is to be put.

Adsorption and elution. Some viruses are known not only to adsorb to red blood cells but also to be able to elute from them. The process may occur too quickly at room or incubator temperatures; hence the adsorption is accomplished at 4°C. This procedure can perhaps best be illustrated by describing it for allantoic fluid containing influenza virus.

A 3 per cent suspension of chicken red-blood cells and the virus preparation are chilled to 4°C. The two materials may then be mixed in equal parts, or the red cells sedimented by centrifuging and then resuspended in the virus preparation. The latter method introduces only insignificant dilution of the virus preparation. After the red cells and virus have been allowed to react for at least 2 hr at 4°C, the mixture is centrifuged, preferably at the same low temperature and employing precooled receptacles. The agglutinated cells are sedimented readily at low speeds, 5 min at 2,500 rpm in a bucket centrifuge, and the supernatant fluid is decanted and saved. The sedimented red cells are resuspended in warm (37°C) saline or other menstruums. The amount of concentration desired dictates the amount of menstruum to use. Ten times concentration can be realized quite efficiently. The resuspended cells are incubated at 37°C for at least 2 hr, and an extra 30 min to 1 hr at this temperature will increase the yield. Occasionally, say once every 30 min throughout the 37°C incubation period, the mixture should be agitated gently.

After the virus has eluted, the red blood cells do not pack so easily by centrifugation. Employing higher speeds on the centrifuge or using a conical-bottom tube will facilitate depositing the red cells securely so the supernatant fluid can be decanted cell-free. Since inactive virus may adsorb but not elute, yet can agglutinate red cells, it is important to be aware of how much this might affect the computation of concentration. Determining the infectivity titer of each fraction aids in determining this error. The adsorption-elution process can concentrate active virus, while some soluble antigens apparently are not concentrated by this method. Perfect concentration is probably never achieved, since the law of mass action comes into play, leaving a certain percentage unadsorbed or adsorbed and eluted. In the attempt to determine the amount of concentration achieved, a sample of the original fluid, the fluid from which the virus was adsorbed, and the concentrated eluate should be examined. If the sample of original fluid is chilled, heated, and centrifuged along with the specimen being concentrated, some conclusion may be drawn as to the extent of virus inactivation caused by the procedure itself. The sum of the unadsorbed virus and the adsorbed and eluted virus should equal the original fluid control. Some potential errors arise in attempting to make this determination. The unadsorbed virus in the supernatant fluid would be tested at relatively low dilutions in anticipation of a low titer. The concentrated virus, if started at the same low dilutions, will appear to have a greater titer than possible. This is probably due to excess virus carried as a film on the inside of the pipet which is gradually washed out during successive serial transfers. This can be obviated by starting the concentrated virus off at high dilutions. The amount of deviation permitted by the coarseness of the test used to quantitate the virus makes precise appraisals almost impossible. As long as exact titers are not taken too literally, this procedure, as well as those described above, is useful in concentrating virus.

Chemical methods of precipitation used alone or in combination with centrifugation are sufficiently diverse that original articles should be consulted to learn about them (Warren, 1950; Sharp, 1953).

STORAGE

A number of methods are available for storing viruses and rickettsiae. The optimum method varies with the agent, equipment available, and convenience.

One of the most satisfactory methods involves lyophilization, and this method is most commonly employed for long-term storage, such as periods of over a year. Generally speaking, the method of lyophilization used for viruses is the same as that used for bacteria discussed elsewhere. Viruses and particularly rickettsiae may be stored frozen on solid carbon dioxide for variable periods up to a year and beyond. Any insulated container, such as a silvered-glass vacuum jug (thermos), that will delay the dry-ice evaporation is satisfactory. Aside from arranging for a reliable and more or less continuous supply of dry ice this method is quite convenient. Most of the more common viruses store well in the deep-freeze boxes if the temperature is maintained between −20 and −40°C. Inasmuch as they are electrically controlled, the preservation of the viruses does not depend on someone's remembering to replenish the dry-ice supply and they may be therefore more reliable.

Viruses stored at 4°C may retain their infectiousness from a few days to a few weeks. This may be especially satisfactory because of its convenience when fresh seed virus may be substituted for the old each week. Storing neurotropic viruses in 50 per cent neutral glycerin at 4°C is usually satisfactory for several months.

The length of time a virus may remain viable depends not only upon the temperature but upon the menstruum employed. As a rule 10 per cent tissue suspensions store satisfactorily. Fluids of low protein content usually store better if the protein content is increased, such as by mixing them with sterile milk. Embryonic membranes may be stored whole. If the infected animal is small, such as a mouse, certain pneumotropic and neurotropic viruses are preserved very well if the whole animal is frozen. When freezing mice it is advisable to use a small airtight container which the mouse or mice will fill sufficiently to exclude practically all the air.

Some evidence is available that repeated freezing and thawing degrade some viruses (Tyrrell and Horsfall, 1954). It would seem advisable then to put away several smaller containers of virus to avoid repeated freezing and thawing of one large sample. Some viral antigens store well at 4°C but are destroyed at freezing temperatures. It is well to be aware of what activity you wish to store before selecting the storage temperature.

REFERENCES

Bernkopf, H. 1950. Study of infectivity and hemagglutination of influenza virus in deembryonated eggs. *J. Immunol.*, **65**, 571–583.

Burnet, F. M., J. F. McCrea, and J. D. Stone. 1946. Modification of human red cells by virus action. I. The receptor gradient for virus action in human red cells. *Brit. J. Exptl. Pathol.*, **27**, 228–236.

Finney, D. J. 1947. "Probit Analysis." Cambridge University Press, New York.

Francis, T. F., Jr. (ed.). 1948. "Diagnostic Procedures for Virus and Rickettsial Diseases," pp. 228–233. American Public Health Association, New York.

Greiff, D., and H. Pinkerton. 1951. Rickettsiostasis in fertile eggs from use of antibiotic residues in poultry feeds. *Proc. Soc. Exptl. Biol. Med.*, **78**, 690–692.

Hartman, F. W., F. L. Horsfall, Jr., and J. G. Kidd (eds.). 1954. "The Dynamics of Virus and Rickettsial Infections," chap. 6. The Blakiston Division, McGraw-Hill Book Company, Inc., New York.

Hultin, J. V., and A. P. McKee. 1951. Fixation of "neutralized" influenza virus by susceptible cells. *J. Bacteriol.*, **63**, 437–447.

Irwin, J. O., and E. A. Cheeseman. 1939. On an appropriate method of determining the median effective dose and its error, in the case of a quantal response. *J. Hyg.*, **39**, 574–586.

Jones, M. 1945. Adaptation of influenza virus to heat. *Proc. Soc. Exptl. Biol. Med.*, **58**, 315–319.

McKee, A. P., and W. M. Hale. 1947. Reactivation of overneutralized influenza virus in chicken embryos. *J. Immunol.*, **58**, 141–152.

Reed, L. J., and H. Muench. 1938. A simple method of estimating fifty per cent endpoints. *Am. J. Hyg.*, **27**, 493–497.

Reitman, M., R. L. Alg, W. S. Miller, and N. H. Gross. 1954. Potential infectious hazards of laboratory techniques. III. Viral techniques. *J. Bacteriol.*, **68**, 549–554.

Salk, J. E. 1944. A simplified procedure for titrating hemagglutinating capacity of influenza-virus and the corresponding antibody. *J. Immunol.*, **49**, 87–98.

Sharp, D. G. 1953. Purification and properties of animal viruses. In "Advances in Virus Research," vol. 1. Academic Press, Inc., New York.

Simms, H. S., and M. Sanders. 1942. Use of serum ultrafiltrate in tissue cultures for studying deposition of fat and for propagation of viruses. *Arch. Pathol.*, **33**, 619–635.

Smadel, J. E. 1951. The hazard of acquiring virus and rickettsial diseases in the laboratory. *Am. J. Public Health*, **41**, 788–795.

Sugg, J. Y. 1951. Further studies on serial passage of a mixture of influenza viruses in embryonated eggs. *Proc. Soc. Exptl. Biol. Med.*, **76**, 199–202.

Tyrrell, D. A. J., and F. L. Horsfall, Jr. 1954. Disruption of influenza virus. Properties of degradation products of the virus particle. *J. Exptl. Med.*, **99**, 321–342.

Warren, J. 1950. Symposium on viral and rickettsial diseases. Part I, 3. Progress in the purification of viruses of animals. *Bacteriol. Revs.*, **14**, 200–206.

CHAPTER XII

Inoculations with Bacteria Causing Plant Disease

A. J. RIKER IN COLLABORATION WITH O. N. ALLEN, P. A. ARK, A. C. HILDEBRANDT, AND E. M. HILDEBRAND

INTRODUCTION

The methods for studying the pathogenicity of bacteria in plants and for making a few selected cognate investigations are briefly treated. The procedures for handling certain organisms and for studying the diseases they induce vary so widely that no given directions apply to the group as a whole. The selected representative methods included are given primarily as guides to the beginner and need modification according to circumstances. Two excellent reviews have appeared recently by Allen and Allen (1950) and Hildebrandt (1950).

Difficulty in interpretation frequently is encountered from variations in results that depend on the methods used. A given bacterial characteristic may sometimes be positive when measured by one method and be negative when measured by a slightly different technique. Students should employ a known positive and a known negative as controls when making critical determinations. The method used should always be given or cited when a characteristic is listed, so that its validity can be estimated by the reader.

A number of topics discussed in Chap. X regarding bacteria pathogenic on animals are applicable to bacteria pathogenic on plants. These include particularly (1) identification of the active agent as the bacterial cell or its products; (2) distinction between invasion and the power to cause disease after entry; (3) variability in virulence of the pathogen, which requires single-cell cultures, and in susceptibility of the host, which frequently calls for plants with known genetic constitution, when critical studies are involved; and (4) relations between reactions induced in the test tube and in the host.

The pathogenicity of a microorganism may be proved by fulfilling Koch's postulates, which have been stated and modified in various ways and which are so important that they are repeated here. One summarized statement follows: (1) The causal agent must be associated in every case with the disease as it occurs naturally, and conversely the disease must not appear without this agent. (2) The causal agent must be isolated in pure culture, and its specific characteristics determined. (3) When the host is inoculated under favorable conditions with suitable controls, the characteristic symptoms of the disease must develop. (4) The causal agent must be reisolated, usually by means of the technique employed for the first isolation, and identified as that first isolated. Obviously, the demonstration of pathogenicity is made only after repeated trials, preferably with a number of different isolates which are of unquestioned purity. While the technique for cultivating causal agents on artificial media has not yet been worked out, their pathogenicity is established in other ways (e.g., Rivers, 1937). When causal relations are being worked out, one may well differentiate between predisposing, inciting, and continuing causes. Various factors that influence the physiology of the plant may also affect pathogenicity.

The ability to induce disease often has been considered a single characteristic. In reality, pathogenicity is a combination of many characteristics, of which we can consider only a few. The microorganism must have a susceptible host that is growing in a suitable environment (especially temperature, moisture, light, and mineral nutrition). It must enter the host plant. It must establish itself inside by finding suitable nutrients and by overcoming any antagonistic factors encountered. It must come out of the plant again; it must find a means of transfer by one way or another to a new host; it must withstand the winters. Variations may occur independently in any one of these and comparable items. Without a suitable combination of such factors, no epidemic[1] will develop.

Unusual succulence of the host plant and certain other abnormalities sometimes may enable saprophytes to invade the tissue and to give the appearance of pathogens. Such conditions have led to erroneous conclusions. To ensure against such conclusions, one should hold the environmental conditions for experimental inoculations as nearly as possible like those occurring in nature at the time of natural infection. When difficulty occurs with artificial inoculations, careful continued

[1] "Epidemic," in the original Greek meaning "on the people," was early applied to plant diseases, together with many other medical terms. It is an old and common word in plant pathology, although on etymological grounds its use for human disease alone is preferred by some authorities. This chapter follows the broad definition from Gould's Medical Dictionary, "Epidemic: of a disease affecting large numbers or spreading over a wide area."

observation of the host plant at the time of natural infection may reveal the cause of the trouble.

The simpler methods for preparing and using both ordinary and differential media, for making isolations, and for studying the morphology and physiology of such bacteria have been adequately described in other chapters. This chapter, therefore, is concerned primarily with methods of inoculation.

In advanced research, investigators working with pathogens, whether against plants, animals, or men, have many common interests. These include, for example, (1) life cycles, referring to changes in the morphology of individual cells and the relation of these different forms to virulence; (2) changes in colony characters and physiology, including particularly changes in pathogenicity; (3) factors attending changes, such as the time, frequency, and conditions of origin, as well as the influence of environment, and relations to earlier and succeeding generations; (4) statistical analyses to classify the origin and frequency of the variations observed; and (5) life histories of the pathogens in relation to entrance into the host, location, exit, and transmission to a new host.

Certain characteristics of plants not possessed by animals facilitate basic research on pathogenicity. These include the following: (1) Large numbers of hosts are easily available. The number used, whether 10 or 10,000, is selected on the basis of experimental needs. (2) The initial costs and expenses of maintaining plants are relatively low. (3) The species of plants studied frequently contain varieties or selections possessing several degrees of resistance and susceptibility. (4) Plants are suited to a wide range of experimental procedures, such as regulation of internal temperature and moisture, that are not feasible with animals. (5) Epidemics are induced with relative ease and without concern for the health of the technician or the public. (6) The genetic purity of the host can be assured. Seeds from long lines of successively self-fertilized parents often are available. When this is not sufficient, one can commonly find or develop experimental units all genetically identical through vegetative propagation on their own roots. With such material, any variations secured can be studied without concern that the genetics of the host may have been obscuring pathogenicity. (7) Certain plant materials are being cultivated *in vitro* on media containing only nutrients for which the chemical formulas are known. This technique offers many interesting possibilities for investigations on the interactions of hosts and pathogens.

SIMPLE REPRESENTATIVE INOCULATION METHODS

The actual procedures for making inoculations vary with circumstances. Some simpler methods are considered briefly by way of illustration. Ways for testing the relative efficiency of several techniques are considered in a later section.

Soil "inoculation." The introduction of large numbers of pathogenic bacteria into the soil depends upon growing sufficient quantities in cultures, either on agar or in liquid media. Special flasks, bottles, and other containers having adequate flat surfaces are useful. Most plant pathogens are aerobic and need incubation under pronounced aerobic conditions for the best growth. When agar is used, the surface growth is washed or scraped off after sufficient growth has appeared, and a suspension is made. When a liquid medium is employed, a satisfactory bacterial count per cubic milliliter develops, with an organism like *Agrobacterium tumefaciens* (Smith and Town) Conn., with a medium less than 2 cm deep or with one well aerated by shaking or by some other means. Satisfactory aeration may be secured in deep liquid cultures by bubbling sterile air through a sintered glass or other aerator placed in the medium. In large containers aeration can be improved by a few pounds of pressure. This forces more air to dissolve in the liquid. Maintaining such pressure also reduces contamination from leaky valves. Chemicals that poise the oxidation-reduction potential may be helpful. The highest count of active bacterial cells may occur somewhat before the maximum turbidity is attained. Bacterial gum may cause considerable turbidity. Usually the whole culture is employed for soil treatment, but one should avoid adding too much extraneous matter with the inoculum. Such aerated liquid cultures also work well with many fungi.

Soil may be "inoculated" by pouring liquid suspensions on relatively dry soil, by allowing the water to be absorbed long enough to avoid puddling, and by mixing. The quantity of culture used for each plant varies. One might begin with 1 part of culture to 10 parts of soil and use a handful of this mixture about the roots of each plant.

Inoculations through the soil are considerably more difficult than those with various other methods.

Seed inoculation. Perhaps the easiest way to infect a large population is through treatment of the seed. Legume root nodule bacteria from a fresh, active culture grown on agar are shaken into a water suspension and are commonly spread on the seed just before planting. Many commercial inocula are prepared by mixing the culture with some moisture absorbing powder, such as autoclaved ground peat. Wood flour is also absorbent and contains almost no bacteria. If the seed is drill-sown, it is made only moist enough to "fix" the bacteria on the seed and then dried sufficiently so as not to clog the drill. To secure uniform results it is best to use plenty of bacterial culture. Fred, Baldwin, and McCoy (1932) have reviewed this general subject.

Spray inoculation. Spraying is the method most commonly used to inoculate growing plants. It is particularly useful with bacteria that enter the host plants through natural openings such as stomata, water pores, and nectaries. For many simple tests, suspensions of bacteria are merely sprayed on the surfaces of susceptible leaves, stems, flowers, fruits, etc. For more exact tests, as for comparative virulence, one suspends the growth from an agar culture in water, saline solution (0.9 per cent NaCl), or a selected buffer (such as suitable mixtures of dilute K_2HPO_4 and KH_2PO_4, and standardizes the concentration according to a selected and measured turbidity. If the bacteria have been grown in liquid culture, the entire culture may be used. This procedure, however, is often unsatisfactory because, after spraying, secondary organisms may grow in the nutrient medium. It is frequently better to separate the bacteria from the medium by means of a centrifuge and to resuspend the cells as with the growth from agar media.

The number of bacteria in a suspension may be determined, for example, (1) by a Breed count or by direct examination in a Petroff-Hausser counting chamber or (2) by mixing a known volume of the bacteria with previously counted suspensions of yeast or red blood cells, by making smears, and by determining the relative number of bacteria and cells. Bacterial suspensions often are duplicated by comparing their turbidity with that of a graded series of barium sulfate standards (described by Riker and Riker, 1936). A common density for a bacterial suspension has the turbidity of a solution obtained by mixing 1 ml of a 1 per cent solution of barium chloride with 99 ml of dilute sulfuric acid. Turbidity can be measured accurately and rapidly with a suitable instrument.

The prepared bacterial suspension is filtered through cheesecloth to remove small pieces of agar or other materials which might clog the spray nozzle and is placed in the spraying device. The plants are sprayed so that good coverage is given, especially to the lower sides of leaves which commonly have more stomata. A strong spray may force the bacteria through the stomata into the leaves. The plants are placed in an environment where they will not dry off for a number of hours.

Certain additional precautions may be necessary for best results. For example: (1) The relative humidity of the air surrounding the host plant is maintained at saturation before as well as after inoculation. This and suitable light help to provide wide-open stomata. The length of time necessary varies with the host plant and the parasite. A saturated atmosphere for 6 to 18 hr in both instances favors infection with many leaf parasites. Various kinds of moist chambers, e.g., that described by Keitt *et al.* (1937), can be used in the greenhouse. Small outdoor plantings can be covered for a short time with a cloth tent (Keitt, 1918), and water sprayed over the exterior. The amount of mo.sture in the air apparently influences the intercellular humidity and, correspondingly, the susceptibility of the host. (2) If the plant parts are difficult to wet because of a waxy covering, the surface can be gently rubbed with a moist cloth. For work on a large scale, the suspension of the organism can be made in a solution of a spreader (e.g., castile soap, 1:1,000) to reduce surface tension. The concentration varies according to requirements. Some spreaders, however, are toxic. (3) A reduced oxygen supply may be important if the pathogen is a facultative anaerobe. For example, the protective wound-cork formation in potato tubers requires abundant oxygen, while certain bacterial pathogens, such as *Erwinia carotovora* (Jones) Holland, grow well with little oxygen. (4) Water pressure, suction, prolonged spraying, and other means can be used to saturate the intercellular spaces below the stomata and thus to improve the penetration of bacterial suspensions into these regions. This is particularly important with a pathogen, like that causing black fire of tobacco (Johnson, 1937), which is often not aggressive. With this method it is possible to induce necrotic areas on plants not ordinarily considered hosts of the microorganism used. Since bacteria that are usually considered saprophytes have caused damage under these circumstances, care is necessary while interpreting such results. For example, such saprophytes would hardly fulfill the first of Koch's postulates, as given earlier.

Wound inoculation. Suspensions of bacteria and small portions of culture or of diseased tissue can be introduced into healthy plants through wounds when they do not readily gain entrance through natural openings or when heavier or more rapid infection is desired. The simplest procedure is to smear the point of a dissecting needle with the bacterial mass and to insert the needle into the plant tissue. If large numbers of inoculations are to be made, various instruments are useful. For example, Ivanoff (1934) has described an inoculator in detail. It consists of a hypodermic needle (size varied according to needs) with end closed and smooth-walled opening made above the point, a suitable chamber to hold a bacterial suspension, and a valve to regulate the flow. This needle with a side opening may be used with an ordinary syringe. The common hypodermic needle when pushed into a plant clogs too easily to be practical.

Known small numbers of bacteria may be introduced into microwounds by means of a micromanipulator. Such wounds may resemble those made by insects (Hildebrand, 1942).

Insect inoculation. The translocation of microorganisms causing plant disease and their introduction into susceptible plants by insects are large and relatively undeveloped fields. The simplest technique with active insects like cucumber beetles or leafhoppers is merely to place the plant to be inoculated in the same insect cage with an infested diseased plant (Leach, 1940).

For virus diseases, inoculation with slow-moving insects, like aphids, is accomplished by placing a paper on a caged plant to be inoculated and by laying on this paper a portion of a diseased leaf which carries aphids. As the new leaf tissue dries, the insects crawl over the paper to the fresh leaf below. When insects are involved, a variety of special cages (Leach, 1940) may be necessary.

All stages in the life cycle of the insect employed must be considered because inoculation capabilities often vary in this respect. The insect should be identified by a competent authority, and if significant results are obtained a specimen should be deposited in a permanent reference collection.

A detailed discussion of methods for studying insect transmission has appeared (Leach, 1940). Some knowledge of the mouth parts of insects and of their feeding and breeding habits is necessary if insects are to be used successfully for inoculating bacterial plant pathogens. Sometimes they merely open infection courts for the microorganisms to enter. Aphids and leafhoppers are particularly important as carriers of virus diseases.

Before claims are made about insect transmission of a plant disease, demonstrations of the following (Leach, 1940) seem a minimum for proof: (1) close but not necessarily constant association of the insect with diseased plants, (2) regular visits by the insect to healthy plants under conditions suitable for the transmission of disease, (3) presence of the pathogen or virus in or on the insect in nature or after visiting a diseased plant, (4) experimental production of the disease by insect visitation under controlled conditions and with adequate checks.

Fungus inoculation. In general, inoculations with the spores or mycelia of fungi differ only in detail from those made with bacteria. For pathogenic fungi, variations in the mode of entrance and in other important characters require modified procedures. Some of the more common methods are discussed by Riker and Riker (1936).

Virus inoculation. Brief mention is given to inoculations with viruses without implication that they are microorganisms. Experimental inoculations are more commonly accomplished by mechanical processes, insects (see "Insect Inoculation"), and grafting.

Mechanical inoculation of a virus frequently is made by grinding diseased tissue in a mortar with a little water and by rubbing the juice lightly over leaves of the host plant. With some viruses, the following modifications may be helpful. A favorable reaction between pH 7.0 and 8.5 may be obtained by placing a little 10 M K_2HPO_4 in the mortar before the leaves are triturated. Sometimes viruses have to be protected from rapid oxidation by means of 0.5 per cent anhydrous Na_2SO_3. Just enough friction by a finger, cheesecloth, or similar agent is employed to injure the leaf hairs. With viruses difficult to transmit, better infection may be induced if a fine abrasive material (e.g., carborundum powder, 600 mesh) is lightly dusted on the leaf before it is rubbed. Some plant viruses are highly infectious. (Usually washing with soap and water is sufficient to remove infectious material from the technician's hands.) When the mechanical methods and insect vectors fail, two possibilities are left.

Budding or another form of grafting may be employed and is sometimes the only successful means of virus transmission. When grafts are made, special precautions are necessary to prevent desiccation of the grafted parts before union has been accomplished. This may be achieved by providing high air humidity, by suitable wrappers, or by spraying the scions with one of the commercial wax emulsions.

Some investigators use dodder to carry certain viruses not otherwise transmitted from one plant to another.

TREATMENT WITH BACTERIAL PRODUCTS

The metabolic products found in bacterial cultures are prepared and employed in a variety of ways which are not yet well worked out. Perhaps the least change occurs in the bacterial cells if they are centrifuged from a liquid culture and dried while frozen. The culture filtrate may be concentrated under reduced pressure at a little above room temperature and then "lyophilized" if desired.

A fermented culture or an aqueous extract may be sterilized and placed in a small container. If leaves with petioles or growing tops are removed from the host plant and are placed with the cut surfaces in such liquids, they commonly show injury within one day or two if much toxic material is present. Bacterial contamination may be reduced if the pH is approximately 4.0. A rigid control of temperature and relative humidity is essential to repeat the results. Care is necessary while interpreting such injury because many constituents of media may be toxic, e.g., ammonia in alkaline material or mineral salts. Likewise, many nonparasitic as well as parasitic fungi produce toxic substances in culture that are not necessarily the reason for pathogenicity.

The metabolites are sometimes applied either in liquid form or in a paste made with inert material, like lanolin, a polymer of ethylene glycol, or flour. The paste has the advantage of furnishing a continuous supply of material over a longer period with relatively less desiccation. It is commonly applied to a wound. The liquid can be introduced into the vascular system of a potted plant by placing cut roots extending from the base of the pot or a cut petiole into a container of the material. Likewise, a cup can be made from a rubber stopper and sealed on a plant stem with vaseline. The cup is filled with liquid, under which a cut is made into the vascular system, so that the liquid is taken by the plant directly into the transpiration stream. The stem can be opened to form a small cavity which is kept filled by means of a capillary tube and funnel. If an enzyme like pectinase is being tested, thin sections of tissue need merely be immersed in a few drops of the liquid.

So many substances appearing in cultures influence plants in one way or another that rigid controls are necessary in searching for the products responsible for pathogenicity. Whenever feasible, an attenuated culture of the same organism or a closely related nonpathogenic culture is carried in a parallel series of trials.

The methods of testing for plant "hormones" and "vitamins" are being improved so rapidly that one should consult an active investigator for the latest procedure.

ANTIBODY PRODUCTION

Questions on the development of antibodies in plants following inoculation or natural infection are discussed in a considerable literature reviewed by Chester (1933, 1935). A number of controversial points are involved.

The injection of plant bacteria into an experimental animal commonly results in the production of antibodies useful for various investigations. Suitable methods appear in Chap. IX.

COGNATE CONSIDERATIONS

Strain variations. When studies involving strain variations are made, it is well to consider Frobisher's (1933) comment, "Plating and fishing of colonies, while generally useful, is not a sufficiently reliable method of purifying cultures in work involving bacterial variations. It is sometimes extremely difficult, if not impossible, to separate bacterial species by this means. Single-cell methods are much more reliable and, it would seem, furnish the only satisfactory means of solving our problems, but even such procedures as are at our disposal require very expert manipulation and may lead to error." The relative unreliability of the poured-plate technique for such studies has been discussed by Riker and Baldwin (1939). The need for cultures with a known origin from a single cell has stimulated much work on methods for securing them. Literature on this work has been reviewed by several writers, e.g., Hildebrand (1950). Unfortunately, some reports on bacterial variations have appeared in which the cultures were purified merely by several successive dilution plates, and such purified cultures were called "single-cell cultures." This misleading use of a well-established phrase provides both the investigator and the reader with a false sense of security.

Variations may be induced among plant pathogens by procedures very similar to those employed on other bacteria. Some of the considerations involved in such studies are discussed by Riker (1940).

The pathogenicity of crown gall bacteria can be destroyed (Van Lanen, Baldwin, and Riker, 1940) with certain amino acids and related compounds added to common media. Attenuation was commonly secured in 20–30 successive transfers. The rate of attenuation was increased if bacterial growth was retarded by the strength of the compound (e.g., 0.1–0.3 per cent glycine) and by an alkaline reaction (e.g., pH 8.0).

The virulence of partly attenuated cultures was restored by long cultivation on suitable media and by ultraviolet irradiation (Duggar and Riker, 1940). Likewise, when a virulent culture was inoculated into a tomato stem above an inoculation with an attenuated culture, the gall about the attenuated culture was approximately as large as that about the virulent culture. A gall induced by a plant hormone served as well as that from a virulent culture (Riker, 1942).

Pathogens acting together. Combinations of microorganisms sometimes induce symptoms different from those caused by any one alone. As long as the pathogens can be cultivated on artificial media, the principles in Koch's postulates can be applied with two or more causal agents. For example, a simple inoculation with one organism may involve a series of susceptible plants growing in a suitable environment with the living causal agent and a parallel control series. With two causal agents, however, there should be four series of plants as follows: (1) with both living pathogens, (2) with only one living pathogen, (3) with only the other living pathogen, and (4) with neither living pathogen. Correspondingly, three causal agents would require eight series of plants.

One should not overlook the fact that in nature pure cultures seldom exist except in the most advanced margin of the lesion.

Cultures from another locality. The use of a culture of a pathogen not already present on local plants requires critical consideration. The progress of bacteriology calls for reasonable freedom in the movement of cultures. This science, however, has a duty in the protection of local plant populations and requires that cultures or strains brought into a new locality should be handled with proper consideration of all the factors involved. It must be insisted that cultures be secured and studied only after both the investigators and their administrators have fully considered and accepted the responsibilities involved. Younger research workers and particularly graduate students are advised to employ such cultures only after detailed plans have been made in conference with their advisors.

Various laws apply to the shipment of infected material.

Relative efficiency in technic. The best methods of procedure for making inoculations and for recording results have not always been worked out and are not obvious from inspection. If the question is of sufficient importance, the answer may be secured statistically. A doubt may appear, with a leaf-spot organism, for example, whether to spray or to make needle punctures. Likewise, when infection develops, the question may occur whether the results should be recorded in terms of total number of lesions, of total tissue involved, of the effect of the disease on yield, or of some other criterion. Such possibilities may be tested by means of the frequently described "analysis of variance." Thus the best method for making the trials and for recording the results may be determined. In general, the method that gives the greatest value for the variance ratio F is the most desirable. This value indicates a greater uniformity in readings from different trials with the same technique or a greater differentiation of the varieties used or treatments employed without a proportional increase in error.

Antibiotics. Numerous recent reports have shown that various higher plants and saprophytic microorganisms produce substances that adversely affect certain bacteria. Many instances occur in which various higher and lower forms of plant life make chemicals that inhibit successful plant inoculations.

To secure the latest methods for plant diseases, one should consult an active investigator in this rapidly developing field.

RECORDS

Taking notes on plant inoculations presents various problems depending upon the experiment in hand. To assist with such records, a tentative protocol appears in Table 29. For some lines of work it is obviously too complex, while for others it is clearly too simple.

A number of the items listed for records may be critical factors for the success or failure of an experiment. Since each one cannot be discussed, several examples are mentioned. (1) Infection may fail if the incubation temperature is either too low or too high. Many plant pathogens operate best between 18 and 30°C. (2) Plenty of moisture is usually important for disease development, a deficiency of water often being responsible for negative results. (3) The age of the plant or of the part inoculated may influence the result. The relatively young leaves are frequently more susceptible than old leaves to bacterial leaf spots. (4) Some varieties of plants are highly resistant to pathogens which readily attack other varieties. Similarly, different strains of bacteria often vary in pathogenicity.

TABLE 29. TENTATIVE PROTOCOL FOR PLANT INOCULATIONS

Host:	Manner of inoculation:
Variety...........................	Through soil......................
History...........................	Through wounds..................
Age...............................	By sprays.........................
Morphological condition...........	Spreader used................
Stomata open.....................	By insects (name)................
Physiological condition............	Stage in life cycle.............
Susceptibility......................	Other means......................
Environment......................	Incubation:
Treatment before..................	Time..............................
Treatment after...................	Environment:
Pathogen:	Temperature................
Strain.............................	Moisture......................
History...........................	Light..........................
Culture on........................	Intensity....................
at.........................°C	Length of day...............
for.......................days	Soil nutrients...............
Inoculum used:	Symptoms:
Diseased tissue....................	Location.........................
Entire culture.....................	Age of parts affected.............
Bacteria:	Severity..........................
Turbidity....................	Description:
Number per ml...............	Early........................
Filtrate...........................	Medium......................
Products..........................	Final.........................
Amount used per plant............	Effect on yield:
	Quantity.........................
	Quality..........................

INTERPRETATION OF RESULTS

The results of research are valid only in accord with the reliability of the methods employed and the accuracy of their interpretation. After an experiment has been performed, it is insisted that a report of such work must not be published for the use of others until repeated determinations have been made and the results have been satisfactorily analyzed. The simpler experiments are commonly performed with suitable controls at least in duplicate or triplicate and carried through three separate times. A good investigator does not become so enthusiastic about an experiment that he fails to view it impartially and to accept sound evidence against it. On the contrary, he makes every reasonable effort before publishing to find an error in the experiment itself or in the conclusions drawn from it.

REFERENCES

Allen, Ethel K., and O. N. Allen. 1950. Biochemical and symbiotic properties of the rhizobia. *Bacteriol. Revs.*, **14**, 273–330.

Chester, K. S. 1933. The problem of acquired physiological immunity in plants. *Quart. Rev. Biol.*, **8**, 129–154, 275–324.

———. 1935. Serological evidence in plant-virus classification. *Phytopathology*, **25**, 686–701.

Duggar, B. M., and A. J. Riker. 1940. The influence of ultraviolet irradiation on the pathogenicity of *Phytomonas tumefaciens*. (Abstract.) *Pythopathology*, **30**, 6.

Fred, E. B., I. L. Baldwin, and E. McCoy. 1932. Root Nodule Bacteria and Leguminous Plants. *Univ. Wisconsin Studies in Sci.* 5.

Frobisher, M. 1933. Some pitfalls in bacteriology. *J. Bacteriol.*, **25**, 565–571.

Hildebrand, E. M. 1950. Techniques for the isolation of single microörganisms. II. *Bot. Rev.*, **16**, 181–228.

———. 1942. A micrurgical study of crown gall infection in tomato. *J. Agr. Research*, **65**, 45–59, illus.

Hildebrandt, A. C. 1950. Some important galls and wilts of plants and the inciting bacteria. *Bacteriol. Rev.*, **14**, 259–272.

Ivanoff, S. S. 1934. A plant inoculator. *Phytopathology*, **24**, 74–76.

Johnson, J. 1937. Relation of water-soaked tissues to infection by *Bacterium angulatum* and *Bact. tabacum* and other organisms. *J. Agr. Research*, **55**, 599–618.

Keitt, G. W. 1918. Inoculation experiments with species of Coccomyces from stone fruits. *J. Agr. Research*, **13**, 539–569.

———, E. C. Blodgett, E. E. Wilson, and R. O. Magie. 1937. The epidemiology and control of cherry leaf spot. *Wis. Agr. Expt. Sta. Research Bull.* 132.

Leach, J. G. 1940. "Insect Transmission of Plant Diseases." McGraw-Hill Book Company, Inc., New York.

Riker, A. J. 1940. Bacteria pathogenic on plants. In The genetics of pathogenic organisms, *AAAS Publ.* 12, Lancaster, Pa.

———. 1942. The relation of some chemical and physico-chemical factors to the initiation of pathological plant growth. *Growth*, Sup. to v. **6**, 4, Symposium on Develop. and Growth, 105–117, illus.

———. and I. L. Baldwin. 1939. The efficiency of the poured plate technique as applied to bacterial plant pathogens. *Phytopathology*, **29**, 852–863.

———. and R. S. Riker. 1936. "Introduction to Research on Plant Diseases." John S. Swift, St. Louis. (Planographed.)

Rivers, T. M. 1937. Viruses and Koch's postulates. *J. Bacteriol.*, **33**, 1–12.

Van Lanen, J. M., I. L. Baldwin, and A. J. Riker. 1940. Attenuation of cell-stimulating bacteria by specific amino acids. *Science*, **92**, 512–513.

CHAPTER XIII

Glossary of Terms Used on the Charts and in the Manual

Acid curd, precipitated milk protein resulting from coagulation of milk due to acid production.

Adjuvant, a subsidiary ingredient in a bacterial inoculum which modifies the host response.

Aerobic, growing in the presence of free oxygen; *strictly aerobic,* growing *only* in the presence of free oxygen.

Agar stroke, agar slant.

Agglutinin, an antibody that causes particulate antigens to clump and settle out of suspension.

Anaerobic, growing in the absence of free oxygen; *strictly anaerobic,* growing *only* in the absence of free oxygen; *facultatively anaerobic,* growing in either the presence or absence of oxygen.

Antibody, a modified serum globulin which reacts specifically with an antigen.

Antigen, a substance which when injected into the animal body stimulates the synthesis of antibody globulin, which can then react with the substance injected.

Antiserum, serum-containing antibody.

Antitoxin, an antibody which can neutralize a toxin or cause it to flocculate.

Arborescent, branched, treelike growth.

Ataxic, lacking coordination of voluntary muscular movement.

Autotrophic, able to grow in absence of organic carbon and nitrogen, i.e., uses inorganic salts and carbon dioxide.

Bacteremia, the presence of bacteria in the blood stream.

Beaded (in stab or stroke culture), separate or semiconfluent colonies along the line of inoculation.

Brittle, growth fragile, easily broken with the inoculating needle.

Butyrous, growth of butterlike consistency.

Capsule, an envelope surrounding the cells of some kinds of bacteria. (Also see *Sheath.*)

Central, occupying a mid-position, e.g., a spore in the center of a sporangium.

Chains, four or more bacterial cells attached end to end.

Chromogenesis, the production of color.

Circular, round, with smooth edges, and over 1 mm in diameter. (Cf. *Punctiform.*)

Clavate, club-shaped; applied to a bacterial sporangium, indicates one containing a subterminal spore.

Coagulation, formation of a clot or curd; the solidification of a sol into a gelatinous mass.

Coccus (*pl. -ci*), a spherical organism.

Collagenase, enzymelike substances which dissolve collagen.

Colony, a visible group of microbes in a culture, derived from the reproduction of, usually, a single organism.

Comma, a short curved rod; comma-shaped. (Cf. *Spiral.*)

Compact, refers to sediment in the form of single fairly tenacious mass.

Complement, a nonspecific heat-labile component of animal serum which participates in antigen-antibody reactions. It is composed of at least four fractions.

Concave, presenting a depressed or hollow surface.

Concentrically ringed, marked with rings, one inside the other.

Consistency, degree of firmness, density, or solidity of a substance.

Contoured, an irregular, smoothly undulating surface, like that of a relief map.

Crateriform, a bowl-shaped liquefaction of the medium.

Cultural characteristics, distinctive, typical, or distinguishing form of growth in or on culture media.

Cup, same as crateriform.

Curd, precipitated milk protein consisting mostly of casein; may result from coagulation by acid or by rennet.

Curled, composed of parallel chains of cells in wavy strands, as in anthrax colonies.

Cylindrical, applied to a sporangium, indicates one with parallel sides.

Cytolysis, a dissolving action on cells.

Diastatic action, conversion of starch into simpler carbohydrates, such as dextrins and sugars, by means of enzyme(s), e.g., diastase.

Diphtheritic, diphtherialike.

Diplococci, cocci occurring in pairs.

Dissociation, variation of organisms from the parent, particularly in colony form, but in other characteristics as well.

Drumstick (*Capitate*), applied to the shape of bacterial sporangia, indicates that one end (the one containing the terminal spore) is larger than the other

Echinulate, growth with toothed or pointed margins along line of inoculation.

Edema, intercellular accumulation of abnormally large amounts of fluid in a part of an animal body.

Effuse, growth thin, veily, unusually spreading.

Eldredge tube, a double-compartment culture tube for collecting carbon dioxide given off during growth of a culture.

Ellipsoid, ovate or ovoid (egg-shaped); elliptical in longitudinal section.

Elliptical, same as ellipsoid.

Endospore, thick-walled spore formed within the bacterial cell, i.e., typical bacterial spores like those of *Bacillus anthracis* or *B. subtilis.*

Entire, applied to colonies, indicates a smooth margin.

Erose, irregularly notched.

Erythrocyte, red blood corpuscle.

Excentric, slightly to one side of the center, between the positions denoted central and subterminal.

Facultative anaerobe, see *Anaerobic.*

Fibrinolysin, an enzyme which dissolves fibrin.

Filamentous, growth composed of long, irregularly placed or interwoven threads.

Filaments, applied to morphology of bacteria, refers to threadlike forms, generally unsegmented; if segmented, the organisms are enclosed in a sheath.

Filiform, in stroke or stab cultures, a uniform growth along line of inoculation.

Flagellum (pl.-la), a flexible, whiplike attachment used as an organ of locomotion.

Flaky, refers to sediment in the form of numerous separate flakes.

Flat, lying in one plane, as a thin colony.

Flocculent, containing small adherent masses of various shapes floating in the fluid.

Fluorescent, having one color by transmitted light and another by reflected light.

Gonidia, a type of asexual spores.

Gonidial (phase), referring specifically to a bacterial stage producing gonidialike bodies.

Gram (stain), a differential stain for bacteria, based on a method of Christian Gram.

Granular, composed of small particles or granules.

Habitat, the place where an organism normally lives under natural conditions.

Hemolysin, a substance causing hemolysis, either alone or in presence of complement.

Hemolysis, a dissolving action on red blood corpuscles.

Hemorrhage, an escape of blood from the vessels.

Histolysis, breaking down of tissues.

Hyaluronidase, an enzyme which hydrolytically splits hyaluronic acid of tissues.

Hydrolysis of starch, decomposition of starch by the incorporation of water; includes diastatic action but is a more general term.

Immune serum, a serum containing an antibody.

Incubation, keeping a culture at a temperature (usually optimum) to facilitate development.

Indicator, a colored compound which changes color with changes in pH or with changes in oxidation-reduction potential.

Infundibuliform (***Infundibule***), in form of a funnel or inverted cone.

Intraperitoneal, within the peritoneum.

Intravenous, within a vein.

Iridescent, exhibiting various and changing colors in reflected light.

Irregular, applied to colonies, indicates a nonuniform, variable periphery.

Lecithinase, an enzyme which dissolves lecithins, e.g., in egg protein.

Lesion, a local injury or morbid structural change.

Lobate, having lobes, or rounded projections.

Maximum temperature, temperature above which growth does not take place.

Membranous, growth thin, coherent like a membrane.

Methemoglobin, a modified form of oxyhemoglobin; green hemolysis produced by certain bacteria.

Microaerophilic, growing best in presence of small quantities of oxygen.

Micrococci, cocci that divide randomly in three dimensions, resulting in irregular groups.

Minimum temperature, temperature below which growth does not take place.

Morphology, form, shape, structure.

Motility, ability to move spontaneously.

Mucoid, mucuslike, referring specifically to a bacterial phase producing slimy growth.

Napiform, liquefaction in form of a turnip.

Opalescent, milky white with tints of color as in an opal.

Opaque, not allowing light to pass through.

Optimum temperature, that temperature at which some reaction, commonly rate of reproduction, takes place most rapidly.

Papillate, growth beset with small nipplelike processes.

Parasitic, deriving its nourishment from some living animal or plant upon which it lives and which acts as host; not necessarily pathogenic.

Pathogenic, not only parasitic but also causing disease to the host.

Pellicle, bacterial growth forming either a continuous or an interrupted sheet over the culture fluid; a thick membrane.

Peptonization, conversion of (milk) protein into peptone by the action of proteolytic enzymes.

Peritrichic, having flagella distributed over the whole surface.

Per os, through the mouth.

Persistent, lasting, in contrast to transient.

Phase, a recognizable stage in the growth of bacteria.

Phase variation, separation of a species into strains having somewhat different characters. (See *Dissociation.*)

Photic characters, characteristics relating to light.

Photogenic, glowing in the dark, phosphorescent.

Physiology, study of the functioning of organisms.

Polar, at the end or pole of the bacterial cell.

Precipitin, an antibody which causes the precipitation of soluble antigens.

Pulvinate, cushion-shaped.

Punctiform, point-shaped, i.e., very small, but visible to naked eye; under 1 mm in diameter.

Radiate, to diverge or spread from a common point.

Radiately ridged, ridges extending out from a center.

Raised, growth thick, with abrupt or terraced edges.

Reaction, as used here, refers to the hydrogen-ion concentration, i.e., pH of a solution.

Reduction, basically, the gain of electrons by a substance. May be accomplished by removal of oxygen from or the addition of hydrogen to the substance. A corresponding oxidation always accompanies reduction. Reduction refers here specifically to the conversion of nitrate to nitrite, ammonia or free nitrogen, also to the decolorization of litmus and other indicators.

Reduction of indicators, reduction of certain colored compounds by bacteria, resulting in loss of color.

Rennet curd, coagulation of protein of milk due to rennet or rennetlike enzymes; distinguished from acid curd by the absence of acid.

Rhizoid, growth of an irregular branched or rootlike character, as colonies of *Bacillus mycoides.*

Ring, growth at the surface of a liquid culture, often adhering to the glass.

Rod, a bacterial cell having a straight central axis which is longer than the diameter of the cross section of the cell.

Rough, colonies with an irregular, nonsmooth surface.

Rounded, circular or semicircular.

Rugose, wrinkled.

Saccate, liquefaction in form of an elongated sac; tubular, cylindrical.

Saprophytic, living on dead organic matter.

***Sarcina* (*pl.-ae*),** a regular cubical packet of cocci, resulting from cell division in three planes. Also, the genus of cocci forming such packets.

Saucer, a concave liquefaction of the medium, shallower than crateriform.

Sheath, an envelope similar to a capsule (see above) but surrounding a filamentous organism.

Smooth, colonies with an even surface.

Spherical, globular.

***Spindle*(*d*),** larger at the middle than at the ends. Applied to sporangia, refers to the forms called clostridia. Also describes colonies, usually subsurface forms.

Spiral, a long, curved rod form of bacterial cell; a coiled cell. (Cf. *Comma.*)

***Sporangium* (*pl.-ia*),** cell containing an endospore.

Spore, a reproductive cell of lower organisms; in bacteria, usually same as endospore.

Spreading, growth extending much beyond the point or line of inoculation, i.e., several millimeters or more.

Stratiform, liquefying to the walls of the tube at the top and then proceeding downwards horizontally.

Streptococci, cocci in the form of a chain.

Strict aerobe, see *aerobic.*

Strict anaerobe, see *anaerobic.*

Subcutaneous, under the skin.

Subterminal, situated toward the end of the cell but not at the end, i.e., intermediate between the positions denoted excentric (see above) and terminal.

Syncope, loss of consciousness due to cerebral anemia.

Terminal, at the end of a cell, as a spore.

***Thermal death point* (*TDP*),** temperature at which an organism is killed after exposure for 10 min.

***Thermal death time* (*TDT*),** time required to kill an organism at a specific temperature.

Toxin, a poisonous substance of plant or animal origin.

Transient, lasting a short time, in contrast to persistent.

Translucent, allowing light to pass through without allowing complete visibility of objects seen through the substance in question.

Truncate, ends abrupt, square.

Turbid, cloudy.

Ulcer, an open sore.

Umbonate, having a raised center; knoblike.

Undulate, wavy.

Vegetative cells, bacterial cells not containing spores, i.e., cells primarily concerned with nutrition and growth.

Villous, having short, thick, hairlike processes on the surface; intermediate in meaning between papillate and filamentous.

Viscid, sticky; growth follows the needle when touched and withdrawn; sediment on shaking rises in a coherent swirl.

Index